System Innovation for Sustainability 3

System Innovation for Sustainability 3

CASE STUDIES IN SUSTAINABLE CONSUMPTION AND PRODUCTION — FOOD AND AGRICULTURE

EDITED BY URSULA TISCHNER, EIVIND STØ, UNNI KJÆRNES
AND ARNOLD TUKKER

Published by Greenleaf Publishing Limited
Aizlewood's Mill
Nursery Street
Sheffield S3 8GG
UK
www.greenleaf-publishing.com

Printed in Great Britain on acid-free paper by Antony Rowe Ltd, Chippenham and Eastbourne

Cover by LaliAbril.com.

British Library Cataloguing in Publication Data:
Case studies in sustainable consumption and production :
food and agriculture. -- (System innovation for
sustainability ; 3)
1. Sustainable agriculture--Case studies. 2. Food supply--
Environmental aspects--Case studies.
I. Series II. Tischner, Ursula.
338.1-dc22

ISBN-13: 9781906093242

Contents

Acknowledgements

The SCORE! core team managing the activities within the project on Sustainable Consumption and Production in the agriculture and food domain—Ursula Tischner from econcept, Germany, Eivind Stø and Unni Kjærnes from SIFO, Norway—supported by the manager of the SCORE! project, Arnold Tukker from TNO, The Netherlands, and his team would like to thank all experts in the closer and wider SCORE! team for their excellent and valuable contributions during the past three years. These are, first, the contributing authors of this book and all experts who have provided papers for the SCORE! workshops and conferences.[1] Second, we would like to thank all participants in the SCORE! sessions on SCP in the agri-food domain for their active participation and interesting contributions to discussions. The table overleaf lists the active experts on SCP in agri-food domain who were involved in SCORE!, though the list is not and cannot be exhaustive. Furthermore, the EU is acknowledged for providing financial support to the SCORE! project via its Sixth Framework Programme.

As the SCORE! project will have a follow-up and, as activities will be continued, the organisers hope for further fruitful cooperation not only with the established group of experts but also with new members. It is recommended to visit the SCORE! website regularly for information updates about the continuation of activities on SCP research, both in Europe and internationally.

1 All papers that could not be included in this publication can be downloaded from the SCORE! website: www.score-network.org.

Experts active in the Sustainable Agriculture and Food group of SCORE!

(note: this list is not exhaustive)

Expert	Institution	Country
Ursula Tischner	econcept, Agency for Sustainable Design	Germany
Eivind Stø Unni Kjaernes Ingrid Kjørstad Ingri Osmundsvåg	National Institute for Consumer Research (SIFO)	Norway
Carlo Vezzoli Fabrizio Ceschin Anna Meroni	Politecnico Milano	Italy
Chris Foster Paul Dewick	Manchester Business School	UK
Tim Cooper	Nottingham Trent University	UK
Steve Webster	Delta-innovation Ltd	UK
Michael Søgaard Jørgensen	Technical University of Denmark	Denmark
Oksana Mont Helen Nilsson	IIIE E, University of Lund	Sweden
Edina Vadovics	Central European University	Hungary
Matthew Hayes	Szent István University	Hungary
Burcu Tunçer	UNEP/Wuppertal Institute Centre for Sustainable Consumption and Production	Germany
Patrick Schroeder	University of Wellington	New Zealand
Chris Wille Joke Aerts	Rainforest Alliance	USA
Bernward Geier	Colabora	Germany
Martina Schäfer Adina Herde	Technical University Berlin	Germany
Cordula Kropp	Munich Institute for Social and Sustainability Research	Germany
Gerd Scholl	IÖW	Germany
Frank Belz Sandra Silvertant	Technical University of Munich	Germany
Irmgard Schulz Doris Hayn	ISOE	Germany
Ulrike Eberle	Öko-Institut Freiburg	Germany

Expert	Institution	Country
Carolin Baedeker	Wuppertal Institute for Climate, Environment and Energy	Germany
John Thøgersen	Aarhus School of Business	Denmark
François Jégou Sara Girardi	SDS/Dalt	Belgium
Fritz Reusswig	Potsdam Institute for Climate Impact Research	Germany
Bruno Scaltriti	University of Gastronomic Science/Slow Food	Italy
Benjamin Nölting	Technische Universität Berlin	Germany
C. Anna Catania	University of Palermo	Italy
Patrizia Ranzo Maria Antonietta Sbordone Rosanna Veneziano	Seconda Università degli Studi di Napoli (SUN)	Italy
Conny Bakker	Info-Eco	The Netherlands
Angelija Bučienė Martynas Šlažas Aušra Steponavičiūtė Marija Eidukevičienė	Klaipeda University	Lithuania
Brigitte Biermann	Fernuniversität Hagen	Germany
Doris Fuchs	University of Twente	The Netherlands
Elizabeth Sargant Gerd Spaargaren	University of Wageningen	The Netherlands
Simone Maase	Technical University Eindhoven	The Netherlands
Inge Røpke	Technical University of Denmark	Denmark
Karl-Werner Brand	Munich Project Group on Social Research	Germany
Benny Leong	Hong Kong Polytechnic University	Hong Kong
Toshisuke Ozawa	National Institute of Advanced Industrial Science and Technology (AIST)	Japan
Atsushi Inaba	University of Tokyo	Japan

Part I
Overview and introduction

1
Introduction

Ursula Tischner
econcept, Agency for Sustainable Design, Germany

Eivind Stø and Unni Kjærnes
National Institute for Consumer Research (SIFO), Norway

Arnold Tukker
TNO Built Environment and Geosciences, The Netherlands;
Norwegian University of Science and Technology (NTNU), Department of Product Design

This publication is a result of the European project Sustainable Consumption Research Exchanges (SCORE!)[1] and summarises the findings of the working group on Sustainable Consumption and Production (SCP) for the domain 'food and agriculture'. Food is of great significance in people's social life and welfare—on the supply side as well as on the consumption side. At the same time, food and its connected (agricultural) production systems have received considerable attention in recent years regarding their sustainability. Food and the domains 'mobility' and 'energy and housing' are consumption areas associated with the greatest negative impacts on the environment (Hertwich 2005; Tukker and Jansen 2006; Tukker *et al.* 2006). This is why these three consumption domains were selected in the SCORE! project as the areas to be tackled in case-study research and expert exchange activities to understand how a shift to more sustainable consumption and production can be organised and supported in the three domains.

Experts estimate that close to half of human impact on the environment is directly or indirectly related to food production and consumption (Jongen and Meerdink 1998). Indeed, food production, distribution, consumption and disposal are important in terms

1 See Section 1.1 and www.score-network.org, accessed 17 July 2009.

of the use of land and resources, pollution and emissions, biodiversity, landscape design and so on. Consequences are to be found in the external environment as well as 'internally', in health hazards and in the satisfaction of the basic needs of citizens. More than 200 million adults in the European Union are overweight or even obese (CEC 2005) as a result of unhealthy diets and too little exercise.

Sustainability issues are on the agenda for food producers and market actors, for politicians and regulators, and for ordinary people as citizens—in public discourse and as a subject of collective mobilisation—as buyers and eaters of food. The issue has also received considerable attention from experts and scientists of various kinds. Many attempts in a variety of forms to reduce environmental threats and increase sustainability have been instigated, and numerous studies, research programmes and publications have addressed such issues. Agri-food issues have also been prominent in the evolving definition of what 'sustainability' means, including what should go under the umbrella terms 'sustainability' and 'sustainable development'.

One significant difference between the area of food and other consumption domains is that, whereas people are able to live without private cars or holidays, people cannot live without food. In this domain it is more a question of what we eat and how much of it. The availability of sufficient healthy food constitutes a basic human need still not met for all people worldwide. In 2008, the overall number of undernourished people in the world increased to 963 million, primarily because of higher food prices, compared with 923 million in 2007. The vast majority of the world's 907 million undernourished people live in developing countries.[2] Thus for the field of agriculture and food under the sustainability umbrella we need to discuss two basic questions: (a) how can we secure access to enough healthy food for all inhabitants of planet Earth even with a growing world population and (b) how can the systems of food production and consumption stay within the limits of the carrying capacity of our natural environment (e.g. in relation to availability of land, resources, water, global warming issues and so on)? This book focuses more on question (b) as it summarises the results of a European project with a perspective on sustainable consumption and production (SCP) in Europe, but the challenges of food security and nutrition for all is of high importance and has to be dealt with urgently. Solutions cannot be suggested that overlook such needs.

1.1 The SCORE! project

SCORE! (Sustainable Consumption Research Exchanges) is a network initiated with EC funding supporting the development of the UN's 10 Year Framework of Programmes on SCP. The mission of SCORE! is to organise a leading science network that provides input to this framework. The EU funding for SCORE! ran between 2005 and 2008, engaging 28 institutions; however, through the organisation of various major workshops and conferences the project engaged and structured a larger community of a few hundred professionals in the EU and beyond.

2 Source: www.fao.org/news/story/en/item/8836/icode, accessed 17 July 2009.

The SCORE! philosophy assumes that SCP structures can be realised only if experts that understand (1) business development, (2) (sustainable) solution design, (3) consumer behaviour and (4) system innovation policy work together in shaping those structures. Furthermore, this should be linked with the experiences of actors (industry, consumer groups, eco-labelling organisations) in real-life consumption areas: mobility, agri-food, and energy and housing. Broadly, this gives the following approach to the project.

The first phase of the project (marked by a workshop co-organised with the European Energy Agency [EEA] in Copenhagen, April 2006) aimed to arrange a positive confrontation of conceptual insights developed in the four aforementioned science communities of how 'radical' change to SCP can be governed and realised. The results of this phase have been published in the first book in the 'System Innovation for Sustainability' series (Tukker *et al.* 2008). The second phase put the three consumption areas centre stage. SCORE! work package leaders inventoried cases 'that work' with examples of successful switches to SCP in their field. In a series of conferences and workshops, these cases were analysed in terms of their 'implementability', adapted where needed, and policy suggestions were worked out that could support their implementation. The results of this phase are published in this book and in two parallel books covering the areas of mobility and housing and energy-using products (Geerken *et al.* 2009; Lahlou *et al.* 2010).

1.2 The structure of this book

After this introduction first Tischner and Kjærnes summarise the state of the art of sustainable agriculture and food in Europe. Then ten case-study chapters present different approaches, strategies, concepts and opinions about SCP in the agricultural and food system, as follows. In Chapter 3 Dewick, Foster and Webster describe strategies for the 'greening' of mainstream agri-food business and policy, focusing especially on the dairy industry in the UK. In Chapter 4 Cooper discusses the issues of self-sufficiency and localisation, reflecting on sustainability and ambiguity in Britain's food policy.

The question of how organic farming and food can be found in niches in Europe is discussed in Chapter 5 by Jørgensen, who deals with the development of production and consumption of organic food in Denmark.

After this, different local food production and distribution systems are presented. In Chapter 6 Nilsson and Mont discuss the socioeconomic aspects of farmers' markets in Sweden, in Chapter 7 Vadovics and Hayes discuss the Open Garden concept (a local organic producer–consumer network in Hungary), and in Chapter 8 Kjørstad presents the state of the art of Slow Food in Europe, especially in Poland.

The global–local and fair trade perspective of the food and agricultural system is discussed in the next three chapters. In Chapter 9 Tunçer and Schroeder introduce the Sambazon concept of marketing the açaí berry from the Brazilian Amazon. In Chapter 10 Osmundsvåg discusses the Fairtrade concept of Max Havelaar in Norway. In Chapter 11

Wille, Aerts and Geier present the 'verified sustainable agricultural system' of the Rainforest Alliance.

In the final case study, by Schäfer, Herde and Kropp, in Chapter 12, the authors discuss how 'life events' can serve as turning points in consumer behaviour towards more sustainable nutrition.

The book is concluded by the summary of Tischner, Stø and Tukker, who merge the learning of the food working group in the SCORE! project with the findings of the case studies in this book. This final chapter includes the summary by Tukker of overall conclusions of the SCORE! project about SCP, cross referencing the outcomes of the other two SCORE! working groups on SCP in the mobility and energy and housing domains.

References

CEC (Commission of the European Communities) (2005) 'Kampf gegen Fettsucht in Europa', *Ernährung im Fokus* 5.5: 145.

Geerken, T., and M. Borup (eds.) (2009) *System Innovation for Sustainability. 2: Case Studies in Sustainable Consumption and Production—Mobility* (Sheffield, UK: Greenleaf Publishing; www.greenleaf-publishing.com/SCP2).

Hertwich, E. (2005) 'Life-cycle Approaches to Sustainable Consumption: A Critical Review', *Environmental Science and Technology* 39.13: 4673.

Jongen, W., and G. Meerdink (1998) *Food Product Innovation: How to Link Sustainability and the Market* (Wageningen, Germany: Wageningen Agricultural University).

Lahlou, S., M. Charter and T. Woolman (eds.) (2010) *Case Studies in Sustainable Consumption and Production. 4: Housing/Energy Using Products* (Sheffield, UK: Greenleaf Publishing; www.greenleaf-publishing.com/SCP4).

Tukker, A., and B. Jansen (2006) 'Environmental Impacts of Products: A Detailed Review of Studies', *Journal of Industrial Ecology* 10.3: 159-82.

——, G. Huppes, S. Suh, R. Heijungs, J. Guinee, A. de Koning, T. Geerken, B. Jansen, M. van Holderbeke and P. Nielsen (2006) *Environmental Impacts of Products* (Seville, Spain: European Science and Technology Observatory/Institute for Prospective Technological Studies [ESTO/IPTS]).

——, M. Charter, C. Vezzoli, E. Stø and M. Munch Andersen (eds.) (2008) *System Innovation for Sustainability. 1: Perspectives on Radical Change to Sustainable Consumption and Production* (Sheffield, UK: Greenleaf Publishing; www.greenleaf-publishing.com/SCP1).

2
Sustainable consumption and production in the agriculture and food domain

Ursula Tischner
econcept, Agency for Sustainable Design, Germany

Unni Kjærnes
National Institute for Consumer Research (SIFO), Norway

This chapter summarises the state of the art and research on consumption and production in the agriculture and food domain.

The subject of food is connected as much to individual well-being and economic issues as it is to North–South divide and sustainable development topics. Agriculture and the food industry are still very important sectors for most countries. Furthermore, food is important in societal and political issues such as public health or the self-sufficiency of individual nations.

As we will see below this also makes the agri-food domain a very diverse field to discuss, with a lot of different influences and complex relationships between several actors. Thus we can give only a broad overview in this chapter, concentrating on the most important issues related to the agri-food production and consumption system and sustainability issues.

In Section 2.1 of this chapter we describe trends and sustainability issues of the agri-food domain; in Section 2.2 we introduce the agri-food production–consumption system with its main actors and conditions; Section 2.3 draws preliminary conclusions about typical sustainable production and consumption (SCP) strategies in the agri-food domain.

2.1 Trends and sustainability issues in the agri-food domain

Which trends are influencing the agri-food domain and what are the sustainability issues and problems connected with the production and consumption of food? These questions are discussed below; first we introduce the larger context of the agri-food system, including societal megatrends (Section 2.1.1), then more system-specific boundary conditions that emerge from these megatrends (Section 2.1.2), followed by sustainability problems and issues (Section 2.1.3).

2.1.1 Context factors and megatrends for the agri-food production–consumption system

There are some general societal, economic and technological developments influencing the food production and consumption system.

2.1.1.1 General societal trends

2.1.1.1.1 An ageing population in Europe

Today about a third of the European population is senior and the population in Europe is ageing faster than in any other continent (*Drudge Report* 2006). The average life expectancy in Europe is 82 years for women and 76 years for men (Eurostat 2007). This has implications for the agri-food domain in two ways: (a) the population of farmers is ageing as well, and few young people have the desire to become farmers,[1] and (b) the consumer group of elderly people has specific food and nutrition demands, including more food services.

2.1.1.1.2 Decreasing household size and individualisation

Fewer couples in Europe are having children. Since 1965 the overall fertility rate in the EU-25 has been decreasing constantly: currently, on average it is 1.5 children per woman (Eurostat 2007). The share of single households is increasing; in some countries it is up to 46 % of the population.[2] This has consequences for eating habits as well; trends are towards smaller meals and portions of food, more eating out, and buying quickly prepared convenience food. To a certain extent, we also see more individualised eating.

2.1.1.1.3 A more multicultural society and migration within and from outside Europe

The multicultural society creates cross-cultural influences and a rich and inspiring diversity of different restaurants and food offers. But at the same time there is the challenge

1 For example, the average age for farmers in Sweden is 53 years, and one in six are over 65 years old. Only 6% of farmers are under 35 years old (see Jordbruksverket 2003).

2 Source: German Social Science Infrastructure Services eV (GESIS): www.gesis.org, accessed 18 July 2009.

of integrating diverse cultures, habits, values and religions; for example, Christian, Islamic or Jewish cultures may have very different dietary habits and norms about food and eating.

2.1.1.1.4 Increasing awareness of global warming and its consequences

The presence of this topic in all news media has increased consumer awareness of global warming and created a sense of urgency to act. Thus most consumers are aware that their personal behaviour and consumer choices are connected to this global phenomenon. The EU member states agreed on a common goal: by 2020 to have reduced carbon dioxide (CO_2) emissions by 20% compared with 1990 levels.[3]

2.1.1.2 General economic trends

2.1.1.2.1 Between financial crisis and climate change

On average, Europeans have seen an increase in wealth (CIAA 2002). After a period of stagnation that started in 2001 (associated with the attacks of September 11th, business scandals such as Enron, the introduction of the euro and so on) and decreased consumer expenditure, European economists and consumers were cautiously optimistic again and growth rates were on the rise. This came to an end through the recent US real estate crisis and its consequences and the international financial crisis. Financial markets have been severely shocked and they will take time to recover. Experts see the global imbalances as the roots of this crisis: excessive deficit in the USA and excessive savings in East Asia. The specificity of this recession is that credit conditions and liquidity constraints are its drivers. Politicians agree that it needs new regulations for the banking sector and at the world level to avoid these kinds of problems in the future: the G20 meeting on 15 November 2008 went in that direction. In Europe governments are acting together, recapitalising first and combating recession and stimulating the economy. Most European politicians furthermore agree that there are two crises at same time: the financial crisis that concerns the short term and the energy and climate change challenge that concerns the long term. One should not try solving one or the other as both are linked (DG Research 2008).

At the consumer level the 'scissors phenomenon' of the rich getting richer and the poor getting poorer is gaining more and more importance. The RIVM[4] expects a growing dif-

3 The EU Climate and Energy Package has been confirmed in 2009 and aims at a 20–20–20 formula: a 20% cut in emissions of greenhouse gases by 2020, compared with 1990 levels; a 20% increase in the share of renewables in the energy mix; and a 20% cut in energy consumption. Source: ec.europa.eu/environment/climat/climate_action.htm, accessed 15 April 2010. Especially relevant for agriculture is the integration of climate change into the EU's Rural Development Policy: e.g. goals for the carbon sequestration potential of afforestation and reforestation measures, forest management and natural forest expansion in the EU-15 Member States by 2010: 33 Mt CO_2 equivalent; or the support scheme for energy crops under the EU's Common Agricultural Policy (Regulation 795/2004/EC)—the Regulation makes available €45 per hectare in aid to producers of energy crops, i.e. crops intended for the production of biofuels or electric and thermal energy. Source: The European Climate Change Program, EU Action against Climate Change, European Communities, 2006; ec.europa.eu/environment/climat/pdf/eu_climate_change_progr.pdf, accessed 15 April 2010.

4 See www.rivm.nl/en, accessed 18 July 2009.

ference in income per capita, the continuing rise of the 24-hour economy, an increasing number of one-and-a-half and double-income households and stronger reliance on market forces through increased privatisation. That has implications for the food market as well: there are very price-sensitive consumer groups and products, and there are products and consumer groups where high quality, health and convenience are more decisive factors for food buying decisions.

2.1.1.2.2 Global versus local economy

In the past decade, global integration of markets has been one major economic trend. The promise of economies of scale has been one of the driving forces behind this trend, along with the development of new, cheap production facilities and the opening of new markets. Information technology has been the most important enabler for this. However, it has become clear that there are disadvantages to globalisation as well, such as difficulties in controlling quality in global value chains, the opening of doors for the illegal copying of patented knowledge or copyrighted designs, social problems because of a 'sweatshop' mentality and the exploitation of poor workers in developing countries resulting in companies moving production back to their original countries, the formulation of social standards such as SA8000 and public criticism of globalisation (seen, for example, through protests at meetings of the World Trade Organisation, and in Naomi Klein's publication *No Logo* [2000]).

Global branding, marketing and advertising contribute to the development of a common image of a desirable lifestyle and consumption levels. Complementary to this globalisation, a new local–regional self-consciousness is emerging in several European regions, such as in Italy, Spain, France and Germany where local culture and local specialities (e.g. special food products and recipes), supported by stronger European Labels of Origin, are getting more and more attention and are rediscovered as an alternative to the unified global (Western) culture.

2.1.1.2.3 Peak oil discussion and rapidly increasing oil prices

Another important factor for a move back to more local production are the fast-rising prices for fuel and thus transportation. The prices for kerosene increased by 90.5% from June 2007 to June 2008 (Rubin 2008) and the cost of shipping a standard 40-foot container from Shanghai to the US eastern seaboard (including inland costs) rose from only $3,000 in 2000 to $8,000 in 2008; it is predicted soon to be $15,000 (Rubin and Tal 2008). As intensive agriculture generally depends heavily on the availability of oil, the rapidly rising oil prices, supported by global warming effects such as water scarcity, are an important factor for the rising prices of food products. This tendency is supporting the shift to more localised and less industrialised agriculture and food production.

2.1.1.3 General technological developments important for food

2.1.1.3.1 New production technologies: biotechnology and genetic engineering

Technologies applied in primary food production and processing are developing rapidly, involving highly diverse and complex methodologies at all stages of food provisioning, from breeding and seeds production to functional foods, packaging and chilling techniques. These new technologies may both challenge and support efforts to improve the sustainability of food. The technology that has received most attention regarding sustainability and risk has been genetic modification. Biotechnology has been applied for centuries in the agri-food domain to increase efficiency. More recently the modification of genetic codes of species became possible and is applied to create, for example, systems of crops designed to harmonise with a specific pesticide to gain better agricultural output. While companies active in genetic modification and production of pesticides see these practices as steps towards food security for the whole world population, media and the majority of consumers are concerned that setting genetically modified (GM) crops free in the fields threatens biodiversity, can have unknown and very negative consequences for the ecosystem and, ultimately, via the food chain, for human health. The development of systems of GM crops and fertilisers and pesticides by one single company is particularly problematic because it gives that company too much power to control whole markets and causes total dependence of farmers on these systems. The risks of free field production with GM crops are documented by cases such as Monsanto's GM corn, MON 810.[5] Although the European Commission allowed the cultivation of this GM crop in Europe, several European nations such as France and Austria prohibited its use in their countries in 2008 because studies showed damage to the organs of rats and reduced fertility of mice fed with the corn, and because of a suspected connection of the GM corn with widespread extinction of bees in Europe. Austria now leads European countries with seven bans on GM crops such as corn and rape.[6]

The majority (58%) of Europeans declare that they are opposed to the use of genetically modified organisms (GMOs) whereas around a fifth (21%) support their use. A further 9% say they have never heard of GMOs. At the country level we see that the absolute majority in most countries are opposed to the use of GMOs. This is particularly the case in Slovenia (82%) and Cyprus (81%). Respondents in Malta, Portugal and Spain hold the mildest opinions in this respect, which is explained mainly by the high share of respondents in these countries spontaneously admitting that they have never heard of the concept or do not form an opinion for or against (CEC 2008).

2.1.1.3.2 Information and communication technology and the internet

Information and communication technology (ICT) is everywhere, even in our kitchens. From the full electrification of the household in the 20th century we move to 'intelligent' kitchens that are equipped with ICT-supported appliances: intelligent refrigerators that are connected to an ordering system of a local supermarket and order food stuff that has run out of stock, intelligent microwave ovens that heat up the prefabricated meals to exactly the right temperature and time 'à la minute' and so on. Most of these innovations in an 'intelligent home' are technology-driven and company-driven—consumers are not (yet) convinced that they need all this.

5 Monsanto has developed insect-protected YieldGard corn, MON 810, which produces the naturally occurring *Bacillus thuringiensis* (Bt) protein Cry1Ab. YieldGard corn is protected from feeding damage by the European corn borer (ECB, *Ostrinia nubilalis*), the southwestern corn borer (SWCB, *Diatraea grandiosella*) and the pink borer (*Sesamia cretica*).

6 See The Glocalist news website: www.glocalist.com/index.php?id=20&tx_ttnews%5Btt_news%5D=3789&tx_ttnews%5Bcat%5D=4&cHash=e47ed5753b, accessed 18 July 2009.

One of the developments made possible by ICT that is already changing, and will continue to change even more the way we do things and live, are internet, the use of online media and functions and the possibility of living a 'second life' in virtual worlds. There is a huge growth in online shopping, including for food.[7] Learning and exchange of knowledge (such as recipes) takes place in internet communities.[8] Individual consumers can join online purchasing communities and thus enjoy economy-of-scale effects. The internet is also playing an increasingly important role in the social mobilisation and exchange of information about food issues. Problems and scandals are becoming more and more difficult to keep away from public scrutiny.

2.1.1.3.3 Food distribution systems and logistics: tracking and tracing technologies

Food distribution chains are becoming more and more integrated. This is helped by new and much more efficient forms of logistics. This is about organisation, ownership and power, but technology also plays an important, facilitating role. Developments in ICT and global positioning systems (GPS) make it possible to track the origin of products, record aspects throughout life-cycles, transfer such information online or by specific media and locate products wherever they are: for example, by using radio frequency identification (RFID) tags, intelligent labels and so on. Moreover, quality assessment and hazard analysis and critical control point (HACCP) systems, often linked to external audits, have had a major impact by influencing risk management as well as standardisation and predictability. This allows increasing control over the products and their use along complex provisioning chains, improved maintenance and life-cycle management, improved storage and ordering management, precision farming methods and so on. Such techniques also play a significant role by improving the transparency of complex systems.

2.1.2 The agri-food 'landscape': system-specific factors in the domain

Trends and factors more directly connected to the agri-food domain and often emerging from the context factors and megatrends introduced above are discussed in the following section. We distinguish trends on the production side, in food offers and consumption, on the consumption side and in policy.

2.1.2.1 Production side

2.1.2.1.1 Industrialised food, globalisation of and power concentration in the food chain

Food production has become much more commercialised. Additional processing steps have been added between farmer and consumer. The food industry, as all other indus-

7 For example, with growth rates around 27% in the UK, the figures for buying online were six times better than for the traditional retail market in 2004, according to a study by Verdict: ds.datastarweb.com/ds/products/datastar/sheets/verd.htm, accessed 18 July 2009.

8 For example, see the blogs at www.kiplog.com/food, accessed 18 July 2009.

try sectors, has increased its economic efficiency (higher outputs) over the past century. Conventional agriculture worldwide is still becoming more intensified, characterised by greater use of synthetic fertilisers, pesticides and technical devices and by an increasing average farm size. Materials are being transported over large distances, being processed in countries with low labour cost, with growing amounts of packaging and transportation, more and more product diversification and so on. The input of energy and water has increased dramatically, as has the global spread of diseases and plagues. Furthermore, a lot of efficiencies in the system are delivered at the cost of exploitation of people, soil and resources and by compromising the 'naturalness', quality and taste of the food products.[9]

Globalisation of the food-processing industry has led to much more diversity in the food products available. Thus seasonality of food is losing importance (Marshall 2001). Whether this also leads to homogenisation of food supplies across regions and nations is more disputed (e.g. see Harvey *et al.* 2004). Consumers have greater choice at lower prices and more convenience. However, as the production systems have become more complex and the distance between producers and consumers increases, direct control becomes more difficult. A major trend in current provisioning systems in Europe is the growing centralisation of power, control and information at the retail level, with integrated supply chains and highly targeted marketing. Consumers have more problems knowing where and how food is produced, and producers do not know who is buying their produce. While these tendencies have challenged transparency as well as accountability, overall it has also produced predictability.

2.1.2.1.2 Farmers under pressure

More and more farmers are aware of the problems of intensive agriculture: increasing dependence on energy and fertiliser inputs as well as on transportation, fewer jobs in the countryside, increasing levels of environmental pollution and soil erosion and so on. Instead of selling food to their neighbours, farmers sell into a long and complex marketing chain of which they are a tiny part and are paid accordingly (see Chapter 7).

This could be a fruitful climate in which to discuss changes—for example, towards more sustainable agricultural systems—if only so many farmers did not suffer from tremendous economic pressures. The causes of this pressure are production quotas under the Common Agricultural Policy (CAP) combined with ongoing centralisation at all steps along the food chain. For instance, big food processors (especially of milk and meat) dictate the prices, and big supermarkets and discounter chains retailing the food products compete mainly by lowering prices.

2.1.2.2 Consumption side

2.1.2.2.1 Declining trust in food safety

The changing provisioning and audit systems have generally improved food safety standards, but the situation is highly variable. In the UK, food quality and safety has never

9 One famous example concerns the nicely shaped and coloured Dutch tomatoes that lacked taste and that were refused especially by German consumers, pushing the industry to change back to tastier types (Braun 2009).

been so high and this is a core strength of the food industry (CIAA 2002), but the situation is much more problematic in many countries in the east and south of Europe. Consumer trust varies accordingly. In the late 1990s consumer distrust came on the public agenda, associated with food scandals and scares such as salmonella, bovine spongiform encephalitis (BSE); later came bird flu and foot and mouth disease. A European study of trust shows that in many, but far from all, cases efforts to cope with these problems have served to reinstall confidence in food safety (Kjærnes *et al.* 2007). That has not, however, removed criticism of globalisation and industrialised food production from the public agenda, also raising other issues besides food safety, such as problems of quality, animal welfare and sustainability. Scares and problems with control of food safety seem to have driven many consumers to become more aware and active as political consumers.

2.1.2.2.2 Consumers' expenditure on food

Consumer spending on food as a percentage of total household expenditure has steadily declined in most countries, sharpening competition in the food processing and retail sectors and leading to an explosion in the number of food products and services offered to the consumer (OECD 2001). This is likely to change as a result of the 'world food crisis', as recent developments in increasing food prices are called. The recent sharp increase in food prices has served to raise a number of questions regarding food distribution systems, social inequalities and the regulation of food prices.

2.1.2.2.3 Increasing awareness about hunger and about obesity and the health aspects of food

Twelve years after the 1996 Rome World Food Summit (WFS) the number of undernourished people in the world remains high. There has been no progress towards the WFS goal—to reduce hunger by half by 2015. While some countries, such as Latin America and the Caribbean or sub-Saharan Africa, were well on track towards reaching the summit's target, before the food prices increased drastically, in 2008 primarily because of higher food prices the overall number of undernourished people in the world increased to 963 million, compared with 923 million in 2007. The vast majority of the world's undernourished people—907 million people—live in developing countries. Nearly two-thirds of the world's hungry live in Asia (583 million in 2007).[10]

At the same time it is becoming obvious that obesity is one of the greatest public health challenges of the 21st century. Its prevalence has tripled in many countries in the WHO European Region since the 1980s, and the numbers affected continue to rise at an alarming rate, particularly among children and particularly in lower-class families. Obesity is already responsible for 2–8% of health costs and 10–13% of deaths in different parts of the EU. To facilitate region-wide action, in November 2006 WHO Europe organised a Ministerial Conference on Counteracting Obesity. At the Conference, member states adopted a European Charter on Counteracting Obesity.[11]

10 Source: www.fao.org/news/story/en/item/8836/icode, accessed 18 July 2009.

11 WHO Regional Office for Europe: www.euro.who.int/obesity/import/20060217_1, accessed 18 July 2009.

2.1.2.3 Trends in food consumption

2.1.2.3.1 Trends in nutrition and eating patterns

The general trend in OECD countries is one of too high an intake of calories leading to overweight and obesity becoming a serious health problem. People in these countries tend to have a higher consumption of meat, cheese, fruits, vegetables and bottled drinks compared with developing and emerging countries (CEC 2005; Michaelis and Lorek 2004). A considerable number of people are becoming aware of health-related issues and are increasingly concerned about the nutritional content and functional value of their food; they demand increasingly 'light' products with reduced fat and calories, some even vegetarian alternatives to meat products (Michaelis and Lorek 2004; OECD 2001). On the other hand, convenience, fast food and 'finger foods' are still on the rise. There is a tendency toward the consumption of highly processed foods (fast and convenience foods) and more appliances in the kitchen, accompanied by decreasing knowledge about nutrition and food (Davies 2001; Swoboda and Morschett 2001). The average preparation time for food products in the household has been reduced to around 18 minutes per meal. Consumers use mainly two to five ingredients to cook a complete meal, and the market share for ready-to-heat meals has increased by 50% since 1997 (Tempelman *et al.* 2006).[12]

In many countries we see more eating away from home. This is reflected also in a significant growth in the catering industry. Furthermore, fast-food chains are still successful in Europe, although fast food has received broad public criticism because of its negative health effects, packaging waste, low quality and so on. 'Finger food', as a subcategory of fast food, is ready-to-eat food purchased and eaten in all sorts of different places: the car, the office, while travelling in trains or planes and so on. It is prepared to be eaten from hand to mouth, thus making it very simple to consume.

2.1.2.3.2 More sustainable and high-quality food on the rise

Alternatives to conventional and processed food offers are on the rise: the growth rates of organic, fair trade or slow food are higher than the growth rates in most conventional food offers; for example, for organic food the growth rate is around 30% per year. Parallel to this, and partly in conjunction with, the supply of high-quality products is also increasing. But still they are niche products. Organic agricultural production in 2001 had, on average, a market share of between 0.3% (organic pork and poultry) and 1.8% (organic cereals) in Europe. There are big differences between the EU countries, ranging from 0.1% in the Czech Republic to 3.5% in Denmark and 3.7% in Switzerland. The range includes cereals, potatoes, vegetables, fruit, dairy products, eggs, meat, wine and processed ingredients. In countries with high sales of organic food, organic products are sold not only in specialised shops but also via general food shops such as multiple retailers. The promotion of organic food in most countries takes place via a label. However, the recognition of the label is very different in the EU countries: from 1% recognition in Greece to 94% in Sweden. Most people have relatively undifferentiated motives for buying organic food, referring to combinations of concerns for own health, environmental

12 See www.cbl.nl, accessed 18 July 2009.

and social sustainability and wishes for freshness and taste (Torjusen *et al.* 2004). Consumer price premiums for organic food varied from 28% for baby food to 163% for organic cucumbers (figures for 2001 are from Hamm and Gronefeld 2004; Zanoli *et al.* 2004).

2.1.2.4 Trends in agricultural and food related policies

2.1.2.4.1 The European Common Agricultural Policy

The CAP is a system of EU agricultural subsidies and programmes. It represents about 44% of the EU budget (e.g. €43 billion scheduled spending for 2005).[13] These subsidies work by guaranteeing a minimum price to producers and by direct payment of a subsidy for crops planted. Reforms of the system are currently under way, including a phased transfer of subsidy to land stewardship rather than specific crop production. Detailed implementation of the scheme varies in different member countries of the EU. Subsidies were generally paid on the area of land growing a particular crop rather than on the total amount of crop produced. Current reforms of the system now under way are phasing out specific crop subsidies in favour of flat-rate subsidies based only on the area of land in cultivation and for adopting environmentally beneficial farming methods. This will reduce, but not eliminate, the economic incentive to over-produce.[14]

The EU agricultural policy encounters strong criticism—from conservative and neoliberal economists arguing that subsidies are against the will of the free market, and from a developing-country perspective arguing that the CAP promotes poverty in developing countries, by artificially driving down world crop prices. Agriculture is one of the few areas where developing countries have a competitive advantage and could actually develop their own economy and therefore more self-sufficiency. Furthermore, many economists believe that the CAP is unsustainable in an enlarged EU. Criticism has also been directed towards its support for particular sectors, such as meat, dairy and sugar production, in part for its unsustainability and in part for nutritional concerns. The inclusion of 10 additional countries in 2004 has obliged the EU to take measures to limit CAP expenditure. Even before expansion, the CAP consumed a very large proportion of the EU budget. Considering that a small proportion of employment, and a relatively small proportion of the GDP comes from farms, many consider this expense excessive.

2.1.2.4.2 From agricultural to consumer-oriented policy

For decades, policy related to the nutrition of the population was of much less importance than agricultural politics. Postwar food security concerns allowed the economic interests of the agricultural sector to dominate for a long time, more than consumers' interests. A fundamental reorientation is happening in many countries, also in reaction to the BSE crisis in 2001 (e.g. see Halkier and Holm 2006). Food safety issues and consumer protection are becoming more relevant. For example, one of the most important goals of the 'new agricultural politics' in Germany was to increase the share of organi-

13 See ews.bbc.co.uk/2/hi/europe/4407792.stm, accessed 18 July 2009.

14 See ec.europa.eu/agriculture/healthcheck/before_after_en.pdf, accessed 15 April 2010; and Wikipedia: en.wikipedia.org/wiki/Common_Agricultural_Policy, accessed 18 July 2009.

cally cultivated area to 20% of all farmland by the year 2010. The state also aimed policy measures at new actors: from the 'iron triangle' of agricultural politics, agricultural interest groups and agricultural administration to the 'magic hexagon', comprising consumers, farmers, the fodder industry, the food industry, retailers and the state. In the UK, as a response to the food crises, the Strategy for Sustainable Farming and Food was set up by government to increase not only the economic but also the environmental and social sustainability of the sector (see Chapter 3). Policies have been open to question, and new groups (such as organic farmers and retailers) have become involved. Even though references to consumers have become rhetorically much more prominent, it is, however, doubtful how much this shift really represents a shift in policy-making towards more consumer power (see Bergeaud-Blackler and Ferretti 2006; Kjærnes *et al.* 2007; McMeekin *et al.* 2002). In that regard, the situation is also highly variable across Europe.

2.1.3 Key systemic sustainability issues and problems

No common definition and internationally accepted criteria system for sustainability of food exists. There are many different definitions mentioning often contradictory issues, the biggest contradiction being that food and eating per se should be *enjoyable* and not primarily efficient or eco-efficient. Thus it is by no means sufficient to define, for instance, a functional unit of intake of calories and nutritional factors, then use life-cycle assessment (LCA) to work out which provision system and which food products deliver the functional unit in the most efficient and environmentally friendliest way. Taste, health, culture and well-being are important factors in food consumption too.

In line with the definition of sustainable development, most definitions of sustainable food mention the three dimensions of sustainability—people, planet, profit—as well as issues such as:

- **Social sustainability**: food security, health and food safety, hunger and obesity, local versus global production and consumption, quality of life, skills and knowledge of citizens, protection of local culture and wisdom, fair trade and fair wages and labour conditions
- **Environmental sustainability**: land use, emissions, pesticides and artificial fertilisers, hormones and antibiotics, energy use and CO_2 and methane emissions, diversity and GMO issues, animal welfare
- **Economic sustainability**: feasibility of agricultural and production systems, subsidies and production quotas, profit and power distribution in the food chain, efficiency and quality, value for money

Accordingly, **sustainable agriculture** is a way of producing or raising food that is healthy for consumers, does not harm the environment, is humane to workers, respects animal welfare, provides fair wages to farmers and supports and enhances rural communities (see also the pragmatic definition of the Rainforest Alliance, presented in Chapter 11, Table 11.1, pages 196-97).

According to von Koerber *et al.* (2004) **sustainable nutrition** is defined by the following aspects:

- Enjoyable and easily digestible foods
- Preferably plant-based foods
- Preferably minimally processed foods
- Organically produced foods
- Regional and seasonal products
- Products with environmentally sound packaging
- Fair-trade products

In conventional agriculture often the definition of sustainability is a little different compared with definitions originating from alternative sustainable agri-food movements: first, the economic demands of commercial efficiency of the industry are driven by the market (and do not depend on subsidies); second, social sustainability more generally refers to the *acceptability* of the industry and 'overall human welfare', including education and training, equal opportunities, the health and safety of workers and consumers, and animal welfare; last, environmental sustainability may be interpreted as *improving* environmental performance, including the landscape, biodiversity, soil, air quality, water quality and waste management, and, increasingly, the connections between climate change and agriculture (see Chapter 3).

In fact, production, processing, transportation, packaging, preparation, consumption and disposal of food today cause major negative environmental impacts. Experts estimate that almost half of humankind's impacts on the natural environment are related to food production and consumption (e.g. see Jongen and Meerdink 1998, see also Chapter 12).

In the following sections a selection of sustainability factors related to the agri-food domain are discussed more in detail.

2.1.3.1 Environmental aspects

2.1.3.1.1 Exploitation of natural resources

This includes the exploitation of natural resources up to the extinction of species, including the diminishing fish population through over-fishing, land erosion and soil degradation through intensive farming and so on.

2.1.3.1.2 Life-cycle impacts of food products

Environmental LCAs of food products show that, normally, the main environmental impacts related to food (from pollution to resource consumption and soil erosion, etc.) are caused during the primary production stage (growing crops and raising cattle), but for energy use the most important contributions occur at other stages of the life-cycle, such as production, storage (cooling) and transportation. Packaging contributes around 5% to total energy consumption (although it can be higher with complex packaging units), and transport is responsible for around 3% of total energy demand (though the percentage is higher if air transport is used; see Eberle *et al.* 2006; Krutwagen and Lindeijer 2001). Very important are food losses, which occur at every link in the food chain and which represent significant energy impacts for each functional unit (i.e. for each meal consumed; see Tempelman *et al.* 2006). In the UK, for instance, one-third of the food bought for home consumption is wasted—that is 6.7 million tonnes annually. Food waste costs the average UK family £420 (€530) a year. Elimination of the unnecessary greenhouse gas emissions that this wasted food produces would be equivalent to taking one in five cars off UK roads. Food waste is not just a problem in industrialised countries: in developing countries up to 40% of food harvested can be lost because of problems with storage and distribution (Defra 2002).

2.1.3.1.3 Food products and global warming

Around 15–20% of all greenhouse gas emissions are connected with food (especially meat and cheese) and agriculture, according to the UN Food and Agriculture Organisation (FAO)[15] and other studies (according to the UNFCCC, deforestation on its own [not caused by farming alone] accounts for up to 20% of global greenhouse gas emissions[16]). Food products with high global warming potential are cheese, followed by meat and eggs. Causes for global warming in food production are, especially, the use of fertilisers, transport, cooling, a high degree of processing, highly packaged products, animal-based (cattle and pigs) products (from methane emissions) and nitrous oxide from manure. According to a study by the German Öko-Institute, organic products are a little better than conventional products in this respect, causing around 6–15% less CO_2 and methane emissions, increasing CO_2 absorption in soil, resulting in less transportation and so on (Fritsche *et al.* 2007). The best way to reduce the global warming effects of food would be to change cooking and eating habits towards more dishes with no or little animal products and to stay away from food products from overseas that are transported by air (Thielicke 2007). Buying seasonal and regional products, doing the shopping by bike, buying fresh and unprocessed food and preparing it with little cooking time are also positive options for consumers.

Another connection between global warming and food is that climate change consequences—such as more severe storms, draughts, floods, water scarcity and other natural disasters—cause food production losses and are highly problematic for food production.

2.1.3.1.4 Animal welfare and biodiversity

Animals and nature have become commodities, raw materials for an industrial production system, which has led to a mainly functional and economic perspective of production with regard to the natural environment. How bad we treat the animals that we eat and our natural environment, and how much we reduce biodiversity, which is very important for evolution and the possibility for the world's ecosystem to adapt and react to changes, are increasingly important concerns of consumers and thus policy.[17] The EC regulations on animal transportation in Europe and activist groups speaking out for ani-

15 See www.fao.org, accessed 18 July 2009.

16 unfccc.int/press/fact_sheets/items/4981.php, accessed 15 April 2010.

17 See for example the Convention of Biological Diversity, with their second meeting in April 2010 in New York; www.cbd.int, accessed 15 April 2010.

mal welfare, or the strong rejection of the majority of European consumers of GM substances in their food are signs of this.

2.1.3.2 Social aspects

2.1.3.2.1 Food security

The term 'food security' describes the 'physical and economic access to sufficient, safe and nutritious food' (FAO, cited in Defra 2006: 51), as opposed to 'food safety', which means food that is safe to eat.

Although, over the past 30 years, worldwide agricultural output has increased faster than the world's population (FAO 2002), the WWF states that, according to the Living Planet Index and the Ecological Footprint, 1.8–2.2 times the Earth's natural biomass productivity would be required for sufficient food production under current consumption patterns. Water will be the major constraint in many developing countries, and the most productive cereal areas in North America, India and China will soon be approaching biophysical limits (Tempelman *et al.* 2006).

Meat products have yields as low as 0.2 tonnes per hectare (t/ha), whereas grain yields range from 4.4 to 6.8 t/ha. Thus around 20 people can be fed from 1 hectare of rice and potatoes, whereas only around 2 people can be fed from beef or lamb per hectare. So the already high demand for meat in industrialised countries and explosive new demand for meat in developing countries will require an exponential growth in land use (WHO/FAO 2002). In this context it is important to point out that the problem is not the growing population but the unsustainable diets in industrialised countries (and increasingly in emerging and developing countries), promoted also by agricultural policies.

Therefore one strong requirement for sustainable agriculture and a sustainable food production and consumption system is to eat less meat, replacing some of the crops grown for livestock feed with cereals and horticultural crops for direct human consumption and to ensure distribution of the foodstuff to regions where food production is threatened.

2.1.3.2.2 Food safety

As already discussed, in a more complex production chain food safety issues are more challenging to regulate and control. A problem may become more difficult to identify and it may affect many more people. Modern complex systems generally employ advanced technological and organisational tools to handle such problems. But failures may happen and there are actors who try to earn quick money by not following the rules. Trust in these systems has therefore become a much more critical issue.

2.1.3.2.3 Fair trade issues: 'trade not aid'

Outsourcing production to wherever it is cheapest creates what is often referred to as a 'race to the bottom'. Globalisation weakens national governments' authority to enforce legislation in industries and introduce new labour laws. The unbalanced trading relationships between poor, small-scale producers in the South and big multinational corporate buyers from the North has created unpredictable and low prices for commodities such as coffee. Commodity prices have in some cases been sinking to less than the level of production costs (FTAO 2005). The unorganised small-scale farmers in the South are dependent on middlemen in the form of buyers, brokers, wholesalers and retailers, who receive most of the profit from end-sales to consumers. Started by Christian organisations in the 1950s as partnerships between non-profit importers and retailers in the North and small-scale producers in developing countries, the fair trade movement has entered mainstream markets. The most common fair trade products are coffee, bananas, roses, orange juice, tea and chocolate. Market shares are around 1% and growth rates have been between 20% and 30% per year since the beginning of the 21st century (see Chapter 10). But fair trade concepts are difficult to control and may still leave too much of the extra payment to middlemen.

2.1.3.2.4 Social accountability and corporate social responsibility

Another way of influencing the behaviour of companies from industrialised countries with respect to low-labour-cost countries is to agree on standards for labour conditions and other social aspects, such as child labour, working hours, minimum wages and so on. A standard for managing ethical workplace conditions throughout global supply chains was established in 1997 by Social Accountability International (SAI) via 'Social Accountability 8000' (SA8000) (www.sa-intl.org, accessed 16 June 2009). A new standard for corporate social responsibility (CSR) from the International Organisation for Standardisation (ISO), Social Responsibility 26000 (ISO 26000), is being developed and will be published in 2010.[18] Also, generic initiatives regarding sustainability issues have been developed at the branch level: for example, with reference to marine resources.

2.1.3.2.5 Cultural diversity

Although less researched, the issue of cultural diversity is gaining in importance. The cultural erosion caused by multinational food producers and retailers offering unified food all around the world causes counter-movements such as Slow Food (Chapter 8), and labels of local food specialities are gaining increasing importance in some countries and with some consumer groups.

2.1.3.2.6 Information and involvement, transparency and trust

With very complex global production and retail chains the transparency of the agri-food system for consumers is lost. Thus it is very difficult for consumers to understand and take responsibility for the consequences of their food choices. There are two major tendencies to counter this, either to produce labels that are independently certified and trustworthy, or by bringing consumers (again) in more direct contact with production (facilities and farmers). Involving consumers more in farming, food processing and preparation reintroduces knowledge about healthy food and nutrition that was lost along the way to industrialisation and convenience food.

18 See www.indianet.nl/pdf/briefing_iso_sr.pdf, accessed 18 July 2009; and isotc.iso.org/livelink/livelink/fetch/2000/2122/830949/3934883/3935096/home.html?nodeid=4451259&vernum=0, accessed 15 April 2010.

2.1.3.3 Economic aspects

2.1.3.3.1 Distribution of income in the food chain

North–South divide

World market prices for coffee, rice and other commodities are highly volatile and often below the costs of production. A stable price, which covers at least production and living costs, is an essential requirement for farmers to escape from poverty and provide themselves and their families with a decent standard of living.[19] At the same time, consumers in the South with low incomes are normally the first to suffer from rising world market prices for food. The recent debate about biofuel production versus food production and consequent increasing food prices reflects this phenomenon. In many European countries there are ongoing debates about the WTO and whether the trade system contributes to relieving or maintaining the poverty of people in the South.

Within Europe

The prices for farm products in Europe are artificial and influenced by agricultural policy and the power distribution in the system, where processors and retailers often have more power in the value chain and are able to dictate prices. This leads to raw material prices (such as for milk) that are below the production cost, to protests by farmers and to instability in the system. Thus long-term business and economic stability in the agri-food system is threatened.

2.1.3.3.2 Value for money, efficiency and quality

Through efficiency, industrialisation and rationalisation, as well as lowering of prices for farm products, the prices for food all over Europe are moderate, with the new member states (the Eastern European states) having the lowest prices for food, Greece, France and Germany being in the middle, and Iceland, Norway, Switzerland and Denmark at the top end of the list (Borchert and Reinecke 2007). Food prices, of course, should be discussed in relation to national levels of purchasing power, but they should also be related to degrees of social inequality. Interestingly, Denmark is also a country with a high consumer acceptance of organic food. In Germany an intense public discussion was started about retail prices being too low to deliver high-quality and safe food products (Thielicke 2006).

2.1.3.4 Conflicts and trade-offs

In the list of sustainability issues discussed above we find conflicting issues and open questions such as:

- Is it more sustainable to prefer local and seasonal products or fair-traded food products from far away even though they imply more food miles?

19 See www.fairtrade.net/impact.html, accessed 18 July 2009.

- Is it more sustainable to buy organic products or fair trade food, and why do they have separate label systems? Why is not all organic food also fair and all fair-traded food also organic?
- Is it more important to 'green' the mainstream food industry or to support small-scale sustainable niche producers?
- Is it better to enforce animal welfare regulations or to concentrate on efficiency and volume of meat production and reduction of methane emissions?
- How could it be possible to reduce the proportion of meat and dairy in our diets despite the common perception that meat consumption is part of quality of life and a sign of wealth?

We will discuss and try to answer some of these issues later on in this book.

2.1.3.5 Cleaner production in mainstream agriculture

The predominant agricultural system in Europe is an industrialised intensive agriculture leading to negative environmental impacts, as described above. Thus, increasing the sustainability of the current mainstream production and consumption system of food is needed urgently. Cleaner production at farms will range from restriction of use of chemicals in favour of shifting to biological pest management, through no-till and low-till agriculture and crop rotation to high-tech forms of farming such as precision farming (which relies on observation, impact assessments and timely strategic responses to fine-scale variation in the components of the agricultural production process).

2.1.3.5.1 No-till and low-till agriculture

Intensive tillage in agriculture degrades the fertility of soils, causes air and water pollution, intensifies drought stress, destroys wildlife habitat, wastes fuel energy and contributes to global warming. Consequently, most tillage-based systems are not sustainable in the long term because of the declining soil quality caused by soil tillage. In contrast, research and farmers' experience indicate that continuous no-till is the most effective and practical approach to restoring and improving soil quality. With continuous no-till, soil organic matter increases, soil structure improves, soil erosion is controlled and, in time, crop yields increase substantially from what they were under tillage management, because of the improved water relations and nutrient availability. A major obstacle that farmers often face with a change to continuous no-till is overcoming yield-limiting factors during the first years of no-till practice, such as residue management and increased weed and disease infestations. However, many problems during the transition are temporary and become less important as the no-till system matures and equilibrates. The use of crop rotations and cover crops may help reduce agronomic risks during the transition years.

Farmers switching to continuous no-till must often seek new knowledge and develop new skills and techniques in order to achieve success with this new and different way of farming. Answers to their questions are urgently needed and research can play an important role here. Research by institutions and farmers continues into developing organic no-till farming methods that utilise the rolling or crimping of cover crops and diverse crop rotations to suppress weeds, insects and diseases. Also, the reintroduction of horses for tilling has been discussed (e.g. Kendell 2005).

One additional positive aspect is the potential for carbon sequestration in the no-till crop fields: by reducing tillage, leaving crop residues to decompose where they lie, and growing winter cover crops such as grains or alfalfa, a farmer can slow down carbon loss from a field while contributing to carbon transfer from the atmosphere to the soil. Already, some large energy corporations are willing to pay farmers to engage in conservation tillage and use their agricultural land as a carbon sink for the power generators' emissions. This helps the farmer in several ways, and it helps the energy companies meet demands for reduction of pollution (Derpsch 2005; Kuepper 2000).[20]

2.1.3.5.2 Precision farming

This computer-aided farming method is applicable to every form of arable farming and is characterised by the 'differential' treatment of field variation as opposed to the 'uniform' treatment that underlies traditional management systems. In precision farming (or precision agriculture) decisions on resource application and agronomic practices are improved to better match soil and crop requirements as they vary in a field and are based on the use of location and crop-monitoring data in the decision-making process.[21] Thus IT is playing an important role in this form of agriculture. The combination of new sensors, on-board computers, electronically controlled machinery, standardised interfaces and both privately and publicly available databases and GPS provides the basis for comprehensive information-driven plant production.

The techniques of precision farming enable field-part specific, location-adjusted crop production, making more efficient use of resources while simultaneously offering potential for increased economic returns. Furthermore, the image of farming and agricultural services and the jobs offered there can be modernised and made more attractive. In addition, precision farming offers potential for the improvement of information flow along the food supply chain.[22]

2.1.3.6 Sustainable niche food concepts: market developments in Europe

2.1.3.6.1 Organic farming and processing in Europe

'Organic agriculture' is often used as a synonym for 'sustainable agriculture'. There are different definitions of 'organic', some stricter, some more pragmatic. According to IFOAM[23] organic agriculture is based on four principles:

20 See also en.wikipedia.org/wiki/No-till_farming, accessed 18 July 2009.

21 See for example the webpage of the Australian Centre for Precision Agriculture at www.usyd.edu.au/agriculture/acpa, accessed 15 April 2010.

22 See also the German research project *pre agro*: www.preagro.de, accessed 18 July 2009.

23 International Federation of Organic Agriculture Movements (IFOAM): www.ifoam.org, accessed 18 July 2009.

- The principle of health: sustain and enhance the health of soil, plant, animal, human and planet
- The principle of ecology: base agriculture on living ecological systems and cycles; work with them, emulate them and help sustain them
- The principle of fairness: build on relationships that ensure fairness with regard to the common environment and life opportunities
- The principle of care: manage agriculture in a precautionary and responsible manner to protect the health and well-being of current and future generations and the environment

The benefits of organic agriculture are, for instance:

- Reduced environmental impact from agriculture (e.g. reduced risk of pesticide residues and of nitrate in groundwater)
- Healthier food because of the lower pesticide residues and nitrate and the higher content of secondary metabolites
- Improved biodiversity in agricultural fields
- Food that does not include GM ingredients
- Better-tasting products (this point is still being discussed and is difficult to prove)
- More ethical animal husbandry
- The opportunity for food exports
- Strengthened regional development with regional interaction along the product chain from field to table

In Europe the production of organic products is regulated by council regulation EEC Number 2092/91 on organic production of agricultural products and foodstuff.[24] This regulation defines a minimum requirement for cultivation, production and marketing and sales. Products referring to this regulation are able to get the official organic label or Bio-label. There are in addition a large number of national or regional organic labels, but the IFOAM principles, or at least parts of them, often constitute a common reference. Although the EC requirements often are lower than the national requirements of European countries that have a long history in organic farming, most countries have combined pre-existing national labels with the European certification.

Approximately 31 million hectares worldwide are now grown organically. The distribution is shown in Figure 2.1 (see also IFOAM 2007).[25]

Organic food often costs more than conventional equivalents because of the lower yields, expensive materials and more labour-intensive production (Belz and Pobisch 2005; GfK Group 2007). Even if the current market development for organic food prod-

24 europa.eu/legislation_summaries/agriculture/food/l21118_en.htm, accessed 15 April 2010.

25 See also www.organic-world.net, accessed 18 July 2009.

FIGURE 2.1 Distribution of organic agricultural land globally and in Europe

Mha = millions of hectares

Source: FIBL and IFOAM 2009 (top) and FiBL and ZMP 2009 (bottom). Figures for 2007, documented in Willer and Kilcher 2009

ucts is positive, consumers' willingness (and capability) to pay premium prices is limited. One major reason mentioned in surveys is the lack of credibility of organic products and labels; another is that availability is still too limited in most European Countries (Torjusen *et al.* 2004).

As organic farms have to be a specific minimum geographical distance from conventional farms, and especially from farms with GM crops, and as organically grown products have to be processed separately from conventional products to avoid mixing, and should be retailed in another way—for example, more directly between producer and consumer and by communicating the qualities of organic products and receiving a premium price—separate value chains for organic products have been established. Many farmers and producers founded cooperatives and marketed their products under special labels, such as 'Demeter', 'Bioland', 'Naturland' in Germany, and most organic food was retailed directly between producers and consumers or via special health food stores and cooperatives. Now this separation, more or less present in different European countries, is disappearing and organic food products have been established in conventional food retail businesses and even in discount markets with an increasing variety of products (GfK Group 2007). Organic food has taken on a more strategic role in the competition of large retailers; the actors in the organic business, previously taking a more ideological approach, are now becoming more professional in terms of their self-organisation and cooperation within agricultural policy, retailing and marketing. At the same time there has been a consolidation of the industrial processing of organic food. However, parallel to this, the alternative forms of distribution are becoming more common, such as farmers' markets and box schemes, aiming at returning to a more direct relationship between producers and consumers. These distribution forms often sell mostly organic (and local) food.

In the future, as the size of organic farms continues to increase, a new set of large-scale considerations will eventually have to be tackled, especially by large organic farms that rely on machinery and automation and purchased inputs. One fundamental question is: will organic farming lose its sustainability when it increases in scale, following the same mechanisms as conventional agriculture, or will it be possible to keep the benefits of a small-scale diverse regional organic agricultural system in Europe while increasing market shares and going mainstream? There are many critical voices, claiming that standards are being lowered and delimited. For example, while small shops and markets often get their supplies locally, large supermarkets may use their global supply chains even for their organic lines (Guthman 2004).

2.1.3.6.2 Local and seasonal versus global food

Seasonality of food almost disappeared from European retailing through complex systems of global sourcing. To fight the disconnection of producers and consumers of food, regional and seasonal food provisioning systems and concepts have been re-established in many different European regions, such as community-supported agriculture or Slow Food (see Chapter 8). Furthermore, discussion on 'food miles' (that is, the transportation) in connection with globalised food chains is taking place (see Chapter 4). In particular, transportation of food by air is criticised, but studies have found that the most significant contributor to CO_2 emissions is domestic transport on roads, from imports or suppliers via hubs and out to the shops. The second most important contribution comes from people using cars for shopping, exacerbated by supermarkets being situated at the margins of cities and residential areas (Defra 2006).

The subject of local food is also connected to efforts to promote food produced locally, from a special region or a country with labels of authenticity (such as Protected Designation of Origin [PDO] and Protected Geographical Indication [PDI]) and traceability schemes. However, this is another dimension of locality of food, as such labels of origin do not guarantee few 'food miles'. Often, these local foodstuffs are traded globally, and production chains can be globalised (for example, fish caught in the North Sea is brought to China for processing because of lower labour costs and are then transported back to a Norwegian processor and out to European supermarkets).

Consumers' role in this context could simply be to substitute products from afar with products from their region or country, if they are available, by selecting seasonal products from the region or at least the same continent, and by purchasing closer to the local or regional producers by using sustainable means of transportation (for example, by shopping at farmers' markets by bike or by signing up to regional food subscription schemes).

Unfortunately, these kinds of regional sourcing concepts contradict the political, economic and even social agenda of free trade and hinder developing countries from developing their economy by exporting agricultural products to Europe. Moreover, local production is not always more environmentally sustainable, such as year-round production of tomatoes in greenhouses, which depends on large quantities of fossil fuels, or intensive animal farming in The Netherlands compared with meat from extensive farming in South America. Finally, local sourcing alone cannot feed the populations in the many regions and countries of Europe or the world. Thus thorough analyses are necessary to understand the environmental, social and economic implications of local and seasonal compared with global food sourcing concepts. All these issues must be studied specifically and locally.

2.1.3.6.3 Fairly traded food products

As a result of broad public criticism of globalisation issues an increasing group of consumers are interested in supporting fair trade through their purchases. Motivated by this movement an increasing number of cooperative initiatives from actors in industrialised countries and production cooperatives in developing countries are now established and successful (see Chapter 10). The most commonly available fair trade products are coffee, bananas, roses, orange juice, tea and chocolate. Market shares are around 1% on average. In 2005 the combined sales turnover of fair trade was more than €1 billion, and growth rates have been between 20% and 30% since the year 2000. The total value of products sold under fair-trade labels in 14 European countries is about €597 million (Krier 2006).

All of these sustainability issues are discussed in more detail in the case study chapters and in the final conclusions of this book but first of all, in the next section, we introduce the agricultural and food production and consumption system with its main actors and opportunities for SCP.

2.2 The agri-food system (regime) and windows of opportunities for sustainable consumption and production

2.2.1 The agri-food production–consumption chain and interlinked practices: the 'regime'

Producers and consumers (together with other actors) share the responsibility for the shift to a more sustainable production–consumption system. Very often in SCP discussions the 'chicken and egg' problem occurs. Producers claim: 'we can produce only what our customers demand!' Consumers say: 'we do not have an option; we can buy only what companies offer and are influenced by advertisements!' These positions are understandable from an individual company's or consumer's perspective. However, if they were true nothing would ever change—but changes happen all the time. In this section we discuss which actors have which roles, scopes and influences in the agricultural and food system, thus trying to identify where the sensitive points in the system are that might lead to changes.

2.2.1.1 The production–consumption chain in the agri-food domain

Figure 2.2 shows in a very simplified way the agricultural and food production–consumption chain, or value chain, with its main actors and activities. All actors in the chain have to comply with food safety regulations, which are maintained by food control organisations with the exception of end-consumers.

Besides the actors directly and indirectly involved in the production and consumption chain of food, there are further actors and organisations with important influences on the agri-food system such as political and governmental institutions, non-governmen-

FIGURE 2.2 Overview of the actors in the food production–consumption chain and their primary roles

Source: Ursula Tischner

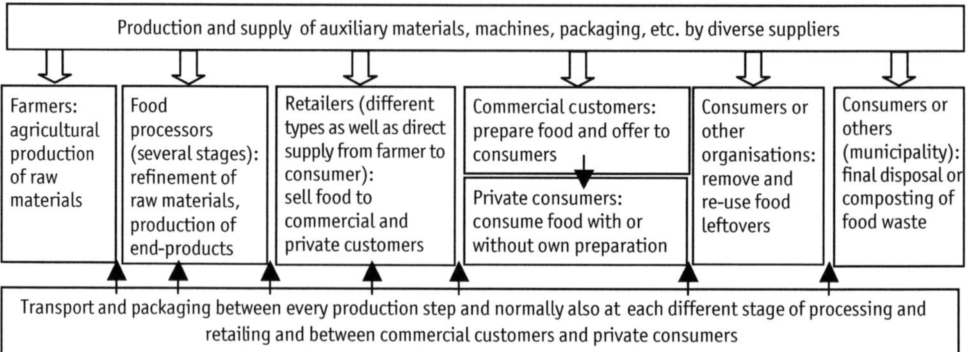

tal organisations (NGOs), media, intermediate organisations (such as lobbying organisations or consumer consultancy groups) and so on.

Thus we can distinguish a triangle of relations and influences of three main groups of actors (see Fig. 2.3), which we will discuss in more detail:

- Actors in the food provisioning system (see above)
- Households and consumers
- State and regulatory authorities and civil society, including policy, intermediary organisations, NGOs, media and research

FIGURE 2.3 Triangle of actors in the agri-food domain

Source: Kjærnes *et al.* 2007

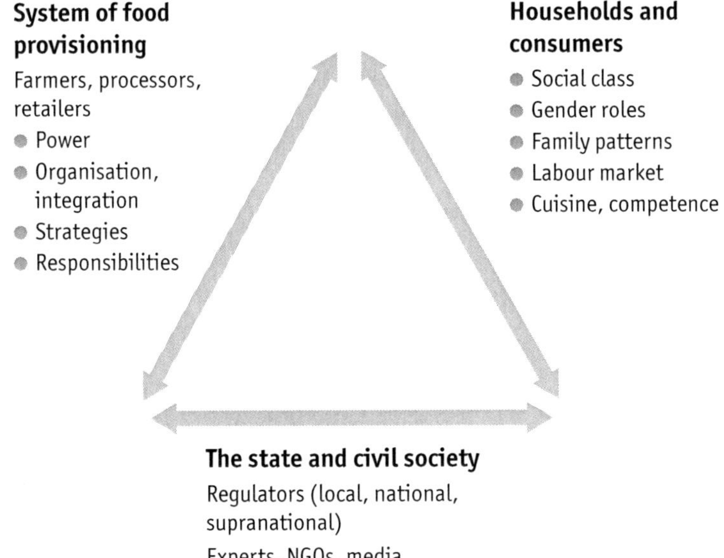

2.2.1.2 Agricultural production

Agriculture is the single biggest user of land in the EU, with farmland covering 41% of the EU-25 landmass. Agri-industries create essential services for society, including the production of food, energy and jobs. Europe's bio-economy (that is, the broad field of agriculture, forestry, fisheries, aquaculture, bio-based handling of resources and rural development) has an estimated annual turnover of more than €1,500 billion and employs 22 million people. Furthermore, farming still remains the most important influence in shaping Europe's rural landscape, as well as being a major determinant of biodiversity in the EU (DG Environment 2008).

As a general trend there is a process of concentration in the agricultural and food industry: fewer and fewer farmers manage larger and larger farms, increasing profitability and efficiency and producing higher yields per hectare. Production is more and more industrialised; for example, there is a strong and direct correlation between the efficiency and profitability of dairy farms and the size of the herd and scale of milk production (Chapter 3).

Historically, there is a strong organisation of individual farmers and farmers' cooperatives taking care of bundling and processing the conventional agricultural products. This makes it difficult for new (organic) farmers and farming concepts to work in the existing infrastructure. For instance, organic producers needed to set up their own cooperatives and infrastructure because organic production must not be mixed with conventional production.

Some 52% of all EU farm holders are over 55 years old and only 7% are under 35 years old, and few young people are interested in becoming farmers (for example see Muller 2005). Dewick, Foster and Webster (Chapter 3) highlight, for instance, that in research on the changes in the UK agricultural system towards more sustainability all responding experts agreed that the changes required in the traditional industry are very great, at an individual (personal) level as well as at a business level.

2.2.1.3 Processing and food industry

The European food processing industry is the third biggest EU industry, employing some 2.7 million people with more than 26,000 companies across the EU. It is the world's leading exporter of food products and has a positive trade balance. This sector is characterised by diversity in its types of activity and in the end-products manufactured. The products covered vary from bakery, pastry, chocolate, spirits, mineral water, beer, confectionery products to modified starches or different food preparations, in addition to the processing of dairy and meat products. More than 70% of the agricultural goods produced in the EU are transformed into food industry products. The most important destinations for exports are the USA, Japan, Switzerland, Russia, Canada and Norway.[26] Also, in food processing there are fewer and fewer companies sharing the market; thus power is concentrated in fewer hands in the system.

2.2.1.4 Retailers

The same situation applies to the retail sector: fewer and fewer larger and larger retail chains share the market and compete mainly through a low price policy. Small stores can survive only if they have a special unique selling point (USP)—that is, special-quality products or services—and are innovative or very much embedded in local culture and traditions. Often, big retailers can dictate prices to agricultural producers and processors in the current system. Studies in the UK showed that only 26% of supermarket checkout prices go to farmers, compared with approximately 50% only 50 years ago (Smith *et al.* 2005).

26 See ec.europa.eu/enterprise/sectors/food/index_en.htm, accessed 15 April 2010.

Although large supermarket chains are gaining increasing proportions of the food market, retailing is still highly diverse across Europe. Not only do the proportions of supermarket shopping vary from nearly 100% in the Scandinavian countries to less than 50% in several southern European countries, the supermarkets are also very different. Although most chains compete on low prices, others compete on quality and a wide selection of goods. In some chains other issues, such as fair trade, animal welfare and organics, have become more important recently, as part of brand management and segmentation strategies. Moreover, whereas some supply chains are characterised by retailer-driven integration, others are dominated by large manufacturers, yet others by fragmentation. All of this has substantial implications for market drives towards sustainability as well as for relationships with consumers (that is, the choices people have and the responses to their demands).

2.2.1.5 Consumers

In several fields of food-related policy consumers are attributed agency and are seen as responsible, through their choices, for a number of societal issues: health, food quality, animal welfare and environmental sustainability. Policy papers explicitly refer to consumer choice and consumers' own responsibility through 'informed choice' and labelling strategies (CEC 2000; Reisch 2004).

We argue that this model is insufficient and even misleading in understanding the roles and responsibilities of consumers. It is evident from several studies of environmental and ethical issues that there are vast variations in consumption practices—across countries and between social groups. An individual, utilitarian approach is of little use to analyse such variations, nor is it helpful in understanding the stability and consistency in food choice often found within national and cultural contexts—or the large-scale shifts in food choice that also occur.

Food consumption must be understood as socially created sets of practices, including purchasing, cooking and eating (and perhaps also storage, gathering, home production, etc.). Food consumption is a form of *social* action and cannot be reduced to a conscious decision at the point of purchase. We have to consider that consumption practices happen within and between social institutions such as the family, work and the marketplace and that these practices are themselves routinised and normalised. There are predictable societal patterns of behaviour related to food provisioning and consumption, emerging from social structures, norms and conventions, and formed by the particular contexts and situations within which consumption takes place. Food represents an intersection between public arenas and the private sphere, the collective and the individual. Meal structure and cuisine affect how people do their food provisioning, but the character of various forms of supply and governance structures also has a significant influence on people's expectations and actions.

Taken-for-granted routines are usually very stable, but Swidler (1986) argues that in some situations routinised practices suddenly become explicit and contested, and routines can intermittently break up. New and alternative, often ideologically justified, sets of practices may be established. These will, however, tend to be gradually routinised, eventually also tacit and taken for granted. Thus it is likely that organic food consumption (in various forms) will also tend to undergo this kind of routinisation and institutionalisation rather than remaining as a new kind of more 'reflexive' consumption. The practice of recycling has been described in this way. Schäfer *et al.* (Chapter 12) argue that life events such as parenthood, disease and retirement can be such points in life of consumers where routines break up and behavioural change in food consumption takes place.

European consumers generally seem positive about sustainable and organic food. Danish consumers might be leading here, with 93% of Danish households buying organic food at least once and 87% at least twice during 2002. However, it is probably less than 1% of the households that buy *only* organic food (see Chapter 5). Most of the consumption of organic food therefore does not take place among highly ideological and motivated groups of the population. Buying and eating organic food has become normalised among the Danes—as one option, but not as a consistent rule (Torjusen *et al.* 2004). Most regular consumers of organic food live close to large cities where the availability of organic food is much higher than in the countryside. Thus **retail structures** seem to be an important factor for sustainable food consumption (see Chapter 5).

2.2.1.6 Consumers' influences on the production side

While consumers are influenced by offers and advertisements, they also have an influence on the production side. The most direct influence of consumers on the food system is the choice of products and services they buy: 'consumers vote with their wallets'. Sales figures directly influence company behaviour. Consumers, however, can hardly influence directly what they want companies to offer. Furthermore, the choice of the point-of-sale, where consumers shop, has consequences for retail structures. These influences are always collective influences: only if a larger number of consumers do the same thing is such behaviour recognised by the production and retail system.

More indirectly, consumers are the 'objects' of market research and trend research by companies and their consultants. Thus trends in consumer behaviour that are made visible by experts or intermediary organisations, media and so on are signals that progressive companies take up and try to incorporate in their business development strategies and product policy and design. Thus, the more consumers prefer sustainable products and services and sustainability-oriented retail channels the more sustainable offers companies will provide.[27]

Recently, more and more companies are offering consumers more active involvement in product design or strategy development and innovation activities such as mass customisation, co-design, focus groups, innovation workshops and so on (see also Belz *et al.* 2007 in the SCORE! proceedings of Workshop 2).

27 This is not always fully correct. In Norway, the demand for organic food has increased significantly over the past couple of years, as recognised also by the retailers. Supplies are, however, far from sufficient to satisfy this demand when it comes to either total volume or type of products demanded. As another example, retailers in the UK are interested only in labelling certain meat product as welfare-friendly: namely, high-cost products where added profits are seen as sufficient (Roe and Marsden 2007).

2.2.1.7 Consumers' influences on policy-making

Most politicians want to be re-elected, and consumers are citizens and voters. Thus the more outspoken they are with their concerns in public, the more consumers are mobilised for a specific subject, the larger the pressure on politicians to act. As a result of the internet and ICT there are many well-organised consumer protest groups and watchdog organisations that channel and mobilise consumer attention on important issues and give them a forum to discuss those issues as well as options to become active. Such issues may relate to GMOs, food security and health aspects or fair trade, where public attention and collective mobilisation are seen as both a precondition for and an outcome of consumers' more politicised or strategic actions in the marketplace. However, the direct influence of consumers on political decisions, such as where tax income should be invested, is rather low.

2.2.1.8 Consumers' influences on sustainability in the agri-food system

Three levels are often outlined for changes to occur in consumption activities induced by sustainability concerns:

- Substitution of conventional with fair trade and/or eco-products without other behavioural change
- Changing the organisation of one's own consumption—for example, one's shopping and cooking behaviour—without affecting one's level of welfare
- Reduced consumption (and reduced welfare or improved welfare in the case of overweight persons)

Even without any special sustainable offers on the food market, consumers' activities can have a strong influence on the environmental impact of food production and consumption. In particular, four areas of consumer activities are relevant: transport (how people go to and from shops and where the shops that they use are located), storage (operation of refrigerators and freezers), preparation (boiling times, etc.) and wastage of consumable food (food purchased but not consumed, re-use of leftovers). Moreover, the composition of the daily diet has an impact. In particular, the reduced consumption of food items of animal origin is of major importance for the sustainability of diet. This may be seen as an organisational matter, but many people consider this as an effect on welfare and well-being as well.

2.2.1.9 Policy and world market influences on power distribution

The influences of the European agricultural policy on farmers and food have already been described in Section 2.1.2.4. In Chapter 3, Dewick, Foster and Webster clearly point out the catastrophic impacts of agricultural policy combined with market mechanisms on the dairy industry in the UK: milk quotas are dictated by policy, and prices are made by processors and retailers and are so low that most farmers cannot produce on a sound economic basis. Thus many farmers farm part-time and have to earn additional income in other jobs or even decide to step out of business.

However, recently things have been changing: the world market price for milk, especially milk powder, is rising because of globally increasing demand (especially from China and India). At the same time, climate change (for example, increasing incidence of droughts) reduces production volumes, and bio-energy production increases competition for land use. EU stocks are all used up, so prices for milk and milk products are rising again. From April 2006 to May 2007 the world market price for low-fat milk powder increased by 50%.[28]

That changes the power distribution in the system: so far, over-production in the EU has held food prices very low, sometimes even below the production price. Currently, scarcity of resource gives milk farmers new opportunities to ask for higher prices from dairies, which ask for higher prices from retailers, which will have to increase end-user prices. In the end, consumers will need to pay more for milk products in the shops (which of course will have an unequal social effect, depending on the purchasing power of various households and nations).

In Germany, the largest producer of milk in Europe, this effect is already changing the market: the shortage started with organic milk and expanded to conventional milk products. Now German milk farmers, to overcome the oligopoly of German dairies, are getting together in new production cooperatives and sell their milk directly to other European countries such as the Netherlands and Italy for much higher prices than they would receive in Germany (*Welt am Sonntag* 2007). At the same time the fight goes on between large discounters (who insist on dictating prices below the production cost), the large agricultural lobby organisations and the policy actors in Brussels (who may or may not raise the production quotas for milk).

The EC agricultural policy originally tried to stabilise prices by stocking products that were over-produced, to reduce over-production by setting production quotas for agricultural products, to increase exports by providing export subsidies to farmers and to protect the EU market against cheap foreign imports by the imposition of import duties. Now, in light of the new world market situation for milk and other food products, many farmers would be willing to live without subsidies from the EC and without production quotas, because they think an unsubsidised and unregulated market economy would offer better opportunities. They will be considering stepping out of the agreement on production quotas in 2015, when the next negotiations at the EC level take place (*Welt am Sonntag* 2007).

These processes seem to continue and even speed up. The outcomes are highly uncertain in terms of effects on sustainability as well as consumer welfare and social inequality. One possible outcome is, for example, claims for higher intensification of agriculture, with more use of artificial fertilisers, GM seeds and higher-yielding animals.

2.2.1.9.1 Food safety regulations

Besides the general function of policy in steering the agri-food system there are food safety regulations formulated by government and controlled by food control organisations. Such regulations are increasingly becoming harmonised across Europe and also globally (via the WTO agreement[29]). The enforcement of food safety regulations has

28 See Zentrale Markt und Preisberichtsstelle (ZMP): www.zmp.de, accessed 19 July 2009.

29 See www.wto.org/english/thewto_e/whatis_e/tif_e/agrm4_e.htm, accessed 15 April 2010.

changed in recent years, with more emphasis now being placed on market self-regulation and public audits. This type of regulation (including, for example, HACCP systems) fits well with the quality management of large and integrated enterprises, but implementation in small businesses and more informal distribution systems has proved much more difficult. This is a challenge for on-farm or small-scale processing, small shops and open food markets.

2.2.1.9.2 Public (green) purchasing and public–private partnerships

Another role of government is to use its purchasing power in the food market to influence the offers of companies: for example, the introduction of sustainable food into the canteens of schools, municipalities and hospitals helps to increase demand for more sustainable food products. There are several green public and private purchasing initiatives throughout Europe, with significant impacts on the strategies of producers and processors.[30]

2.2.1.10 Other societal groups

Other societal groups and their basic roles are described below.

2.2.1.10.1 Consumer organisations

Consumer organisations may be independent or supported by government. They have three functions: to inform and consult consumers, to connect consumers with government and even with companies and to raise public awareness of social and political issues related to consumption. Many such organisations have established activities in the field of food and nutrition consulting for consumers.

2.2.1.10.2 Media

The media generally inform and build public opinion. They report on food scares, social and environmental issues relating to food and are increasing public awareness about these issues.

2.2.1.10.3 Non-governmental organisations

NGOs not only lobby on specific issues but also can form partnerships with companies and inform policy. Many NGOs are active in the area of sustainable agriculture and food. There are NGOs exclusively dedicated to sustainable food issues such as animal welfare, and some of the sustainable food movements (such as the fair trade movement) have even been initiated by NGOs. Consumer organisations, test organisations and even labelling organisations may also engage in such activities, alone or in alliance with single-issue NGOs.

2.2.1.10.4 Test organisations

German Öko-Test and Stiftung Warentest are examples of watchdogs for product quality and company practice and inform consumers, the media and policy. They can be very

30 See for example www.epe.be/workbooks/gpurchasing/contents.html, accessed 19 July 2009.

influential. For instance, when German Öko -Test discovered pesticides in baby food sold in German stores the Spanish producer had to call back the whole range and the brand image was badly damaged.[31] Test organisations play a significant role in some countries, but such tasks of market discipline may be organised in different ways (for example, by public authorities) or be poorly developed in some countries.

2.2.1.10.5 Labelling organisations

Labelling organisations are independent (governmental) or organised by industry, certify specific product qualities, inform consumers, aim to increase credibility and trust, and give incentives to companies to improve production and product quality. For food there are so many different labels on the European market that it is very difficult for consumers to understand and to trust them. As well as the European Bio-label several national and regional eco-labels exist. In addition, there are different fair trade labels and nutritional labels, labels for functional food or the health aspects of food, general controlled quality labels (such as for meat) and so on. Currently under discussion are labels concerning air freight, food miles or carbon labels. Major issues regarding trust are the independence of the labelling organisation, especially from commercial interests, and the transparency of the label regarding the standards on which the labels are based as well as the monitoring of performance.

2.2.1.10.6 Research

(Sustainable) agriculture and food are traditional fields of extensive research. Important functions of research are to increase knowledge and understanding and to support policy decisions and develop (technological) innovations that improve the production–consumption system if taken up by companies and accepted by consumers. Regarding sustainable food consumption, social scientific research is also important, addressing issues such as social and political conditions for change in consumption practices, political consumerism, and power, interests and responsibilities in the food provisioning system.

2.2.2 Stability and leverage points in the system

The most important actors in the agri-food domain belong to one of the three groups: (a) the consumers and households, (b) the production and distribution system, including farmers, the food industry and retailing and (c) civil society, including public policy institutions within the area of food and agriculture, and NGOs, media and intermediate organisations.

When there is a kind of equilibrium between these three areas and types of institutions, the situation in the domain is relatively stable. Instability and change may have a number of different sources, such as shifts in power relations, technological innovation or new political and economic frames. Such changes will often be uneven, producing tension and dissatisfaction among one or more of the actors involved.

31 See www.oekotest.de/cgi/ot/otgs.cgi?suchtext=&doc=3875&pos=0&splits=0:1438:2660:4205:5850:6908:8682:9170, accessed 19 July 2009.

This dissatisfaction can be individual, such as a consumer's distrust of food safety, a farmer's dissatisfaction with the price he or she gets paid for his or her products, a parent's dissatisfaction that there is no healthy baby food on the market and so on. However, it is only if this dissatisfaction turns from an individual and private problem to become a social problem arousing public attention and collective mobilisation, such as several individuals jointly expressing their dissatisfaction or acting to overcome it, that this will have an influence on the system. This has happened in the past, with farmers dumping manure in front of the European Parliament, or with consumers starting 'green' food purchasing groups. How extensive this mobilisation needs to be to induce system changes depends on several factors, such as the size and power position of the actors and also how the social problem is associated with more general political conflict alignments. It needs considerable consumer mobilisation and public attention to convince policy or commercial institutions to change, but it is enough if one huge retail chain starts listing organic products for other retailers to follow. For instance, the announcement in April 2007 by Tesco—the UK's largest and most profitable supermarket—that it intends to establish direct contracts with 850 dairy farmers at a starting price of 22 pence per litre (about €0.28 per litre) milk has encouraged other supermarkets and milk purchasers to do the same (in 2009 Tesco formalised its sustainable dairy activities in the TESCO Sustainable Dairy Project [TSDP]). One important amplifier for these impulses is the public domain: for example, when mass media and leading persons in society take issues up and publicly talk or report about them. Role models such as celebrities, speaking out for animal rights, social rights or environmental issues, setting up their own foundations and projects are influential in this context.

ICT plays an important role in this mechanism because it enables 'normal' consumers to set up or join internet communities and start discussion and activism that can grow quickly and can have considerable influence (examples are blogs, flash mobs, activist groups organising demonstrations, etc.).

Labelling is one tool for consumers to act strategically in the market and is necessary also to organise information exchange along supply chains or enable better control mechanisms. However, they have limited impact with consumers. A large number of consumers do not know or do not trust eco-labels, and they need other more political and emotional communication around sustainable solutions to get excited about and 'fond' of them.

As with conventional products the special qualities of more sustainable products and the associated benefits for consumers have to be communicated in an understandable and transparent way. Unlike in pure and often counterproductive 'greenwashing' campaigns, the genuinely more sustainable products usually have a good story to tell and this can be expressed by designers and marketing experts in product communication and packaging design to help consumers understand and 'feel' the difference.

So-called 'cultural creatives' or LOHAS (lifestyles of health and sustainability)[32] can, furthermore, be very important trend-setting sub-groups in society that also influence

32 See for example www.culturalcreatives.org for the US view (accessed 19 July 2009) and www.creativiculturali.it/laszlo.htm (accessed 19 July 2009) for what Erwin Laszlo writes about the 'cultural creatives' in Europe. LOHAS is a title recently developed in the US for a cluster of consumers particularly interested in 'lifestyles of health and sustainability', discovered by marketing experts as an interesting target group with high incomes; see www.lohas.com (accessed 19 July 2009).

'normal' consumers. They can be addressed by sustainability-oriented companies as first-mover consumer groups, but this must happen in an adequate, intelligent and joyful way.

2.3 Preliminary conclusions: typical sustainable consumption and production strategies in the domain

From the trends and developments, sustainability issues and problems and the structure of the agricultural and food production–consumption system described above we can make preliminarily conclusions that there are several starting points for changes towards more sustainability in the agricultural and food system.

- Change can happen from the bottom up: for example, by consumer groups or individual companies starting an activity that is successful, gets recognised by others and motivates others to follow the example. Activities can be community-based experiments and niches such as community gardens or organic food purchasing cooperatives; they can be business initiatives such as farmers' markets, or sustainable food product innovations such as Bionade®, the highly successful organic lemonade in Germany; or they can be public–private partnership projects such as organic food for schools and so on. So, these bottom-up initiatives can start from conventional or mainstream commercial actors, from alternative commercial actors such as organic producers, from non-commercial actors such as schools or community groups and from alliances between commercial and non-commercial actors
- Change can happen from the top down: for example, when governments formulate legislation and regulations to restrict and control use of chemicals in agriculture, or when the European CAP includes a preference for more sustainable agriculture
- Change can be induced at the intermediate market level, when large powerful companies such as big European supermarket chains decide to source sustainable food products, to create a new food label, buy preferably from local suppliers or introduce advanced supply-chain guidelines on, for example, sustainable packaging

Table 2.1 gives an overview of typical strategies of sustainable production and consumption present in the agriculture and food domain today.

Important elements for up-scaling and multiplying the niche concepts and experiments or for amplifying the changes towards more sustainability of the system are described below.

Table 2.1 Typical sustainable consumption and production (SCP) strategies in the agri-food domain

SCP strategy	Example	Chapter in this book[a]
'Greening' mainstream industrial agriculture	Reduce and phase out those chemicals (pesticides, herbicides, fertilisers) with negative environmental impacts, cleaner production in food processing, low- and no-till agriculture . . .	Chapter 3
Localisation of food production and consumption, national self-sufficiency in food	Setting up and supporting local food production–consumption networks and activities, increasing local food production, reducing imports of food, reducing food miles, etc.	Chapter 4
Organic food going mainstream	Growing consumer demand and thus business success for organic food products, increasing availability of organic food in big supermarket chains and even discounters, supporting conversions of farms to organic by the new version of the Common Agriculture Policy	Chapter 5
Bottom-up localised production–consumption initiatives	Community-supported agriculture, purchasing cooperatives, organic and local food delivery services, farmers' markets, Slow Food, etc.	Chapters 6–8
Fair trade (and organic) solutions	Set up and support cooperative systems or initiatives by actors from industrialised nations with developing countries and regions, with production and trade taking place in a socially acceptable way and supporting the socioeconomic development of the region and protecting the environment	Chapters 9–11
Address especially first-mover target groups open to sustainability	Start with target groups in life-changing events, or so-called LOHAS, and offer them sustainable options that fit into their lifestyles	Chapter 12

a The item may be the main subject within the chapter cited or may be discussed as a part of that chapter.

2.3.1 Networking

Different actor groups should form new partnerships to create greater momentum for change. Consumers can legitimise stronger policy measures; NGOs can work with companies and policy on consumer information and motivation campaigns; research, together with companies and consumers, can develop new sustainable technologies and their applications; consumers, designers and producers together can co-design more sustainable products, services and even new business models. This may, however, challenge trust relations, and care has to be taken in order to keep such processes open and accountable. For example, a lot of criticism has been directed towards so-called 'conventionalisation' processes, the argument being that scaling up and networking means that important values and aims are lost on the way.

2.3.2 Right pace

The agriculture and food system is one of the most traditional and one of the oldest business sectors of humankind. Although it has undergone many changes over time, with massive industrialisation, product innovation and increased power concentration in a few big players in the system, the agricultural system overall still is relatively conservative and certainly needs long-term planning. It is not possible for farmers to react immediately to short product cycles, retail-driven product innovation and quickly changing consumer demands. Sometimes even the election cycles leading to new governments with changing directions in nutritional and agricultural policy can be too short for farmers to react in the way they do agriculture and the type of products they produce.

2.3.3 Strong regulations still needed

The globalised food industry with very complex supply chains needs strong national and international regulations and agreements to secure food safety, reduce the negative environmental impacts and exploitation of the workforce and so on. It is not possible to expect companies and consumers alone to carry all these responsibilities. Regulations are needed at two levels. First, there are basic standards for, for example, food safety, animal welfare and environmental impacts (such as pollution). Such standards have traditionally been developed nationally, but, gradually, international legislation has been put in place, especially regarding food safety, whereas other issues are more weakly developed (or even counteracted with arguments of representing barriers to trade). Second, regulation is needed to ensure accountability in the system of food production and distribution, so that consumers (and businesses) are not deceived. Many aspects of sustainability are not possible to control by inspecting the product, and complex systems offer opportunities for dishonest actors to try making more money by making unfounded claims. This is first of all a challenge in terms of enforcement.

2.3.4 Consumer education

European consumers need to realise that cheap food has its price—that is, an externalised cost—and they have to accept they will need to pay a bit more for healthy and sustainable food. The reasons for this have to be communicated more honestly by governments, companies, intermediate organisations, educators, media and communication experts and better customised to specific target groups than is done today. But this means that efficient information has to be accompanied by measures that ensure that these higher payments really are taken back to improve sustainability and do not end up mainly in the pockets of the 'middlemen'. Moreover, as food is a basic need and central to welfare, social inequalities in purchasing power as well as outright poverty have to be considered in the formulation of policies on sustainable food consumption.

More refined conclusions are discussed in Chapter 13, where the findings of the ten case studies in Chapters 3–12 are summarised.

Further reading

Banks, J., and T. Marsden (2001) 'The Nature of Rural Development: The Organic Potential', *Journal of Environmental Policy and Planning* 3: 103-21.

Brand, K.-W. (ed.) (2006a) 'Die neue Dynamik des Bio-Markts' ('The New Dynamic of the Organic Marketplace'), *Ergebnisband* 1 (Munich, Germany: oekom-Verlag).

—— (ed.) (2006b) 'Von der Agrarwende zur Konsumwende: Die Kettenperspektive' ("From Agricultural Turn to Consumption Turn: The Chain Perspective'), *Ergebnisband* 2 (Munich, Germany: oekom-Verlag).

Gronow, J. (2004) 'Standards of Taste and Varieties of Goodness: The (Un)predictability of Modern Consumption', in M. Harvey, A. McMeekin and A. Warde (eds.), *Qualities of Food* (Manchester, UK: Manchester University Press): 38-60.

Jongen, W., and G. Meerdink (1998) *Food Product Innovation: How to Link Sustainability and the Market* (Wageningen, Germany: Wageningen Agricultural University).

Poppe, C., and U. Kjærnes (2003) *Trust in Food in Europe: A Comparative Analysis* (Professional Report 5; Oslo: National Institute for Consumer Research).

Rosegrant, M.W., M.S. Paisner, S. Meijer and J. Witcover (2001) *Global food Projections to 2020* (Washington, DC: International Food Policy Research Institute).

Tukker, A., and U. Tischner (eds.) (2006) *New Business for Old Europe: Product-Service Development, Competitiveness and Sustainability* (Sheffield, UK: Greenleaf Publishing; www.greenleaf-publishing.com/catalogue/suspro.htm).

——, M. Charter, C. Vezzoli, E. Stø and M. Munch Andersen (eds.) (2008) *System Innovation for Sustainability. 1: Perspectives on Radical Change to Sustainable Consumption and Production* (Sheffield, UK: Greenleaf Publishing; www.greenleaf-publishing.com/SCP1).

Watkiss, P. (2005) *The Validity of Food Miles as an Indicator of Sustainable Development* (London: Department for Environment, Food and Affairs [Defra]).

WHO (World Health Organisation) (1998) *GEMS/Food Regional Diets: Regional Per Capita Consumption of Raw and Semi-processed Agricultural Commodities* (prepared for the Global Environmental Monitoring System [GEMS] Food Contamination Monitoring and Assessment Programme, WHO/FSF/FOS/98.3; Geneva: WHO).

References

Belz, F.-M., and J. Pobisch (2005) 'Shared Responsibility for Sustainable Consumption: The Case of German Food Companies', *Marketing and Management in the Food Industry*, Discussion Paper 4, January 2005.

——, S. Silvertant and J. Pobisch (2007) 'Consumer Integration in Sustainable Product Innovation Processes', in S. Lahlou and S. Emmert (eds.) *TNO SCORE Proceedings WS 2: SCP Cases in the Field of Food, Mobility and Housing*; post-workshop version: 31 May 2007, pages 131-51; www.score-network.org/score/score_module/index.php?doc_id=9594, accessed 28 April 2010.

Bergeaud-Blackler, F., and M.P. Ferretti (2006) 'More Politics, Stronger Consumers? A New Division of Responsibility for Food in the European Union', *Appetite* 47.2: 134-42.

Borchert, E., and S. Reinecke (2007) 'Statistics in Focus: Eating, Drinking, Smoking. Comparative Price Levels in 37 European Countries for 2006', *EUROSTAT* 90/2007.

Braun, C. (2009) 'Das rote Phantom', *Brand eins*, May 2009; www.brandeins.de/archiv/magazin/gegessen-wird-immer/artikel/das-rote-phantom.html, accessed 15 April 2010.

CEC (Commission of the European Communities) (2000) 'White Paper on Food Safety' (COM [1999] 719 final; Brussels: CEC).

—— (2005) 'Kampf gegen Fettsucht in Europa', *Ernährung im Fokus* 5.5: 145f.

—— (2008) 'Attitudes of European Citizens towards the Environment', *Special Eurobarometer* 295/*Wave* 68.2 ('TNS Opinion and Social'); ec.europa.eu/environment/archives/barometer/pdf/report2008_environment_en.pdf, accessed 15 April 2010.

CIAA (Confederation of the Food and Drink Industries of the EU) (2002) 'Industry as a Partner for Sustainable Development: Food and Drink'; www.ciaa.be/documents/brochures/food_and_drink.pdf, accessed 15 April 2010.

Davies, S. (2001) 'Food Choice in Europe: The Consumer Perspective', in L.J. Frewer, E. Risvik and H. Schifferstein (eds.), *Food, People and Society: A European Perspective of Consumers' Food Choices* (Berlin: Springer): 365-80.

Defra (UK Department of Environment, Food and Rural Affairs) (2002) 'The Strategy for Sustainable Farming and Food: Facing the Future' (London: Defra).

—— (2006) 'Food Security and the UK: An Evidence and Analysis Paper' (London: Defra).

Derpsch, R. (2005) 'The Extent of Conservation Agriculture Adoption Worldwide: Implications and Impact', in *Proceedings of the III World Congress on Conservation Agriculture, 3–7 October 2005, Nairobi, Kenya*.

DG Environment (2008) 'LIFE on the Farm: Supporting Environmentally Sustainable Agriculture in Europe' (Luxembourg: Office for Official Publications of the European Communities).

DG Research (2008) 'The Effects of the Financial Crisis on European Research Policy' (Brussels: DG Research, Commission of the European Communities [CEC], 17 November 2008).

Drudge Report (2006) 'Europe Faces Challenge of Aging Population', *Drudge Report*, London, 7 March 2006.

Eberle, U., D. Hayn, R. Rehaag and U. Simshäuser (eds.) (2006) *Ernährungswende. Eine Herausforderung für Politik, Unternehmen und Gesellschaft* (Munich, Germany: oekom-Verlag).

Eurostat (2007) 'Measuring Progress towards a More Sustainable Europe: 2007 Monitoring Report of the EU Sustainable Development Strategy' (Luxembourg: Office for Official Publications of the European Communities).

FAO (Food and Agriculture Organisation) (2002) 'World Agriculture 2030: Main Findings'; www.fao.org/english/newsroom/news/2002/7833-en.html, accessed 16 June 2009.

FTAO (Fair Trade Advocacy Office) (2005) 'Business Unusual: Success and Challenges of Fair Trade' (Brussels: FTAO).

Fritsche, U.R., and U. Eberle (2007) *Treibhausgasemissionen durch Erzeugung und Verarbeitung von Lebensmitteln* (*Greenhouse Gas Emissions from Food Production and Processing*) (Darmstadt/Hamburg, Germany: Öko-Institut eV).

GfK Group (2007) 'Organic Food Products in Germany Surge Ahead', press release, 16 April 2007; www.gfk.com/group/press_information/press_releases/001133/index.en.print.html, accessed 16 June 2009.

Guthman, J. (2004) 'The Trouble with "Organic Lite" in California: A Rejoinder to the "Conventionalisation" Debate', *Sociologia Ruralis* 44.3: 301-16.

Halkier, B., and L. Holm (2006) 'Shifting Responsibilities for Food Safety in Europe: An Introduction', *Appetite* 47.2: 127-33.

Hamm, U., and F. Gronefeld (2004) *The European Market for Organic Food: Revised and Updated Analysis* (OMIaRD publication, Vol. 5; Aberystwyth, UK: School of Management and Business).

Harvey, M., A. McMeekin and A. Warde (eds.) (2004) *Qualities of Food* (Manchester, UK: Manchester University Press).

IFOAM (International Federation of Organic Agriculture Movements) (2007) *The World of Organic Agriculture Statistics and Emerging Trends 2007* (Bonn, Germany: IFOAM).

Jordbruksverket (2003) *Jordbruksföretag, företagare och ägoslag 2003: Statistiska meddelanden (Farms, Businesses and Types of Land in 2003: Statistical Reports)* (JO 34 SM 0401).

Jungbluth, N., and M.F. Emmenegger (2002) 'Ökologische Folgen des Ernährungsverhaltens: Das Beispiel Schweiz', *Ernährung im Fokus* 2.10: 255-58.

Kendell, C. (2005) 'Economics of Horse Farming', *Rural Heritage* 30.3: 71-74.

Kjærnes, U., M. Harvey and A. Warde (2007) *Trust in Food: A Comparative and Institutional Analysis* (London: Palgrave Macmillan).

Klein, N. (2000) *No Logo!* (Toronto: Alfred A. Knopf).

Krier, J.-M. (2006) *Fair Trade in Europe 2005: Facts and Figures on Fair Trade in 25 European Countries* (FLO, IFAT, NEWS! and EFTA).

Krutwagen, B., and E. Lindeijer (2001) 'LCI of Food in the Netherlands', IVAM Environmental Research, paper from *International Conference on LCA in Foods: Proceedings, Swedish Institute for Food and Biotechnology (SIK)*, Gothenburg, Sweden, 26–27 April 2001.

Kuepper, G. (2000) *Conservation Tillage Systems and Management: Crop Residue Management with No-till, Ridge-till, Mulch-till and Strip-till* (Publication #MWPS-45; Ithaca, NY: Northeast Regional Agricultural Engineering Service [NRAES]).

Marshall, D. (2001) 'Food Availability and the European Consumer', in L.J. Frewer, E. Risvik and H. Schifferstein (eds.), *Food, People and Society: A European Perspective of Consumers' Food Choices* (Berlin: Springer): 317-38.

McMeekin, A., K. Green, M. Tomlinson and V. Walsh (2002) *Innovation by Demand: An Interdisciplinary Approach to the Study of Demand and its Role in Innovation* (Manchester, UK: Manchester University Press).

Meier-Ploeger, A. (2005) 'Grundsatzpapier Ernährungspolitik des Wissenschaftlichen Beirats "Verbraucher- und Ernährungspolitik" beim BMVEL' ('Position Paper on the Nutrition Policy of the Scientific Board "Consumer and Nutrition Policy" at BMVEL'); www.uni-kassel.de/fb11cms/nue/img/publication/Grundsatzpapier.pdf, accessed 16 June 2009.

Michaelis, L., and S. Lorek (2004) *Consumption and the Environment in Europe: Trends and Futures* (Copenhagen: Danish Environmental Protection Agency).

Muller, M. (2005) 'Intervention by Martin Muller, Representing the European Council of Young Farmers, CEJA, in the Study Days of the European Association for Rural Development Institutions, to be Realised 20–24 October 2005 in Athens, Greece'; www.agrogi.gr/files/pdf/CEJA%20-AEIER%20Oct2005.pdf, accessed 19 July 2009.

OECD (Organisation for Economic Cooperation and Development) (2001). *Household Food Consumption: Trends, Environmental Impacts and Policy Responses* (Paris: OECD).

Reisch, L.A. (2004) 'Principles and Visions of a New Consumer Policy', *Journal of Consumer Policy* 27: 1-27.

Roe, E.J., and T. Marsden (2007) 'Analysis of the Retail Survey of Products that Carry Welfare-Claims and of Non-retailer Led Assurance Schemes whose Logos Accompany Welfare-Claims' (Cardiff, UK: Welfare Quality; eprints.soton.ac.uk/58688/1/WQR2-p2.pdf, accessed 28 Apeil 2010).

Rubin, J. (2008) 'The New Inflation', *StrategEcon* (CIBC World Markets Economics and Strategy), 27 May 2008; research.cibcwm.com/economic_public/download/smay08.pdf, accessed 15 April 2010.

—— and B. Tal (2008) 'Will Soaring Transport Costs Reverse Globalisation?', *StrategEcon*, 27 May 2008: 4; research.cibcwm.com/economic_public/download/smay08.pdf, accessed 16 June 2009.

Smith, A., P. Watkiss, G. Tweddle, A. McKinnon, M. Browne, A. Hunt, C. Treleven, C. Nash and S. Cross (2005) *The Validity of Food Miles as an Indicator of Sustainable Development* (London: Department for the Environment, Food and Rural Affairs [Defra]).

Swidler, A. (1986). 'Culture in Action: Symbols and Strategies', *American Sociological Review* 51.2: 273-86.

Swoboda, B., and D. Morschett (2001) 'Convenience-Oriented Shopping: A Model from the Perspective of Consumer Research', in L.J. Frewer, E. Risvik and H. Schifferstein (eds.), *Food, People and Society: A European Perspective of Consumers' Food Choices* (Berlin: Springer): 177-96.

Tempelman, E., P. Joore, T. van der Horst and H. Luiten (2006) 'Need Area 4: Food', in A. Tukker and U. Tischner (eds.), *New Business for Old Europe: Product-Service Development, Competitiveness and Sustainability* (Sheffield, UK: Greenleaf Publishing; www.greenleaf-publishing.com/catalogue/suspro): 267-302.

Thielicke, R. (2006) 'Natürlich gut?' ('Naturally good?'), *Focus* 24 (October 2006): 55.

—— (2007) 'Essen gegen die Hitze?' ('Eating against climate change?'), *Focus* 9 (February 2007): 28-30.

Torjusen, H.C., U. Kjærnes, L. Sangstad and K. O'Doherty Jensen (2004) *European Consumers' Conceptions of Organic Food* (Professional Report 4; Oslo, Norway: National Institute for Consumer Research).

Von Koerber, K., T. Männle and C. Leitzmann (2004) *Vollwert-Ernährung: Konzeption einer zeitgemäßen und nachhaltigen Ernährung* (*Whole Foods Diet: Conception of a Contemporary and Sustainable Nutrition*) (Stuttgart, Germany: Karl F. Haug Verlag).

Welt am Sonntag (2007) 'Die neue Lust der Bauern am Markt' ('The New Delight of the Farmer in the Marketplace'), *Welt am Sonntag*, 6 May 2007: 29

WHO/FAO (World Health Organisation/Food and Agriculture Organisation) (2002) 'Diet, Nutrition and the Prevention of Chronic Diseases', draft report of the joint WHO/FAO Expert Consultation, Geneva, Switzerland, 28 January 2001 to 1 February 2002.

Willer, H., and L. Kilcher (eds.) (2009) *The World of Organic Agriculture: Statistics and Emerging Trends 2009* (FiBL–IFOAM Report; Bonn, Germany: IFOAM/Frick, Switzerland: FiBL/Geneva: ITC).

Zanoli, R. (ed.) (2004) *The European Consumer and Organic Food* (OMIaRD publication Vol. 4; Aberystwyth, UK: School of Management and Business).

Part II
Case studies

3

Facilitating a more sustainable food and farming sector in the UK*

Paul Dewick and Chris Foster
Manchester Institute of Innovation Research, UK

Steve Webster
Delta-innovation Ltd, UK

Sustainability can be thought of as 'continuance into the long term' and sustainable development as the means by which we achieve sustainability (Porritt 2005). The triple-bottom-line concept is a useful means to highlight the interconnectedness between the economic, social and environmental dimensions of the food and farming industry. Recent strategy papers on sustainable farming and food identify the consequences *of* farming on measures of sustainable development and also the consequences *for* farming of a changing economic, social and environmental scenario. But what do economic, social and environmental sustainability mean in the context of food and farming?

Economic sustainability may be interpreted as the commercial efficiency of the industry being driven by the market: that is, 'internationally competitive without reliance on subsidy or protection' (HM Treasury and Defra 2005) or, more simply, 'succeeding in the market' (Defra 2006a).[1]

* This chapter draws on two conference papers presented to the 'Cases in Sustainable Consumption and Production: Workshop of the Sustainable Consumption Research Exchange (SCORE!) Network', supported by the EU 6th Framework Program, Paris, 4–5 June 2007: those by Dewick and Foster (2007a) and Webster (2007).

1 Since the mid-term review reforms of the Common Agricultural Policy (CAP) (in 2003 and 2004) a large element of direct production support has been removed, replaced by the single farm payment (SFP); this will continue until 2012, at which point further reform of the CAP will be implemented.

Social sustainability may be interpreted as 'overall human welfare' but more generally refers to the *acceptability* of the industry. Strategy documents refer to a diverse set of issues, including animal welfare, education and training, equal opportunities and the health and safety of workers and consumers of the industry's products, all under the umbrella of 'social sustainability'. These are often set in the context of the broader context of wider social goals such as ensuring access to public services for all or tackling local inequalities (HM Government 2005; Defra 2006a, 2006b).

Environmental sustainability in the context of farming and food may be interpreted as *improving* environmental performance. The Department of Environment, Food and Rural Affairs (Defra)'s interpretation of environmental sustainability is evidenced in phrases such as farming 'maximising its role in the renewal of the natural environment' and 'making a positive net contribution to the environment' (Defra 2006a). Specific environmental objectives for Defra address landscape, biodiversity, soil, air quality, water quality and waste management. The interaction between climate change and agriculture is treated as a separate entity and is a major driver of government environmental policy. Environmental sustainability remains the most complex element of sustainable farming and food because of the possible conflicts between pursuit of different environmental objectives.

In Section 3.1 we focus on the structure of the UK dairy industry and the power relations between actors along the supply chain. By understanding the structure of the production and consumption system and the current environmental impacts, we can consider the environmental implications of opportunities for improving the economic sustainability of the dairy industry. The case provides evidence that transition in the manner described, in the absence of technological change, is likely to lead to increased environmental impacts.

We know that technological change is neither the whole answer nor is it an easy, quick or cheap option. Playing an equally important role is behavioural change in the industry and it is to this we turn our attention in Section 3.2. Important aims of the Policy Commission on Sustainable Farming and Food and of the Strategy for Sustainable Farming and Food (SSFF) (Defra 2002) were to change the attitudes and behaviour of stakeholders in food and farming. The Policy Commission and the SSFF argued that to be economically sustainable farmers would need to become more market-driven, though not at the expense of significant environmental damage—that is, economic sustainability should not come at the expense of environmental sustainability. A 'stick and carrot' approach of legislative and subsidy reform was advocated, to be undertaken centrally but with close liaison between government, farming and environmental bodies at a *regional* level to help gain industry 'buy-in'. The approach was sensitive to the overarching political context of farming and food industries and consistent with the increasing role in economic development given to regional agencies by central government. In Section 3.3 we evaluate the success of one regional initiative. The case demonstrates that a government–industry partnership coordinated at the regional level can have some success, particularly in terms of encouraging 'buy-in' from stakeholders in the industry. It also reveals that changing behaviour and attitudes is very difficult, particularly in a highly political industry where individuals and institutions hold deeply entrenched views, and that triple bottom line sustainability requires very considerable change, not least at the business level but also at the individual level.

Section 3.4 draws general conclusions from the two cases for a more sustainable future for food and farming.

3.1 Improving sustainability by promoting industry structural change: the case of the UK dairy industry

The UK dairy industry is the third largest milk producer in the European Union after Germany and France (and the tenth largest in the world). Dairying accounts for over a fifth of total agricultural production by value in the UK and is the largest single farming sector by value employing around 50,000 people at the farming stage and another 34,000 in the milk industry, driving tankers, pasteurising milk and packing and distributing dairy products (NFU/CPRE 2006). Over the past 20 years or so each stage of the milk production and consumption system has become increasingly concentrated. In 2004, a quarter of milk producers accounted for 50% of total UK dairy output, milk purchasing was dominated by three large firms, and around 60% of milk was processed by five firms that provided as much as 90% of the milk sold by the five largest supermarkets (Foster *et al.* 2007).

The stages have become more concentrated because of a number of different drivers. At the dairy farming stage, for instance, the number of milk producers has been declining for decades and large farms have emerged as the dominant form. Larger herds tend to be more profitable and more efficient because variable costs are similar regardless of the size of operation and because fixed costs decline on a unit output basis as herd size increases. There are strong and direct correlations between efficiency (and profitability) of dairy farms and herd size and the scale of milk production. At the other end of the system, the way in which we buy our milk has changed significantly over the past few decades. In 1980, nearly 90% of milk was delivered direct to the doorstep. Doorstep deliveries now account for less than 25% of total sales and the proportion purchased in retail outlets, predominantly supermarkets, has increased dramatically. This continued trend can perhaps best be explained by the difference in price from the two sources: doorstep deliveries were between 13p and 16p per pint more expensive between 1995 and 2007. However, despite this concentration at each stage, and in contrast to the dairy industries in other European countries, there is little concentration (that is, vertical integration) between stages.

This evolution in the structure of the dairy industry has led to an unequal distribution of value added among the actors in the system and accentuated 'power' distortions across it. Whereas retailers' profits are more resilient, profits remain low for dairy farmers (having fallen by more than 50% in real terms since the 1990s) and processors' profits are not high by food industry standards. The power distortions are highlighted by the steadily increasing difference between farm-gate prices of raw milk and retail prices of milk during the 1990s and 2000s. Analysing this trend, London Economics (2003) estimated that the shift from doorstep delivery to purchase through retail outlets was the most important factor in explaining the price shift between retail and farm gate. In 2003, the farm-gate price of a litre of milk was 18p (including retrospective bonuses), the retail price of a litre of milk was 48p and the doorstep-delivered price was 76p. This analysis also found that price changes at the retail stage were fully transmitted through to the farm gate. However, changes in prices at the farm gate are transmitted only partially through to retail prices. That is, while a retailer can pass the cost of changed circumstances back fully to its suppliers, those suppliers (in this case, farmers) can pass only part of the costs of changing circumstances forward to their customers, the retailers. This helps explain why, in 2000, the industry was worth £2.38 billion at farm-gate prices and £6 billion at the retail level (Defra 2001). It also helps to underline the unsustainable state of the industry in economic terms. With costs of production now averaging more than farm-gate prices (NFU/RABDF 2007), farmers are leaving the industry in their thousands (MDC 2007). Indeed, the low profits and low expectations of future profits evidenced by the survey carried out by the National Farmers' Union (NFU) with the Royal Association of British Dairy Farmers (RADB) and by the Milk Development Council (MDC) report (MDC 2007; NFU/RABDF 2007) undermine investment and support a vicious circle that ultimately calls into question the economic sustainability of the industry.

While all stages of milk production and consumption have some impact on the environment, the activities behind those impacts are very different. The direct impacts of farming practice on animal welfare, the land and landscape are unique to primary production. The production of raw milk is also where energy consumption and greenhouse gases (GHGs; mainly carbon dioxide [CO_2], methane [CH_4] and nitrous oxides [NO_x]) emissions are highest by a large margin (Foster *et al.* 2006). Moreover, the drive to maximise productivity from this stage, coupled with higher resource efficiencies within larger farms, disfavours both the diversity of farm sizes and structures that many would say characterises the UK landscape and, perhaps more importantly, the maintenance of non-productive areas that nurture biodiversity.

The impacts arising from the production of packaging materials (that is, from conversion of natural resources into materials for people to use) depend to a large extent on the type of material produced. For example, production and transport energy costs are lower for plastic than for glass, though not as low as for paperboard cartons (Foster *et al.* 2006); water consumption (and pollution) and air emissions (that is, acidification and photochemical oxidants) are more harmful in the case of oil-based plastics than for paperboard (Sonnesson and Berlin 2002). The efficiency with which the original resources are used in the system as a whole depends strongly on the way in which packaging waste is ultimately managed. For example, the impacts of a glass bottle become much less per unit of milk if the glass bottle is re-used several times; the impacts associated with paperboard cartons are very strongly mitigated if they are incinerated after use in energy-generating plants as a substitute for fossil fuels.

The most significant impacts of milk processing are in the use of energy in different forms—to run electric motors, to produce steam and for cooling and generating compressed air. In addition, milk processing involves a high consumption of water (notably for cleaning) and gives rise to water pollutants (for example, discharge of effluent with high organic loads) and air pollutants (for example, oxides of nitrogen and sulphur and suspended particulate matter). Noise levels can be high, and strong odours can also be emitted in the production process.

Retailing involves further energy use, for refrigeration and for transport. Transport involves burning petrol or diesel fuel, which contributes to climate change, acidification and local deterioration in air quality. The impact of transport at different stages of the system depends on the mode of transport, the utilisation of space in vehicles, the type of product (including packaging format) and the distance travelled.

Waste materials, notably from the processing stage and from households, also give rise to environmental impacts, either directly as noted above or indirectly by creating further energy demand from power treatment plants. Milk wastage within homes is remarkably high: a Swedish study found that dairy products were the biggest components of food wasted from homes (Sonesson *et al.* 2005). The environmental impacts of consumers' activities extend beyond the waste bin: arising from transport (people going to and from shops to buy milk) and milk storage (operation of refrigerators and freezers).

Despite the diversity of environmental impacts across the stages of production and consumption the use of energy provides a common currency that can be tracked across the whole system. Figure 3.1 shows that the primary production stage, packaging stage, processing stage and transport (when considered cumulatively) are all significant in terms of energy use.

FIGURE 3.1 Energy consumption across the milk production and consumption system

Source: data from Foster *et al.* 2006, 2007

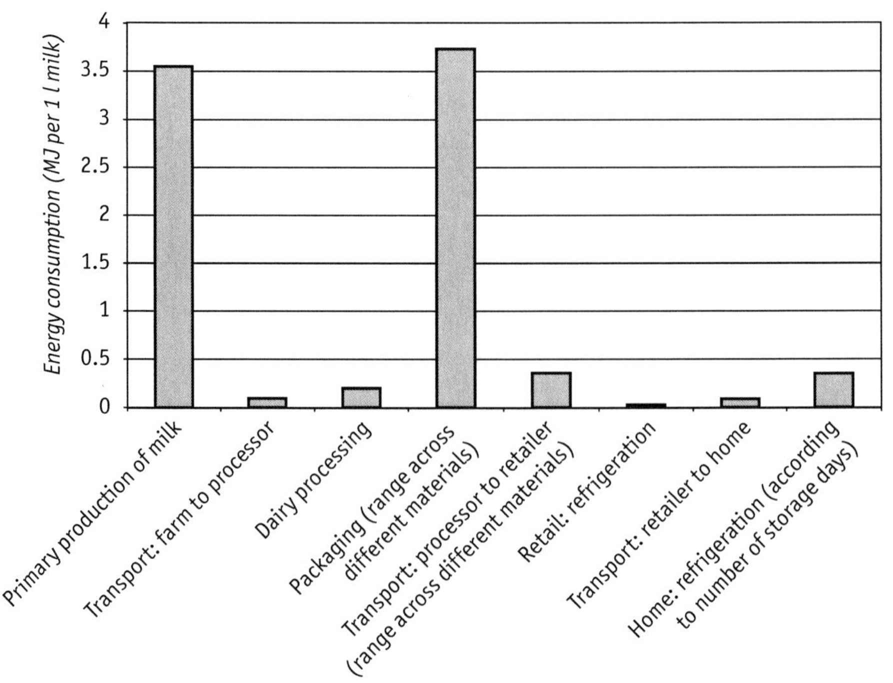

Given the interconnections between economic sustainability and environmental sustainability in the industry, attempts to address the former will impact almost inevitably on the latter. Three opportunities for improving the economic sustainability of milk production and consumption are discussed below: (1) vertical integration between the stages of production and processing, with more producers processing their own milk; (2) further product differentiation at the processing stage; and (3) local production and consumption (that is, milk sold direct to consumers by producers). It is our contention that there may be considerable environmental inefficiencies in a shift away from the current regime along these lines, and the following analysis explores the trade-offs. With these three opportunities, we can assume that the environmental impacts at the milk production and packaging stages will stay the same and, as shown in Figure 3.1, that these two stages are the most energy-intensive. The opportunities are likely to have significant effects at the processing and transport stages.

At the processing stage, there are scale efficiencies and it is our contention that change here in the directions outlined above would lead to more carbon emissions (see Keoleian and Spitzley 1999; Sonnesson and Berlin 2002). CO_2 emissions associated with energy use for liquid milk processing vary from around 0.02 kg per litre milk in the most modern plants, to perhaps twice that in smaller, older installations. Using energy data from Integrated Pollution Prevention and Control (IPPC) permit applications, one can calculate and plot the associated CO_2 emissions (direct from gas or diesel, and indirect CO_2 from electricity production) against raw milk input across four milk processors. Figure 3.2 provides some evidence of scale efficiencies with regard to energy use and CO_2 emissions.

FIGURE 3.2 Specific carbon dioxide (CO_2) emissions from energy use in milk processing

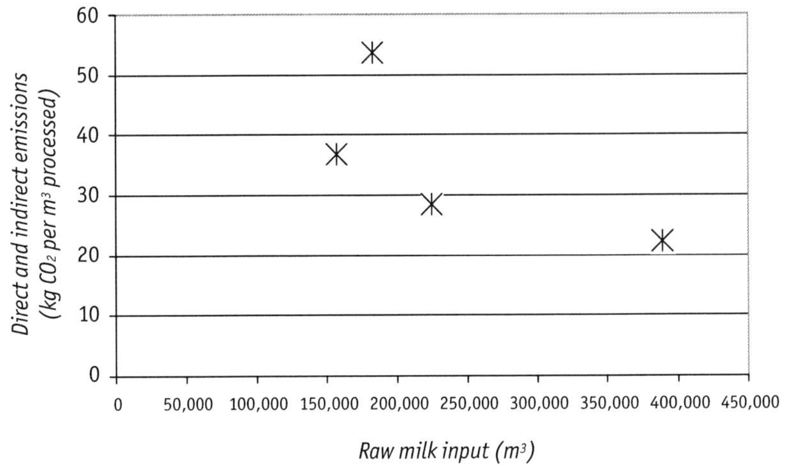

Dairy UK, the industry trade association, is not alone in calling for new product development to improve the margins in dairy processing. About 14 billion litres of raw milk are used for milk and milk-product processing. Just under half of the raw milk is used to produce drinking milk; the remainder is used to produce butter, cheese, yoghurt and milk powder. Further processing associated with a more complex product with more ingredients, or a larger portfolio of products, involves more energy and water consumption and leads to more wastage. For example, compared with milk processing, the processing of yoghurt requires about the same amount of raw milk and fuel but uses twice as much water and six times as much electricity (Feitz *et al.* 2005). CO_2 emissions from UK facilities producing yoghurt and similar products appear to be about six times higher per tonne of raw milk than they are from dairies producing only liquid milk products. Indeed, if we include a major branded yoghurt processor with the analysis presented in Figure 3.2, it suggests that there is a considerable environmental impact associated with producing many, differentiated, value-added products: raw milk input for this particular processor is 260,000 tonnes per year and overall CO_2 is 126 kg per m^3 raw milk, significantly higher than the other processors in the analysis. With more products, more water is required in all stages of the process, particularly for additional equipment cleaning between runs of different products, while smaller batch sizes mean that the residues left in vessels and pipes represent a higher proportion of the batch.

Changes in the way people buy their milk are likely to have implications for transport-related environmental impacts. We know that transporting food to retail outlets involves some 6 billion road vehicle kilometres per year and that transportation of food from retail outlets to the home involves around 10 billion car kilometres per year—so car transport is certainly a significant part of the food logistics system. Moreover, in terms of fuel consumption per weight of food, cars are fuel-intensive.[2] This comparison matters, because changes in the way we buy food might involve more of it being moved further in medium-sized vehicles (home delivery of internet-purchased goods), or more of it being moved further in cars (if consumers travel by car to farmers' markets or a variety of specialist shops that are further away than the nearest supermarket).

3.1.1 Learning experiences

These three opportunities demonstrate the economic and environmental trade-offs associated with transition in the milk production and consumption system. Over the past few decades, technological change has improved incrementally the resource efficiency of production, processing and distribution in the dairy industry (Dewick *et al.* 2007b). Technological trajectories and 'lock-in' have reinforced and strengthened the regime (Dewick *et al.* 2007b). So, in the absence of radical new technologies for efficient small-scale, differentiated processing and new energy and water efficiencies or new transport or fuel technologies to reduce the impact of more localised food distribution to individual consumers (likely, given current retail structures in the UK, to involve more car and

2 Compare, for instance, the approximate amounts of CO_2 associated with moving 1 tonne of goods over 10 km: 200 g on a fully-loaded 10,000 dwt freighter, 2,400 g in a van, carrying 1 tonne, 10,000 g in 50 cars, each carrying 20 kg.

small van transport *vis-à-vis* large lorries), a shift away from the current regime may lead to greater environmental impacts, whatever its economic and social consequences.

Behavioural and attitudinal change may facilitate a more environmentally benign route to improvement of the industry's economic situation. In 2007 we saw that pressure from regulatory bodies (for example, competition authorities) and increased social awareness exerted an influence on the industry to redistribute margins from retailers to farmers, which may lead to improvements in the economics of the industry without compromising environmental limits. The idea of facilitating behavioural and attitudinal change as part of a sustainability agenda has not escaped government. Indeed, it was at the centre of Defra's Sustainable Farming and Food Strategy (Defra 2006a) and the creation of regional implementation groups to help deliver government policies by working in partnership with industry at the regional level. In the Yorkshire and Humber region this action plan (and the steering group set up to guide its delivery) was called Framework for Change (F4C). In the next section we evaluate the success of the F4C scheme and highlight the lessons we can learn in terms of policy initiatives to improve sustainability of food and farming.

3.2 Improving sustainability by facilitating behavioural and attitudinal change: the case of the Framework for Change project

The F4C project was focused on institutional change, moving away from central government 'top-down' control of the farming industry (through regulation and subsidy) towards greater self-regulation and customer focus by the farming industry. It was assumed that if farming became more market-driven it would be more responsive to the demands of consumers for sustainable means of production, providing commercial benefits to the industry alongside environmental and social benefits to the public. Although the explicit aims of the F4C changed over time as the policy backdrop shifted (such as reform to the CAP) and funding from central government for the delivery of projects diminished, the activities of F4C focused on facilitation, influence and guidance, under the organisation's self-description: 'Framework for Change is a partnership between the public and private sector to deliver the vision: Yorkshire and Humber will have an entrepreneurial, dynamic, and sustainable farming and food industry' (KPMG 2005).

F4C drew on a comprehensive network of stakeholders from across industry and government.[3] The initial vague aims and objectives of the F4C may arguably have contributed to encouraging a broad range of stakeholders to participate; providing clear

3 The voluntary F4C forums were led by an independent industry representative and included stakeholders from central and regional government, representatives from regional food and farming industries (such as specialist farming groups, agricultural colleges, levy bodies, etc.), government bodies (for example the Environment Agency and local government) and a number of organisations set up specifically to deliver government initiatives.

objectives prior to the establishment of the group may have removed the sense of ownership that was clearly present in many of the stakeholders.

To gauge the role of F4C in contributing to change in the farming and food sector, a series of questions were asked of key stakeholders directly involved:

- Have projects supported by F4C resulted in long-term behavioural changes within government (for example, with regard to public procurement, or industry liaison)?
- Have projects supported by F4C resulted in long-term behavioural changes within industry (such as in market focus or response to opportunities offered by climate change)?
- Have the networking opportunities offered by F4C resulted in the wider uptake of new ideas and initiatives, by industry and/or government, which would make farming and food more sustainable?
- Have key stakeholders, in general, and farmers, in particular, been more responsive to initiatives and activities as a result of F4C?
- Have farmers more readily changed or responded to changes in the subsidy and legislative regime as a result of F4C?
- Is farming in Yorkshire and Humber more sustainable as a result of F4C? If only a part of farming is more sustainable, which part and why? If F4C had a part in making farming more sustainable, can that part be articulated?
- How well has F4C adapted to changing policy and sustainability priorities?
- Is F4C helping farmers to respond to new challenges?
- What could have made F4C more effective? Is it limited by funds, partners, time-scale or other factors? What has made F4C a success?

The respondents' views of the original purposes of the F4C initiative and the 'key performance indicators' by which different stakeholders judged its overall success varied widely. Respondents generally agreed that F4C had resulted in long-term behavioural changes within government: F4C was seen as helping government bodies to target their support for sustainable behaviour in farming and food and as a route for piloting initiatives involving significant change within industry. For example, F4C was able to help bring together public-sector procurement managers and wholesale meat suppliers. By liaising between them and providing funds for a small pilot project, changes were made to the specifications given to the wholesalers, resulting in an increased use of local (rather than imported, often from outside the UK) foods and a higher specification of nutritional quality of the food product provided to schoolchildren. Although none of the respondents would claim that the F4C had full responsibility for this (given that it brought together a range of initiatives) there was agreement across all of them that F4C had played a central role. Another example given was of a partnership across colleges, skills and business support agencies that had been developed in response to issues raised through F4C. As such, a key role of F4C in this long-term change was identified as one of facilitation and networking.

There was less agreement as to whether the initiative had changed the attitudes and behaviour of farmers. Reasons for this included (a) the overwhelming commercial pressures and financial constraints acting on farmers; (b) the fact that the F4C initiative generally supported projects that would (initially at least) have any impact only on the most forward-looking farmers, the 'pioneers', and relied thereafter on a trickle-down effect; and (c) the deeply entrenched attitudes and behaviour of many farmers. However, the continued and active involvement in the partnership of the two major farming membership groups would indicate that, at least at this level, the initiative is having some influence.

The responsiveness of farmers to initiatives and activities was not thought to have been greatly affected by F4C, principally because it dealt directly with only a small number of farmers but also because it operated to influence the delivery of projects directed at farmers rather than delivering these projects in its own right. While examples of activities emerged from interviews that provided evidence that F4C had influenced the responsiveness of farmers through its role as an 'honest broker', the major changes to which industry must respond (that is, those occurring to the CAP) were not thought to be being addressed in any greater depth as a result of the initiative.

Respondents differed on their view as to whether F4C had made farming in Yorkshire and Humber more sustainable. Some thought that it had not; others thought that it had made a small positive impact. In all cases there was acceptance that the changes required in the industry are very great, at an individual (personal) level as well as at a business level. Therefore, the pace of change induced by the F4C initiative was highlighted by some respondents as 'expected'.

The F4C initiative was seen as responding well to changes in the policy context and is now, for example, working on projects relating to non-food crops, biomass energy and climate change. F4C has also adapted to the changing needs of government partners, with one F4C management group serving a role within the newly established Rural Development Programme for England. This changing context had also reduced the need within Yorkshire and Humber for the types of project that F4C had initially been involved with. There was therefore the possibility that these projects would, over a period of time, diminish in number whereas the networking and influencing role of F4C would continue.

3.2.1 Learning experiences

There were a number of common threads in respondents' views as to what had made F4C a success.

- The fact that F4C started off as a government–industry partnership, with an independent, industry-based chair was seen as critical
- The independent chair allowed other industry organisations and individuals to become involved in the initiative without the need to overcome any disagreements with government in general or with government departments in particular. Thus, F4C was able to achieve an apolitical status within a highly politicised industry

- The facilitators, being seconded from government positions but with a farming industry background, were able to effectively bridge government–industry relations

While the funding provided to F4C (for project work) was seen as useful, the potential value of additional funding was not seen as consistently high among respondents. It appears that this view was partially contingent on attitudes to behavioural change and whether financial support would facilitate it. The one area where this was not the case was in long-term organisational planning of F4C, where longer-term funding was seen by a large proportion of respondents as having definite benefits.

Other issues relating to the longer-term future of F4C were also raised in discussions and one in particular is of note. This is that the ability to specify the benefits of the F4C initiative to stakeholders would allow a different approach to funding core activities, within the principle that the beneficiary organisations should contribute at least some of this funding in proportion to the benefits they receive. This would avoid funding being sought from one organisation only, which was seen as possibly distorting the nature of the partnership.

3.3 Conclusions

The critical situation in milk production and consumption is typical of much of the food and farming industry in the UK. Concentrated stages of production and skewed margins across the supply chain, combined with conservative adversarial stakeholders largely disconnected from consumers, support a vicious circle that ultimately calls into question the economic sustainability of the industry. Pursuit of scale economies and productivity gains in primary milk production, although being perhaps the best short-term response to commercial pressures available to farmers, is detrimental to the environment in certain respects. However, failure to take up the resource efficiencies that accompany larger-scale processing increases other environmental impacts.

Improving economic sustainability in food and farming requires government and industry interventions sensitive to environmental (and social) trade-offs. On the one hand, technological change and industrial structural change offer some potential for improvements in the economics of the industry without compromising environmental limits. The first case has provided some evidence to suggest that a shift away from the current regime towards a more locally based, smaller scale system (based on vertical integration by farmers producing more differentiated products with local processing and distribution) may lead to greater environmental impacts unless it is accompanied by the development, adoption and diffusion of new technologies such as efficient small-scale processing and storage (refrigeration) technologies and by the emergence of local retailing and logistics models that are not reliant on shopping by car. On the other hand, the second case highlighted the important role of behavioural and attitudinal change (on the part of the food and farming stakeholders, and understanding by their consumers) in facilitating triple-bottom-line sustainability. Not only was the F4C successful in identifying and shaping targeted interventions (for example, projects and campaigns), it enabled the arguments about sustainability to be brought into mainstream thinking at the relevant farming and food industry organisations. Moreover, through the development and exploitation of better government–industry relationships it was able to change public-sector food procurement directly.

Overall, it is clear that considerable inertia remains in the food and farming industry. The cases provide some evidence that a more sustainable future is achievable; what is required is a combination of technological and industrial structural change on the one hand and behavioural and attitudinal change on the other. The entrenched views of many stakeholders in all parts of the industry, existing power relationships and the economic fragility of farming, and the generally politicised nature of the dairy sector inevitably afford barriers to both forms of change.

References

Defra (Department for Environment, Food and Rural Affairs) (2001) *Milk Task Force Report* (London: Defra).

—— (2002) 'Strategy for Sustainable Farming and Food'; www.defra.gov.uk/farm/policy/sustain/strategy.htm, accessed 16 June 2009.

—— (2006a) 'SSFS Forward Look'; www.defra.gov.uk/farm/policy/sustain/index.htm, accessed 16 June 2009.

—— (2006b) 'Farming: Food Industry Sustainability Strategy (FISS)'; www.defra.gov.uk/farm/policy/sustain/fiss/index.htm, accessed 16 June 2009.

Dewick, P., and C. Foster (2007a) 'Transition in the UK Dairy Industry: A More Sustainable Alternative?', paper presented at the 'Cases in Sustainable Consumption and Production: Workshop of the Sustainable Consumption Research Exchange (SCORE!) Network', supported by the EU 6th Framework Program, Paris, 4–5 June 2007.

——, C. Foster and K. Green (2007b) 'Technological Change and the Environmental Impacts of Food Production and Consumption: The Case of the UK Yoghurt Industry', *Journal of Industrial Ecology* 11.3: 133-46.

Feitz, A., S. Lundie, G. Dennien, M. Morain and M. Jones (2005) 'Allocating Intra-industry Material and Energy Flows using Physio-chemical Allocation Matrices: Application to the Australian Dairy Industry', paper presented at *Fourth Australian Life-Cycle Assessment Conference*, 23–25 February, Sydney.

Foster, C., K. Green, M. Bleda, P. Dewick, B. Evans, A. Flynn and J. Mylan (2006) 'Environmental Impacts of Food Production and Consumption: A Report to the Department of Environment, Food and Rural Affairs' (Manchester Business School; London: Defra; www.defra.gov.uk/environment/business/scp/evidence/theme2/products0506.htm, accessed February 2010).

——, E. Audsley, A. Williams, S. Webster, P. Dewick and K. Green (2007) 'The Environmental, Social and Economic Impacts Associated with Liquid Milk Consumption in the UK and its Production: A Review of Literature and Evidence' (London: Department of Environment, Food and Rural Affairs; www.defra.gov.uk/foodfarm/food/industry/sectors/milk/pdf/milk-envsocecon-impacts.pdf, accessed February 2010).

HM Government (2005) 'Securing the Future: Delivering UK Sustainable Development Strategy' (Cm 6467); www.defra.gov.uk/sustainable/government/publications/uk-strategy/documents/SecFut_complete.pdf, accessed February 2010.

HM Treasury/Defra (Department for Environment, Food and Rural Affairs) (2005) 'A Vision for the Common Agricultural Policy'; www.defra.gov.uk/farm/policy/capreform/pdf/vision-for-cap.pdf, accessed 16 June 2009.

Keoleian, G.A., and D.V. Spitzley (1999) 'Guidance for Improving Life-cycle Design and Management of Milk Packaging', *Journal of Industrial Ecology* 3.1: 111-26.

KPMG (2005) *Framework for Change: Review Findings and Implementation Plan Overview* (London: KPMG).

London Economics (2003) 'Examination of UK Milk Prices and Financial Returns' (London: London Economics).

MDC (Milk Development Council) (2007) 'Farmers' Intentions Survey'; www.mdcdatum.org.uk/PDF/FIS2007.pdf, accessed 16 June 2009.

NFU (National Farmers' Union) and CPRE (Campaign to Protect Rural England) (2006) 'Living Landscapes: Hidden Costs of Managing the Countryside'; www.nfuonline.com/documents/NFU%20and%20CPRE%20report.pdf, accessed 16 June 2009.

—— and RABDF (Royal Association of British Dairy Farmers) (2007) 'British Milk: What Price 2007?'; www.nfuonline.com/documents/Dairy/British%20Milk%20draft%208.pdf, accessed 16 June 2009.

Porritt, J. (2005) *Capitalism as if the World Matters* (London: Earthscan).

Sonesson, U., and J. Berlin (2002) 'Environmental Impact of Future Milk Supply Chains in Sweden: A Scenario Study', *Journal of Cleaner Production* 11: 253-66.

——, F. Anteson, J. Davis and P.O. Sjödén (2005) 'Home Transport and Wastage: Environmentally Relevant Household Activities in the Life-cycle of Food', *Ambio* 34: 4-5.

Webster. S. (2007) 'Review of a Regional Sustainable Farming and Food Initiative', paper presented at the 'Cases in Sustainable Consumption and Production: Workshop of the Sustainable Consumption Research Exchange (SCORE!) Network', supported by the EU 6th Framework Program, Paris, 4–5 June 2007.

4

Self-sufficiency or localisation?

Sustainability and ambiguity in Britain's food policy

Tim Cooper
Nottingham Trent University, UK

Concern within industrialised nations about future food supplies is not a new phenomenon. In the mid-1970s, rising energy and commodity prices and anxiety about the future voiced in the 'limits to growth' debate prompted British ecologist Kenneth Mellanby to write *Can Britain Feed Itself?*, a book that aimed 'to assess the possibility of Britain becoming reasonably self-sufficient as a food-producing nation' (Mellanby 1975: 2). Although Britain had recently joined the European Economic Community, Mellanby's focus on policy at a national level was shared by the Government's White Paper 'Food from Our Own Resources', which proposed an expansion in domestic agricultural production (MAFF 1975).

By the early 1980s, however, the aim of achieving greater national self-sufficiency in food had been abandoned. The level peaked in 1984, since when it has declined by around 20% to barely 60%. Britain now imports around 90% of its fresh fruit, over half of its ham and bacon, almost one-half of its cheese and well over one-third of its fresh vegetables and butter (Defra 2010a; Living Countryside 2010).

Signs have emerged in recent years of renewed public interest in self-sufficiency prompted by unease about trends in British farming, food security fears relating to climate change and the rising global population, and the growing attraction of locally sourced food to consumers.[1] Nonetheless, despite developing a strategy to encourage

1 An example is the Symposium entitled *Can Britain Feed Itself? Should Britain Feed Itself?* (James Martin 21st Century School, University of Oxford, 15 October 2008; www.21school.ox.ac.uk/downloads/briefings/PFPShould_Britain_Feed_Itself.pdf, accessed 26 March 2010).

sustainable farming and food practices, Britain's government has consistently stated that it does not seek to increase the level of national self-sufficiency in food. This chapter presents a case study of Britain in order to assess the validity of its approach and consider whether, as a general principle, a minimum level of national self-sufficiency in food should be regarded as an essential element of a sustainable farming and food strategy.

4.1 Self-sufficiency in practice

4.1.1 Conceptual overview

Self-sufficiency concerns independence of provision. The focus in this chapter is on the provision of food at a national level, although self-sufficiency could also be considered at a European, bioregional, local or household level.

In Britain, official statistics on self-sufficiency are calculated as the value of domestic production of food as a share of national consumption of food in a given year, with adjustments to take account of imports and exports (Defra 2006).[2] Inputs of livestock feed and seeds are included in the production data, but other elements in the food production process, such as fertiliser, pesticides, machinery and oil, are not.[3] Britain's level of self-sufficiency in 2009 was 59% for all food and 72% for 'indigenous-type food' (i.e. food that can be grown in a temperate climate). If exports are excluded from the equation, however, the statistics reveal that only around 50% of the food consumed in Britain originated here (Defra 2010a).[4] Historical trends and variation in levels of self-sufficiency for different types of food are further explored later in this chapter.

Environmentalists have long associated self-sufficiency with sustainability. Influenced by books such as *Small is Beautiful* (Schumacher 1974) and *The Complete Book of Self-sufficiency* (Seymour 1976), many have regarded a shift towards localised food production, within a context of economic and political decentralisation, as a prerequisite for a more sustainable future (Hines 2000).

In the aftermath of rising oil prices in 1973–74, fears grew that shortages of finite natural resources would result from the world's ever-increasing industrial output, ultimately leading to a global crisis. Some environmentalists responded with a romanticised nostalgia for aspects of the pre-industrial era and this inspired interest in self-sufficiency at a personal level—a desire to go 'back to the land'. Others advocated land value taxation or land nationalisation in order to reduce speculation and redistribute land more fairly, thus making land on which to grow food more accessible to ordinary people. Those who sought change in the food system were by no means all political radicals, however. They included conservatives who were wary of closer political integration with

2 They are therefore based on market value, rather than measures such as volume or calorific requirement.

3 Britain imports 37% of its fertiliser requirement, 69% of pesticides and 63% of primary energy used in agriculture. Government statistics quoted in Lawrence *et al.* 2009.

4 See Charts 7.4 and 7.5 in Defra 2010a.

mainland Europe and critical of any kind of 'dependency culture', whether reflected in social welfare payments or imported food.

Although, strictly speaking, national self-sufficiency concerns crossing predefined geographical boundaries, interest in increasing Britain's level of self-sufficiency has also been prompted by public unease at the distance travelled by foodstuffs from farm gate to consumer—'food miles' (Paxton 1994). Potential environmental and socio-cultural benefits from shortening food supply chains, notably through the development of local food networks, have become widely accepted, and commercial interest in local food has developed, especially in the form of niche markets for regionally produced food.

Recent debate on self-sufficiency has, however, largely been constructed around issues relating to 'food security': physical and economic access to sufficient, safe and nutritious food (FAO 2003). It has grown in response to threats to future food supplies such as climate change, population growth, oil shortage and international terrorism, and was prompted further by the rising global food prices between 2006 and 2008 caused by bad weather, increased energy costs, the use of land for biofuels and commodity speculation.

'Food sovereignty' and 'food patriotism' have also featured in this discussion, serving to highlight how the debate is being constructed around the role of the nation state as well as food security. Public interest in local food and 'buying British' appears to have been partly prompted by people's sense of powerlessness in the face of globalisation. Food sovereignty describes people's right to shape their own food systems and produce food on their own territory to avoid being vulnerable to international market forces (Windfuhr and Jonsén 2005). Although developed in the context of less developed countries, many of its underlying principles are applicable to industrialised countries (Tulip and Michaels 2004).[5] Food patriotism describes consumers' predisposition to favour home-produced food (IGD 2005) and has been used by politicians in the context of proposing clearer country of origin labelling on food.[6]

4.1.2 Historical trends

Britain was largely self-sufficient in food until the Industrial Revolution started in the second half of the 18th century. As the nation industrialised, however, problems in the agricultural sector began to develop. The Corn Laws of 1815, which were intended to encourage food production by restricting grain imports, had to be abolished in 1846 following widespread social unrest in response to rising prices. New developments in economic theory led to a widespread belief that Britain should exploit its comparative strength in manufacturing industry by trading manufactured goods for food and raw materials. This led to growing disinterest in the agriculture sector and its neglect, such that by the start of the Second World War the nation was only 30% self-sufficient in food (Beresford 1975).

5 The concept of food sovereignty has been promoted by *La Via Campesina* (which in Spanish means *the farm worker's way*), an international coalition of NGOs.

6 Notably by the leader of the Conservative Party, David Cameron, in a speech to the annual Oxford Farming Conference in January 2007 (BBC News 2007).

Concern at the evident vulnerability of the nation in wartime to food shortages (to which the response was the 'Dig for Victory' campaign) led to dramatic change after the Second World War. The 1947 Agriculture Act promoted greater self-sufficiency[7] and within five years agricultural output had risen to a level some 50% higher than before the war (Defra 2006). During subsequent decades output continued to increase as a result of greater investment in plant and machinery, drainage and land improvement, and scientific advances in pest control and in animal and plant breeding.

By the early 1970s the nation produced the majority of its food. Although Britain joined the European Economic Community in 1973, domestic agricultural expansion was still regarded as economically important (Bowers 1985). The Labour government's White Paper 'Food from Our Own Resources' proposed increasing agricultural production partly as an insurance against the commodity shortages and fluctuating prices experienced in previous years, but also to reduce the threat posed by an unacceptably large balance of payments deficit (MAFF 1975). A similarly expansionary approach was adopted in a subsequent White Paper, 'Farming and the Nation' (MAFF 1979).

This approach was quietly abandoned by the Conservative government elected in 1979. The level of self-sufficiency in food peaked in 1984, at which point in time the nation produced 83% of indigenous-type food and 63% of all food (95% and 78%, respectively, using the current methodology).[8] The government increasingly interpreted self-sufficiency in a European context. Moreover, the wider political climate had changed, with a growing trend towards trade liberalisation.

Since then, over a 25-year period, the level of self-sufficiency has continually declined, ultimately falling to its present level of barely 60% (Defra 2010a).[9] Various factors have combined to cause this downward trend. Consumers increasingly demand more varied, exotic and out-of-season produce, and the large supermarkets that now dominate food retailing have developed a capacity to source widely in order to meet this demand. Rules governing international trade, the relatively low cost of transport, increased specialisation in the agricultural sector, competition regulations and other factors have favoured the development of international supply chains (Garnett 2003). In addition, from the mid-1990s the high level of sterling, together with the effect of animal health problems such as BSE, curtailed growth in export markets and encouraged imports.

More detailed analysis reveals significant variation in the level of self-sufficiency for different types of food with, for many, a downward trend over the past 15 years. The level of national self-sufficiency in milk and cereals has remained close to 100% and the level for poultry and potatoes has fluctuated around 80–90%, while that of beef has

7 Increased food production was encouraged through an incentive system of 'deficiency payments' to farmers.

8 See asterisked note under Table 4.1, opposite.

9 Despite the revision to the methodology, a sustained downward trend is clear. From 1984 to 1997 the level declined from 63% to 53% (using the previous methodology), while from 1997 to 2009 it continued to decline, from 68% to 59% (using the revised methodology). Equivalent data for indigenous-type food show a decline from 1984 to 1997 of 83% to 69% and from 1997 to 2009 of 82% to 72% (source: Defra; www.defra.gov.uk/evidence/statistics/foodfarm/browsebysubject/documents/selfsuff.xls, accessed 20 May 2010).

TABLE 4.1 Indicative British self-sufficiency ratios (all foods)

Period	Percentage
Pre-1750	~ 100%[a]
1750 to 1830s	~ 90–100%[b]
1870s	~ 60%
1914	~ 40%
1930s	30–40%
1950s	40–50%
1980s	70–80%[c]
2000s	60%

Source: adapted from Defra 2006*

a Temperate produce b Except for poor harvest c Author's estimate†

* The methodology changed in 1998, since when imports and exports of processed food have been revalued such that the data now represents the value of the constituent ingredients (i.e. raw material content). This significantly reduces the estimated value of food consumption. A review of overlapping data for the period 1988–1997 suggests that the effect has been to increase the nominal level of self-sufficiency for all food by around 16-17% and indigenous-type food by around 12–13%. See Annex C in Defra 2006 and the explanation at www.defra.gov.uk/evidence/statistics/foodfarm/browsebysubject/documents/selfsuff.xls (accessed 20 May 2010).

† Defra records the range in the 1980s as 60–70% but this appears to be an underestimate: using the current methodology, in the lowest year during that decade (1988) the level was 71%.

FIGURE 4.1 The UK self-sufficiency ratio, 1956–2007

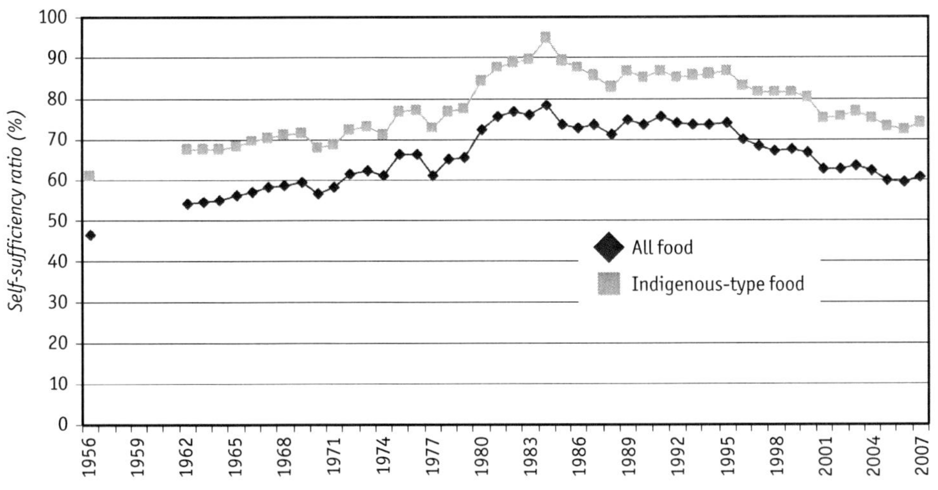

Source: Defra 2008

recovered to around 80% after a decline linked to the BSE crisis. However, Britain has had a consistently low level of self-sufficiency in fruit, with around 90% being imported. In the case of vegetables, the imported proportion has risen from around 25% to more than 40%, while a rising share of ham and bacon (over one-half) and cheese and butter (over one-third) is also being imported. Moreover, imports now account for nearly one-third of the nation's pork and a fifth of its lamb and eggs, whereas around a decade ago it was almost self-sufficient in each (Cabinet Office 2008; Defra 2010a; Living Countryside 2010).

4.1.3 Public policy and its critics

Recent farming and food policy in Britain has been shaped to a large degree by the government-commissioned Curry Report (Policy Commission on the Future of Farming and Food 2002), a key recommendation of which proposed localisation of food chains.[10] Although the report made no direct reference to self-sufficiency, the Commission's chairman, Sir Donald Curry, subsequently revealed that it had considered proposing self-sufficiency targets but concluded that targets for a particular home-produced commodity would attract undue focus and inevitably be difficult to match. His personal judgement was that, as a nation, 'we need to maintain our productive capacity' and 'produce a large and significant share of our own needs' (Maxwell 2005).

The Curry Report was shortly followed by the government's 'Strategy for Sustainable Farming and Food', which, after noting with apparent satisfaction that the prevailing level of self-sufficiency was higher than in the 1950s, declared, 'in an increasingly globalised world the pursuit of self-sufficiency for its own sake is no longer regarded as either necessary or desirable' (Defra 2002: 10). The strategy was later summarised in the following terms: 'profitability matters more than production, sustainability more than size; efficiency more than self-sufficiency' (Defra 2006: 4). It transpired through evidence to the Committee of Public Accounts (2005) that government ministers had been asked directly whether the ability of the country to be self-sufficient in food production should be 'a consideration' in the strategy and replied that it should not. Sir Brian Bender, the senior civil servant in Defra (the Department for Environment, Food and Rural Affairs) appearing before the Committee, said that his personal opinion was that there was no 'bottom line' for self-sufficiency in food.[11] This reflected the view of Margaret Beckett, then Environment Secretary, who, when asked by the Environment, Food and Rural Affairs Committee (EFRAC 2002) whether maintaining a high level of self-sufficiency was government policy, replied that the past decline was 'not an intrinsic concern of mine' and that a 'market approach to British agriculture' was 'incompatible with having a fixed view as to what percentage of what Britain consumes should be produced within the UK'.[12]

10 The Policy Commission was established in the aftermath of a series of animal health problems in the farming sector, including salmonella, BSE and foot and mouth disease.

11 He added that the only exception was the organic sector; the government accepted that consumers wanted more home-produced organic food. See Committee of Public Accounts 2005: Ev. 13.

12 *Minutes of Evidence*, HC550-II, para. 1072-1073.

The Labour government's recent stance has been somewhat ambiguous. It has sought to convince the public that the level of self-sufficiency is satisfactory. Thus the Environment Secretary, Hilary Benn, has highlighted the fact that the level of self-sufficiency in food is higher than in the 1930s and 1950s (Benn 2009),[13] while Defra (2006: 47) has described current levels as 'relatively high by historical standards', presumably because the government considered this to be praiseworthy.[14] At the same time, the government has downplayed the significance of self-sufficiency. Thus the 'Strategy for Sustainable Farming and Food' is critical of pursuing self-sufficiency 'for its own sake' (Defra 2002: 10), self-sufficiency is not among Defra's 'Indicators for a Sustainable Food System' (Defra 2010b), and a paper on food security argues that 'a discourse centred on "UK self-sufficiency" is fundamentally misplaced and unbalanced' (Defra 2006: v). Reluctance to use the concept is also apparent in a key strategy document, 'Food Matters': 'Improving competitiveness in food production, raising sustainable output and building a successful food chain economy are important objectives in their own right. They may result in "positive" movement in self-sufficiency measures but do not need to be justified in those terms' (Cabinet Office 2008: 32).

The government's opposition to targeting an increase in self-sufficiency does not necessarily imply indifference to the level of agricultural output. Defra (2006, 2010c) has noted the potential negative implications of reducing the domestic agricultural base and diminishing the nation's long-term 'capability'. In this respect, its policy is, again, ambiguous: it favours an increase in domestic food production but not on the explicit basis of increasing the nation's level of self-sufficiency.

The government's formal position on self-sufficiency, outlined in a series of reports by Defra on food security (2006, 2008, 2010c), is that increased self-sufficiency 'may offer psychological reassurance' but would not enhance food security because of 'the realities of an interdependent world' (Defra 2006: 47). The fact that its stance has been developed within this specific context may be significant. It is unclear whether Defra has addressed the possibility that a strategy to increase self-sufficiency might help to stimulate a reduction in food miles and thus aid progress towards environmental and social sustainability. By contrast, environmental NGO Friends of the Earth has considered this possibility, arguing in evidence to a Commons enquiry that increased self-sufficiency is 'a desirable policy goal for food security and environmental sustainability' and opposing claims that it will not provide for greater food security: 'High levels of self-sufficiency reflect a strong and resilient food system . . . [and] . . . must be a central plank of a food security policy for the UK' (House of Commons 2009).

There is scope for confusion as the government has sometimes appeared to construct its argumentation around total self-sufficiency rather than increased self-sufficiency. For example: 'Even if it were possible, self-sufficiency would not insulate us against disruptions to our domestic supply chain and retail distribution system' (Defra 2008: 18). More explicitly, the Environment, Food and Rural Affairs Committee (EFRAC) con-

13 Current levels may indeed be regarded as high if compared with the nadir of the 1930s or the recovery period of the 1950s, but they are lower than at any time since the 1970s.

14 As a sign of this ambiguity, Defra (2006) also used the phrase 'pretty normal by historical standards' on page iii of the same report.

cluded, 'The UK should not aim to be self-sufficient, even in indigenous food stuffs. Total self-sufficiency would make the UK's food supplies less secure rather than more secure' (EFRAC 2009: 22). Yet this Committee had already noted that total self-sufficiency is not widely advocated. In other words, Defra and EFRAC have each criticised the straw man of total self-sufficiency without adequately addressing the full range of potential benefits from increased self-sufficiency.

Opinion among farmers concerning self-sufficiency appears to be mixed. Representatives of the National Farmers Union (NFU) support the government's approach of favouring increased agricultural production but rejecting targets for increased self-sufficiency. Former NFU President Tim Bennett argued that targets 'would risk expanding production without market demand, which would inevitably lead to lower prices' (Bennett 2006). Other farming organisations have been critical of the government and expressed support for increased self-sufficiency. A report by the Commercial Farmers Group (2003: 2), for example, argued that the internationalisation of the food chain, combined with the strength of sterling, was leading to 'erosion of our national capacity to produce food and other agricultural products', while another farmer-based lobby group, Food Security (2006), used a national media campaign to 'alert' the public to declining self-sufficiency and the associated threat of disruptions to the food supply through war, terrorism or climate change, the prospect of oil shortages and the 'dubious quality' of imported food.

Consumer opinion on self-sufficiency also appears to be divided, and there is evidence that consumers' purchasing behaviour may sometimes be inconsistent with their stated beliefs. One survey found that 89% of respondents wanted British food to remain widely available, yet only 18% always tried to buy British food even if more expensive, while 29% preferred to buy British food but only if it was the same price and quality as other food; the majority, 51%, indicated that they did not mind where their food came from (IGD 2005).[15] Other research has indicated that consumers who prefer to purchase British food consider the supply to be inadequate: in one survey almost a third of respondents complained that British produce was not readily available (Mintel 2003).

4.2 Increased self-sufficiency as a sustainability strategy

The focus of recent debate in Britain on self-sufficiency has until now been on food security. This chapter adopts a different approach and explores the links between self-sufficiency and sustainable development. Was the former Parliamentary Under-Secretary of State at Defra, Lord Whitty, right to conclude that 'there is no direct relationship between self-sufficiency levels and sustainability'?[16]

In this section arguments for and against increased self-sufficiency in food are considered with reference to each of the three pillars of sustainability (economic, environ-

15 The remaining 2% said that they tried to avoid buying British food.

16 *HL Deb* 2004–05, 22 February 2005, c1,096-1,097.

mental and social).[17] and some underlying themes are identified. The final section assesses whether the need to progress towards sustainable development adds weight to arguments in favour of greater self-sufficiency and, if so, what level would be appropriate and which policy measures might be effective.

4.2.1 The case for increased self-sufficiency

Many advocates of increased agricultural production and, by implication, greater self-sufficiency in food, associate the past decline in self-sufficiency with a lack of *economic sustainability*. They have pointed to a threat to 'critical mass' in the sector, particularly with regard to livestock or at the level of regional economies. For example, in recent years farmers have highlighted acute problems in the pig and dairy sectors (Bennett 2006; Food Security 2006).

Trends in incomes, employment and overseas trade in the agricultural sector appear to indicate a prolonged economic malaise. For example, despite the growth in agricultural output between 1970 and 1990, average income from farming per full-time person employed fell in real terms. Since then they have been volatile, rising to a peak in 1995 and falling to a 30-year low in 2000 (Defra 2010a). Employment in the agricultural sector has long been in decline, with numbers halving over the past 30 years. Less than 2% of the workforce is now employed in agriculture and 29% of employees are over 65 years old (ESRC 2007). The overseas trade deficit in food, feed and drink has widened by over 50% during the past decade, reaching £15 billion in 2007, of which around £9 billion is accounted for by fruit and vegetables and meat; since 1998 imports have risen by 23% and exports have fallen by 2% (Defra 2010a).[18]

Such statistics reflect the downward trend in self-sufficiency over the past 25 years and raise the question of whether there is a floor below which the level should not be allowed to fall. There is a case for government intervention to ensure economic sustainability and preserve a critical mass in vulnerable farming sectors, at the very least. Thus, the decline in fruit and vegetable production has attracted particular concern, and in 2009 the government established a task force 'to increase consumption and production of domestic fruit and vegetables' in response to a report by its Council of Food Policy Advisors (Defra 2010d). Significantly, however, it did not propose a measurable target against which to assess the aims or judge success.[19]

Increased self-sufficiency may also enhance the prospect of greater *environmental sustainability* because a strategy designed to this end should result in shorter food supply chains and less transportation. Research for Defra by AEA Technology (2005) revealed that the amount of food freight has increased substantially over many years. Most is

17 The arguments presented here relate directly to sustainability. Others, such as whether domestically produced food is fresher and (through easier traceability) safer, together with issues relating to consumer rights and 'choice editing', are beyond the scope of this chapter.

18 The increase in imports was greatest for highly processed foods.

19 It thus followed the advice of EFRAC (2009: 39), which also proposed an increase in the production of fruit and vegetables but described targets as 'a crude and, in most cases, impractical way of increasing food production'.

being transported by road; the food industry now accounts for 25% of all HGV vehicle-kilometres. The full cost of this is not included in the price of products but externalised to the general public: additional costs relating to (for example) congestion, accidents and pollution were estimated at £9 billion per annum. The hidden environmental costs of food have been calculated at around 12% of food expenditure (Pretty *et al.* 2005). By encouraging shorter supply chains, a strategy of increased self-sufficiency could reduce these costs significantly.

Estimates suggest that 18–20 million tonnes of food is wasted each year in different parts of the food supply chain. The 'Love Food, Hate Waste' campaign revealed that over 8 million tonnes of this is household waste, equivalent to around one-third of the food that households purchase (WRAP 2010). One means of achieving increased self-sufficiency is to reduce food waste. This would reduce the overall amount of food required to meet consumers' needs and thus the associated environmental impacts: resources used in unnecessary food production and waste that ends up in landfill.

Food policy also affects *social sustainability*.[20] Poor people are especially vulnerable to the food supply problems that are liable to arise from increased global demand (e.g. caused by population growth or increasingly meat-based diets) and threats to supply (e.g. from climate change or competition for land from biofuels). Although Britain's imports account for only a small share of global trade in food, increasing domestic production in order to achieve greater self-sufficiency would contribute towards easing long-term pressures on the global food supply. Adopting a more discerning approach to trade could also enable reduced involvement in global food networks where these are unjust.

In Britain, the poor are liable to feel especially insecure as a result of declining self-sufficiency because they are most vulnerable to higher prices in the event of a disruption to the food supply. In today's interconnected world no nation would be immune from such disruption, but increasing its level of self-sufficiency, particularly if based on a less energy-intensive form of agriculture, would make people feel less susceptible to the threat. Such a strategy would certainly help rural farming communities in Britain, many of whom feel marginalised, exploited and vulnerable (Fearne *et al.* 2005). A significant proportion of small rural settlements have experienced the closure of their last general store or food shop during the past half-century (DETR 2000). Higher farm incomes from increased agricultural output would help to revive such areas and thereby reduce social exclusion.

As a result of today's lengthy, international supply chains, many consumers have only very tenuous links with producers and are largely unaware of how their food has been produced. By encouraging food localisation, increased self-sufficiency could foster local networks of producers and consumers and so create stronger regional economies and an increased sense of community. Jha (2007: 7) concludes: 'The rise in "food patriotism" could be one response to disappearing national boundaries in an amalgamating Europe. Local food can offer a way in which to rediscover valuable community structures and identity.' An example of the social significance of food provenance is seen in the success

20 Social sustainability is concerned with justice, protecting vulnerable groups, overcoming social exclusion and promoting public engagement with environmental problems.

of farmers' markets, which have exposed strong latent public interest in food origins and an appreciation of local 'cultural distinctiveness': 'the richness of difference between places which reflects meaning back to us through the particular accumulations of story upon history upon natural history' (Common Ground 2005).

4.2.2 The case against increased self-sufficiency

An increase in national self-sufficiency in food achieved through market forces would be largely uncontroversial. Criticisms may arise, however, when government intervention to this end is proposed.

One of the fundamental objections to such a strategy is concerned with *economic sustainability*. According to economic theory (the law of comparative advantage), nations gain mutual benefit if they specialise in areas of production in which they have relatively lower costs (due to factors such as climate, location, skills or expertise) and trade with other nations in order to meet consumer demand. It follows from this that Britain should import food if it can be produced more cheaply overseas, because this would reflect the fact that the exporting nation has a comparative advantage in food (perhaps owing to Britain's relatively dense population and variable climate). It would be wrong for the British government to encourage resources to be switched away from other types of economic activity to food production as this would lead to inefficiency, higher costs and, ultimately, less overall prosperity.

The case for increased self-sufficiency in order to reduce the trade deficit in food may similarly be rejected on the grounds that in an open trading system such a deficit is of no real consequence because, in due course, it will be corrected through a decline in the exchange rate. In any case, the contribution of food to Britain's overall trade deficit may be considered low by historical standards (Defra 2006).

An assumption that greater self-sufficiency would enhance *environmental sustainability* may also be challenged. Increasing agricultural production could imply intensification that would harm the environment. Writing as the level of self-sufficiency in Britain peaked in the early 1980s, Bowers (1985: 76) argued that post-war agricultural expansion had a disastrous impact on the British landscape and wildlife habitats, and concluded: 'the threat to the environment in all its aspects would benefit from an abandonment of the wearying dash for growth in agricultural production and still more from some reversal.'

It is certainly the case that, although a strategy designed to increase self-sufficiency would be expected to encourage food localisation, the twin goals of increasing self-sufficiency and reducing food miles may occasionally conflict. In the south-east of England, for example, the latter could involve importing food from northern France rather than transporting it from the north of England or Scotland.

A further tension concerns organic farming, which is often supported on environmental grounds. Preliminary research by Fairlie (2007) compared a range of production systems according to different diets in order to explore the feasibility of total self-sufficiency. His findings indicated that organic farming makes self-sufficiency harder to achieve because 'conventional' farming is more productive in items of land use.

Lastly, far from promising *social sustainability*, a strategy to increase self-sufficiency may threaten to disturb Britain's relationship with other countries. Advocates of trade liberalisation associate open markets with healthy foreign relations. Thus Defra has argued that 'trade makes conflict more costly and less likely' (Defra 2006: 25), whereas promoting self-sufficiency 'fosters reciprocal protection, isolationism and nationalism' (Defra 2006: 45).

A multicultural nation such as Britain also needs to be sensitive to the likelihood that people whose cultural origins lie in other parts of the world will want to purchase food that cannot be grown in this country's temperate climate and any implied restraint on their freedom to do so might create social tension. A report on the ethics of food distribution has warned of the dangers of insularity: 'Local food initiatives and campaigning can be parochial, downplaying the development benefits of international trade and alienating ethnic communities' (Food Ethics Council 2008: 5).

A final concern is that increased self-sufficiency would reduce food security, as Defra argue, and the poor, who are least able to pay higher prices for food, would suffer most in the event of disruptions to the food supply. Britain currently sources food imports widely, mostly from stable countries: 34 countries supply at least 0.5% of its imports and eight European Union countries account for almost two-thirds of total imports. Thus 'serious trade disruption in food would require concerted action by a large number of competitive exporters, which is highly improbable' (Defra 2006: 40). By contrast, a strategy to increase self-sufficiency would make the nation vulnerable to 'risks of adverse weather events, crop failure and animal disease outbreaks' (Defra 2008: 18).

4.2.3 Underlying themes

4.2.3.1 The significance of the nation state

The debate on self-sufficiency in food brings into question the future significance of the nation state. How important are national boundaries when industrial economies are locked into the forces of globalisation and environmental impacts from climate change and pollution cross all such frontiers?

Britain, as an island state, is especially sensitive to the role of the nation in relation to supranational bodies such as the European Union. In the context of importing a vital product such as food, for example, it is faced with distinctive logistical implications, as virtually all imports have to be shipped to a limited number of ports.[21] This may make the nation's inhabitants feel especially vulnerable to external threats, whether or not these are a reality.

Another consideration is whether food should be treated as a special case in international discussions aimed at trade liberalisation. Advocates of food sovereignty argue that food should not be treated as a commodity: 'Food is first and foremost a source of nutrition and only secondarily an item of trade. National agricultural policies must prioritize production for domestic consumption and food self-sufficiency' (Windfuhr and Jonsén 2005: 27). At the same time, any strategy to increase self-sufficiency in an affluent

21 Only a small fraction of food imports is transported by air or through the Channel Tunnel.

nation such as Britain should seek to avoid harming exporters from poor nations by, for example, focusing on indigenous-type foods.

While free trade may well be widely supported in principle, some people appear wary of allowing the nation's food supply to be governed entirely by market forces. The theory that nations should specialise in types of production in which they have a comparative advantage over others attracts only conditional support. One survey found that only 17% agreed with the statement 'If we stopped trying to farm in Britain we could focus our attention on things we're better at doing', whereas 65% disagreed (IGD 2005).

4.2.3.2 Land use priorities

The prospect of food production in Britain being increased in order to achieve greater self-sufficiency raises questions concerning land use priorities. Although global agricultural output has increased faster than the world's population over the past 30 years (FAO 2002), recent trends suggest that self-sufficiency in food at a global level cannot be taken for granted. Increased meat consumption in countries with rapidly expanding economies (notably China) is already having a significant impact on international grain markets and, as indicated above, climate change threatens to reduce food production, particularly (and most disturbingly) in countries that already suffer from serious poverty.

One of the most critical issues concerns the competing demands of food security and energy security. In Britain, the knowledge that North Sea oil production had peaked in 1999 increased interest in biofuels. The United Nations (2007) has warned that rising demand for biofuel crops could push up the price of food, and the increase in world grain prices from 2006 suggested that it might already be having an effect. Having previously been supportive, environmental campaign groups began to express concern about biofuel crops, reasoning that a considerable amount of land would need to be used for biofuels to make a significant contribution towards meeting transport fuel requirements. The OECD (2006) estimated that over 70% of the land currently cultivated in the EU for cereals, oilseeds and sugar would be needed to produce biofuels equivalent to a mere 10% of its transport fuel requirement.

This trade-off between using land for food or fuel has been highlighted by Bill McKelvey, Chief Executive of the Scottish Agricultural College, who, while supporting the development of biofuels, has argued that their production represents a greater challenge to the availability of food in Britain than global population growth, emerging economies or climate change. He attracted national media attention when he warned, 'I don't believe that if we look 50 years ahead we will be in a food-secure situation. It's possible in the next 25 to 50 years that there will be food shortages in the UK' (Clover 2007). His conclusion was that the solution is to be found in more intensive agriculture, including the use of genetically modified crops (von Radowitz 2007).

Other observers have adopted a different stance. The Soil Association (2007), for example, has concluded that the case for mass production of biofuels is flawed because intensive agricultural methods would be required and these demand nitrogen-based fertiliser, which is energy-intensive to produce. It proposes organic farming and a more localised food system as the means of resolving the food–energy trade-off, on the basis that local and organic food production would require less energy than conventional systems.

4.2.3.3 Globalisation and localisation

A strategic decision by government to encourage greater self-sufficiency would need to take account of global politics. The contemporary political culture in Europe is very different from that which prevailed immediately following the Second World War, for example, when centralised planning, quotas, production-linked subsidies and import tariffs were regarded as acceptable means of expanding domestic agricultural production. Over the past half-century international negotiations have sought to 'liberalise' trade by ending such practices, albeit with only partial success. Any government action to encourage greater self-sufficiency would represent a challenge to this free trade ideal.

Trade liberalisation may sometimes be problematic when nations have different values and traditions as standards can often only be agreed at the level of the lowest common denominator. In such a situation, countries that adopt higher ethical standards may disadvantage their domestic producers if required by international agreements to allow access to imports that have been produced with lower standards. For example, British farmers have expressed concern that in other countries 'enforcement of food production standards, in particular controls over inputs such as antibiotics and pesticides, may not be as stringently enforced' (Policy Commission 2002: 38) and have complained about being required to meet stricter animal welfare standards than those that apply in other European countries.[22] A requirement to label goods with the production standards adopted would appear fairer to domestic producers.

In the debate over food security Defra (2006) has largely avoided the trend towards globalisation and the growing concentration of market power among food processors and retailers. By contrast, anti-globalisation campaigners take a more critical view, linking this market concentration with increased risk and insecurity, particularly for poorer social groups. A report by Corporate Watch has proposed that measures be introduced to 'dismantle' food corporations and curb the power of supermarkets in order to promote food localisation. This would enable a shift in the power balance 'away from corporations in favour of a more democratic food system' and the development of 'new food networks . . . outside the corporate controlled food system' (Tulip and Michaels 2004: 38). There has already been a significant growth of interest in localised food networks in recent years, but whether they will progress beyond the fringe remains uncertain. Contrasting the government's goals of supporting trade liberalisation and promoting British agriculture, Garnett (2003: 7) speculates that 'what we may see in future is the co-existence of separate, parallel supply chains: one for niche local and regional food; and another, international one, for the vast majority of the goods we eat.'

22 According to the British Pig Executive (BPEX 2006), over two-thirds of imported pork is from production systems that would be illegal in Britain.

4.3 Increased self-sufficiency as policy

4.3.1 Would greater self-sufficiency in food increase progress to sustainability?

The strongest case in favour of greater national self-sufficiency in food is that it would add impetus to the socio-cultural trend towards food localisation that has already been established because of perceived environmental and social benefits. An added focus on reducing the amount of imported food would further help to shorten food supply chains, resulting in less transportation and reduced environmental impacts.

Defra has argued that 'the links between food transport and environmental damage are weak, complex and differentiated' and that most environmental damage is caused by internal road traffic rather than imports (Defra 2006: 71). The limited data available indeed suggest that, in general, shipping is less harmful to the environment than road transportation. However, food is often transported significant distances by road in other countries before being shipped to Britain and then forwarded from port to final destination. A report on carbon dioxide emissions from transport relating to the food supply chain concluded that 'a reduction in overseas imports is perhaps the most significant challenge we have to address' (Garnett 2003: 10).

A strategy of increasing self-sufficiency would not guarantee progress towards environmental sustainability. Attention would need to be paid to factors such as the mode and efficiency of transport (e.g. loading), retail distribution structures and earlier stages in the life-cycle of food in order to ensure that environmental impacts were reduced. For example, although criticism is sometimes targeted at supermarkets' centralised distribution systems, their logistical efficiency has been carefully developed. Nor can it be assumed that small grocery stores are necessarily more 'local': most of them source widely, if not globally, through wholesalers (Garnett 2003). In addition, it would be important to scrutinise any strategy to increase domestic food production at the level of individual produce. For example, the use of polytunnels or heated greenhouses for growing tomatoes in Britain may consume more energy than importing produce grown outdoors in a warmer climate (AEA Technology 2005).

Arguments that greater national self-sufficiency in food would enhance social sustainability, notably by reviving rural areas, also appear persuasive, although more research is needed to provide empirical evidence. The social benefits of a more localised food system have been seen in the growth of county-based 'food links' initiatives, which are generating prosperity within regional economies and strengthening communities through thriving farmers' markets. In addition, increased production would enable Britain to contribute more towards feeding the ever-rising global population, which would be particularly important if there were to be a global shift towards organic food.

The economic implications of increasing self-sufficiency are more uncertain and shaped by political considerations. Increased agricultural production could lead to higher farm incomes, greater rural prosperity and a reduced trade deficit. On the other hand, internalising social costs would result in higher-priced food for consumers, at least in the short term. Opinions on economic sustainability will be influenced by political judgements on the appropriate model of development (centralised or decentralised, liberal or radical) and distribution of income and wealth between different social groups.

4.3.2 What level of self-sufficiency in food is appropriate?

If the case for increased national self-sufficiency in food is judged to be persuasive, it is necessary to consider what might be an appropriate level of overall self-sufficiency and whether there should be a minimum threshold.

In Britain, the government's stated position is that it would not be appropriate to identify and aim for a specific level of self-sufficiency and neither is there a minimum threshold: 'there is little logic in having fixed minimum targets for self-sufficiency' (Defra 2006: 48). By contrast, a survey showed that the general public believe that there is a minimum level below which the nation's food production should not fall. Only 21% agreed with the statement 'It doesn't matter if Britain produces very little food because we can buy it from elsewhere', whereas 61% disagreed (IGD 2005).

A preliminary question to ask is what level of self-sufficiency is feasible. Although Britain's population density is relatively high by global standards, recent studies have suggested that just enough land is available for the nation to become totally self-sufficient in food, if this was considered necessary, under certain conditions. Cowell and Parkinson (2003) concluded that total self-sufficiency would demand a modest increase in the amount of land utilised by agriculture, alternative crops to be grown in place of non-indigenous crops and agricultural products with relatively low yields to be substituted with others that offer higher yields. Crucially, it would also require a switch to a less meat-based diet, because rearing livestock is a relatively inefficient way of meeting the population's nutritional requirements: meat products have yields as low as 0.2 tonnes per hectare (t/ha), whereas grain yields range from 4.4–6.8 t/ha. Reducing meat consumption would enable crops grown for livestock feed to be replaced by cereals and horticultural crops for direct human consumption. Similar conclusions were drawn by Mellanby (1975) and, more recently, Helms (2004), Watts *et al.* (2006) and Fairlie (2007). The latter two studies considered whether Britain could be self-sufficient in both food *and* energy and concluded that this was feasible as long as people tolerated significant changes in lifestyle.

During the 1980s the level of self-sufficiency in indigenous-type food appears to have regularly exceeded 80% (and, in 1984, 90%), while the average for all food appears to have been well over 70%.[23] An increase from the current levels of 72% (indigenous-type food) and 59% (all food) appears justified, although further research on potential and desirable levels of self-sufficiency should be undertaken. The need for more detailed analysis on potential self-sufficiency was acknowledged by Fairlie (2007) in his update of Mellanby's 1975 study. Assessing the desirable level of self-sufficiency is an even more complex task: the effects of increased agricultural output and a more localised food supply system on, for example, wildlife, food processing and distribution, the retail sector and waste need to be carefully considered.

23 These figures are based on the current methodology. Only estimates are possible due to the change in methodology introduced in 1998.

The susceptibility of food production to climatic and other variables means that self-sufficiency targets will rarely be met precisely, but this is not a persuasive argument for rejecting the idea of an acceptable range or a critical threshold. More research is needed but, on the basis of existing knowledge, a reasonable medium-term aim would be for the nation to produce between two-thirds and three-quarters of its overall food requirement and be no less than 80% self-sufficient in indigenous-type food.

4.3.3 What policy measures might be effective?

Policies to increase the level of national self-sufficiency in food will need to be different to those adopted previously, in the aftermath of the Second World War. The starting point for any contemporary policy discussion is the free market. Thus, Lord Whitty observed that following the 2003 reforms to the Common Agricultural Policy British farmers would need 'to produce what the market wants, rather than what subsidy dictates or what any artificial target for self-sufficiency might dictate' and, in any case, 'there are restrictions on what government can do under the European rules on state aid for promotion of British products within Britain'.[24] The EFRAC report stated bluntly: 'Fixing minimum production levels is incompatible with the unfettered operation of the marketplace' (EFRAC 2002: HC 550-I, para. 93). On the other hand, governments have scope for promoting increased self-sufficiency in ways that do not involve setting 'artificial' targets or 'fixing' production levels, such as adopting measures that enable consumers to identify domestically produced food more readily.

In Britain, a survey by a leading consumer organisation found that around three-quarters of the respondents considered it important for the country of origin to be labelled on products such as fruit and vegetables, dairy products, fish, and meat and poultry used in processed foods (*Which?* 2010). At present, however, European Union rules require manufacturers to provide country of origin information for only a limited number of products,[25] or if it would be misleading not to provide it. Most food labelling is voluntary and this would remain so under recent reforms proposed by the European Commission. One of the most effective policies to increase self-sufficiency might therefore be to improve labelling relating to the country of origin in order to enable consumers to choose more effectively. This was advocated in the Curry Report:

> Compulsory country of origin labelling should be introduced for as wide a range of foods as possible. Such labelling must be straightforward and honest, without legalistic trickery. Country of origin on labels should have its simple English meaning, and food should not qualify as being from a particular country merely on the grounds of having been processed there (Policy Commission 2002: 97-98).

International regulations currently allow 'country of origin' to be interpreted as the 'place of last substantial change', which means that labels may refer to where food has been processed rather than the source of the main ingredients. Evidence suggests that

24 *HL Deb* 2004–05, 22 February 2005, c1,096-1,097.

25 Beef, eggs and poultry (if imported from outside the EU), fish in some circumstances, some fresh fruit and vegetables, honey and olive oil.

this results in considerable confusion among consumers (FSA 2010). There have been two attempts to clarify the law in Britain in recent years but without success.[26] Another problem to be addressed is the use of images carrying 'implications of product origin' rather than wording which indicates more precisely the country of origin of individual ingredients.

Much of the food that people eat has not been prepared in the home from primary ingredients. In these circumstances they are even less well informed about the source of ingredients and less able to identify domestically produced food, which may be their preference. One food critic noted,

> Almost all of the 16 million fresh chickens we buy from supermarkets each week are British. However, many ready meals are made with imported chicken, mostly from the EU, as well as Brazil and Thailand. It's perfectly legal for a chicken curry ready meal to say it's produced in the UK if it's made in a British factory with chicken from Thailand (Stacey 2010).

The need to improve country of origin labelling thus extends to ready meals and food sold in restaurants, cafes, canteens and takeaways.

Improved labelling should be effective in exploiting latent consumer demand for domestically produced food. Another approach would be to stimulate consumer demand by drawing attention to the quality of British food. In Britain, as in other countries, domestically produced food tends to be perceived as of higher quality than imported food (FSA 2010). Research by the IGD (2005: 32) concluded that to increase demand 'people must be given better reasons to buy British food' and recommended that it be better marketed, through, for example, references to seasonality, heritage and the countryside. The National Farmers Union concurred: 'we need to give consumers a reason to prefer British products' (Bennett 2006). Other farming groups have advocated a more interventionist approach. The Commercial Farmers Group (2003), for example, has argued that the import of agricultural products involving unacceptable standards of animal welfare or environmental care should be subject to appropriate labelling or banned, while Food Security (2006) has advocated a similar policy on imports and has urged government action to raise prices for dairy farmers, provide grants to young entrants into farming and offer greater support to the fruit and vegetable sector.

Trade over long distances and, by implication, the level of self-sufficiency will be influenced by energy costs, which government policy can influence through fuel duties. The relatively low price of oil between the early 1980s and 2002, typically under $30 a barrel, made long-distance transportation relatively cost-effective and facilitated the creation of centralised distribution and delivery systems (Garnett 2003). Since then, oil prices have risen substantially, reaching around $80 a barrel in 2010. This increase should encourage shorter, more localised, food supply chains, as long as the public pressure to reduce fuel duties is resisted by the government and industry treats this higher level as the future long-term norm. The government thus needs to signal that it would use its ability to raise fuel duties to prevent energy prices from falling below the current

26 Richard Bacon MP has introduced Food Labelling Bills, in 2004 and 2008, but neither obtained government support and passed into law (*HC Deb* 2003–04, 22 March 2004, c580-82; *HC Deb* 2007–08, 29 October 2008, c897-99).

level, a policy that is fully justifiable in the light of declining domestic reserves of oil and gas.

Finally, the level of self-sufficiency that is achievable will depend in part on the nation's diet, which may be influenced by the government's public health policy. In Britain, the policy of encouraging people to eat more fresh fruit and vegetables is liable to reduce overall self-sufficiency because, even now, domestically produced fruit and vegetables account for only 37% of consumption (Defra 2010d). Some fruit cannot be grown in Britain's temperate climate, but there is considerable potential to reverse the decline that has taken place: for example, Britain has an ideal climate for growing apples and yet around two-thirds are imported (Defra 2009). Another public health policy that would increase the nation's self-sufficiency is to promote lower meat consumption because, in addition to the potential health benefits, rearing livestock is (as noted earlier) relatively inefficient in terms of land use (Fairlie 2007).

4.4 Conclusion

Recent public debate in Britain on the level of self-sufficiency in food has primarily been undertaken in the context of food security. This chapter has adopted a different approach, focusing on issues relating to economic, environmental and social sustainability, in order to reframe and broaden the debate. It has presented a case study of Britain to explore whether, as a general principle, a minimum level of national self-sufficiency in food should be considered an essential element of a sustainable farming and food strategy.

The chapter has raised issues such as vulnerability to the domestic weather, possible negative impacts from more intensive production, conflict with the lower yields of organic production, the benefits of healthy trading relationships, the possibility of higher prices and an implied reduction in consumer choice. It has nonetheless concluded that, on balance, the case for increased self-sufficiency on grounds of increased environmental and social sustainability is persuasive.

While there are clear benefits to be gained from the European Union as a whole being broadly self-sufficient in food, a considerable amount of trade in food takes place over unnecessarily long distances.[27] There is increasing support for food localisation and a key means of achieving such change is for action towards this end to take place at the level of individual member states.

This chapter has demonstrated that there are clear linkages between self-sufficiency and sustainability. The reduction in self-sufficiency in food that has taken place in Britain over the past 25 years appears to be at conflict with the need for environmental and social sustainability. A strategy to reverse this would add further impetus to the trend, already established, towards localisation of the food system. Whether the same would be true for other European countries will depend on their individual circum-

27 A particular concern is the extent to which virtually identical products are traded between nations. See Lucas 2001.

stances. This chapter has presented a simple framework that could be used to assess whether their levels of national self-sufficiency in food are satisfactory.

In Britain, the government has opposed setting target levels for self-sufficiency in food but has failed to indicate whether it believes that there is a threshold below which a reduction would be unacceptable. Its stance is, at best, ambiguous. It has sought to reassure the public that the level of self-sufficiency is 'relatively high by historical standards' while, at the same time, downplaying the significance of self-sufficiency. It evidently favours an increase in domestic food production but not on the basis of increasing the level of self-sufficiency in food.

A minimum level of national self-sufficiency in food ought to be identified and strategies developed to enable it to be achieved. As food production cannot be predicted in the same way as industrial production, due to factors such as the weather, broad and flexible targets should be used. A reasonable target for the medium term would be for Britain to produce between two-thirds and three-quarters of its overall food requirement and be no less than 80% self-sufficient in indigenous-type food.

The necessary policies to encourage increased national self-sufficiency would complement those that are already in place to promote regional and local food. They ought primarily to be market-led, exploiting existing consumer preference for domestically produced food and stimulating further demand. Initiatives to improve country of origin labelling, promote the high quality of domestically produced food, ensure that energy prices are at stable and realistic levels, and encourage less meat consumption are examples of policies that should be introduced in Britain and may well be equally appropriate elsewhere.

References

AEA Technology (2005) 'The Validity of Food Miles as an Indicator of Sustainable Development' (final report produced for Defra; https://statistics.defra.gov.uk/esg/reports/foodmiles/final.pdf, accessed 15 March 2010).

BBC News (2007) 'Cameron's "Buy British Food" Call', BBC News; news.bbc.co.uk/1/hi/6226703.stm, accessed 19 March 2010.

Benn, H. (2009) Speech by Rt Hon Hilary Benn MP, *Chatham House Conference on Food Security 2009*, London, 3 November 2009; webarchive.nationalarchives.gov.uk/20100401103043/http://www.defra.gov.uk/corporate/about/who/ministers/speeches/hilary-benn/hb091103.htm, accessed 20 May 2010.

Bennett, T. (2006) 'UK Farming's Place in World Markets', *Sentry Conference*, Linton, Cambridgeshire, UK, 15 February 2006; www.sarpn.org.za/documents/d0001948/UK_NFU_speech_Feb2006.pdf, accessed 15 March 2010.

Beresford, T. (1975) *We Plough the Fields* (Harmondsworth, UK: Penguin).

Bowers, J.K. (1985) 'British Agricultural Policy since the Second World War', *Agricultural History Review* 33.1: 66-76.

BPEX (British Pig Executive) (2006) 'An Analysis of Pork and Pork Products Imported into the United Kingdom' (Milton Keynes, UK: BPEX).

Cabinet Office (2008) 'Food Matters: Towards a Strategy for the 21st Century' (The Strategy Unit; www.cabinetoffice.gov.uk/media/cabinetoffice/strategy/assets/food/food_matters1.pdf, accessed 22 March 2010).

Clover, C. (2007) 'Growing demand for biofuels "could lead to fuel shortages" ', *Daily Telegraph,* 19 April 2007; www.telegraph.co.uk/news/uknews/1548917/Growing-demand-for-biofuels-could-lead-to-food-shortages.html, accessed 12 March 2010.

Commercial Farmers Group (2003) 'The Case for a Sustainable UK Agricultural Industry and National Food Security'; www.commercialfarmers.co.uk/CFGDoc2003.pdf, accessed 12 March 2010.

Committee of Public Accounts (2005) 'Helping Farm Businesses in England' (12th Report of Session 2004-05, Minutes of Evidence 13 October 2004; HC 441).

Common Ground (2005) *Producing the Goods: Goods that Reflect and Sustain Locality, Nature and Culture* (Shaftesbury, UK: Common Ground).

Cowell, S.J., and S. Parkinson (2003) 'Localisation of UK Food Production: An Analysis Using Land Area and Energy as Indicators', *Agriculture Ecosystems and Environment* 94: 221-36.

Defra (UK Department for Environment, Food and Rural Affairs) (2002) 'The Strategy for Sustainable Farming and Food: Facing the Future' (London: Defra).

—— (2006) 'Food Security and the UK: An Evidence and Analysis Paper' (London: Defra).

—— (2008) 'Ensuring Food Security in a Changing World' (Discussion Paper; www.ifr.ac.uk/waste/Reports/DEFRA-Ensuring-UK-Food-Security-in-a-changing-world-170708.pdf, accessed 3 March 2010).

—— (2009) 'Basic Horticultural Statistics'; https://statistics.defra.gov.uk/esg/publications/bhs/2009/Basic%20Horticultural%20Statisitcs%202009.pdf, accessed 26 March 2010.

—— (2010a) 'Agriculture in the UK 2009'; www.defra.gov.uk/evidence/statistics/foodfarm/general/auk/latest/documents/AUK-2009.pdf, accessed 20 May 2010.

—— (2010b) 'Indicators for a Sustainable Food System'; www.defra.gov.uk/evidence/statistics/foodfarm/general/foodsystemindicators/documents/foodsystemindicators.pdf, accessed 20 May 2010.

—— (2010c) 'UK Food Security Assessment: Detailed Analysis', August 2009, updated January 2010; www.defra.gov.uk/foodfarm/food/pdf/food-assess100105.pdf, accessed 3 March 2010.

—— (2010d) 'First Report of the Council of Food Policy Advisors'; www.defra.gov.uk/foodfarm/food/policy/council/pdf/cfpa-rpt-090914.pdf, accessed 22 March 2010.

DETR (UK Department for the Environment, Transport and the Regions) (2000) 'Our Countryside: The Future' (Cm 4909; London: DETR).

EFRAC (Environment, Food and Rural Affairs Committee) (2002) 'The Future of UK Agriculture in a Changing World' (Ninth Report of Session 2001-02, 'Report', HC 550-I and 'Minutes of Evidence', HC 550-II).

—— (2009) 'Securing Food Supplies up to 2050: The Challenges Faced by the UK' (Fourth Report of Session 2008-09, 'Report', HC 213-I).

ESRC (Economic and Social Research Council) (2007) 'Agriculture, Farming and Rural Policy in the UK' (ESRC Fact Sheet; www.esrc.ac.uk/ESRCInfoCentre/facts/UK/index46.aspx?ComponentId=12663&SourcePageId=18130, accessed 21 May 2007).

Fairlie, S. (2007) 'Can Britain Feed Itself?', *The Land* 4 (Winter 2007/8): 18-26; www.thelandmagazine.org.uk/sites/default/files/The%20Land%204.pdf, accessed 20 May 2010.

FAO (Food and Agriculture Organisation) (2002) 'World Agriculture 2030'; www.fao.org/english/newsroom/news/2002/7833-en.html, accessed 16 March 2010.

—— (2003) 'Trade Reforms and Food Security'; www.fao.org/docrep/005/y4671e/y4671e00.htm#Contents, accessed 16 March 2010.

Fearne, A., R. Duffy and S. Hornibrook (2005) 'Justice in UK Supermarket Buyer–Supplier Relationships: An Empirical Analysis', *International Journal of Retail and Distribution Management* 33.8: 570-82.

Food Ethics Council (2008) 'Food Distribution: An Ethical Agenda' (Report; www.foodethicscouncil.org/system/files/fooddistribution.pdf, accessed 22 March 2010).

Food Security (2006) *The Real Facts, The Real Problems, The Real Solutions, The Action* (Epping, UK: Food Security Ltd).

FSA (Food Standards Agency) (2010) 'Country of Origin Labelling: A Synthesis of Research' (Executive Summary, January 2010; www.food.gov.uk/multimedia/pdfs/coolsynsummary.pdf, accessed 20 May 2010).

Garnett, T. (2003) 'Wise Moves: Exploring the Relationship between Food, Transport and CO_2' (London: Transport 2000).

Helms, M. (2004) 'Food Sustainability, Food Security and the Environment', *British Food Journal* 106.5: 380-87.

Hines, C. (2000) *Localization: A Global Manifesto* (London: Earthscan).

House of Commons (2009) Memorandum SFS 70 submitted to EFRAC by Friends of the Earth; www.publications.parliament.uk/pa/cm200809/cmselect/cmenvfru/213/213we65.htm, accessed 8 March 2010.

IGD (Institute for Grocery Distribution) (2005) 'Connecting Consumers with Farming and Farm Produce' (Watford, UK: IGD).

Jha, P. (2007) 'Global v local: How to Choose', in 'The Ethics of Food', *New Statesman*, 22 January 2007 (supplement): 9.

Lawrence, G., K. Lyons and T. Wallington (eds.) (2009) *Food Security, Nutrition and Sustainability* (London: Earthscan).

Living Countryside (2010) 'Farming Today'; www.ukagriculture.com/farming_today/livestock_self_sufficiency.cfm and www.ukagriculture.com/farming_today/crops_self_sufficiency.cfm, accessed 16 March 2010.

Lucas, C. (2001) 'Stopping the Great Food Swap' (Brussels: The Greens/European Free Alliance, the European Parliament).

MAFF (UK Ministry of Agriculture, Fisheries and Food) (1975) 'Food from Our Own Resources' (Cmnd 6020; London: HMSO).

—— (1979) 'Farming and the Nation' (Cmnd 7458; London: HMSO).

Maxwell, F. (2005) 'Food Self-sufficiency Low on the Political Agenda', *The Scotsman*, 2 February 2005.

Mellanby, K. (1975) *Can Britain Feed Itself?* (London: Merlin).

Mintel (2003) 'Attitudes towards Buying Local Produce: UK', Mintel Market Intelligence, January 2003.

OECD (Organisation for Economic Cooperation and Development) (2006) 'Agricultural Market Impacts of Future Growth in the Production of Biofuels' (Directorate for Food, Agriculture and Fisheries; Paris: OECD).

Paxton, A. (1994) 'The Food Miles Report: The Dangers of Long Distance Food Transport' (London: Sustainable Agriculture, Food and Environment [SAFE] Alliance).

Policy Commission on the Future of Farming and Food (2002) 'Farming and Food: A Sustainable Future' (London: Cabinet Office).

Pretty, J., A.S. Ball, T. Lang and J.I.L. Morison (2005) 'Food Miles and Farm Costs: The Full Cost of the British Food Basket', *Food Policy* 30.1: 1-19.

Schumacher, E.F. (1974) *Small is Beautiful* (London: Abacus).

Seymour, J. (1976) *The Complete Book of Self-sufficiency* (London: Dorling Kindersley).

Soil Association (2007) 'Peak Oil: The Threat to our Food Security'; www.soilassociation.org/peakoil, accessed 21 May 2007.

Stacey, C. (2010) 'England's Food Renaissance', BBC Food; www.bbc.co.uk/food/food_matters/britishfood.shtml#truly_english?, accessed 16 March 2010.

Tulip, K., and L. Michaels (2004) 'A Rough Guide to the UK Farming Crisis' (Oxford, UK: Corporate Watch).

United Nations (2007) 'Sustainable Bioenergy: A Framework for Decision Makers'; www.fao.org/docrep/010/a1094e/a1094e00.htm, accessed 16 March 2010.

Von Radowitz, J. (2007) 'Intensive Farming and Food Production: What are the issues?', *Scottish Agricultural College News*; www.sac.ac.uk/news/newsletter/may2007/?page=intensivvefarming, accessed 13 May 2007.

Watts, B., S. Howard, L. Markham, A. Elmualim and K. Hutchinson (2006) 'A Framework for UK Self Sufficiency for Food and Energy from Renewable Resources', *Association for Environment Conscious Building Conference*; www.aecb.net/conference2006.php, accessed 16 March 2010.

Which? (2010) 'Country of origin rules should be expanded, says *Which?*', press release, 14 January 2010; www.which.co.uk/about-which/press/press-releases/campaign-press-releases/food-and-health/2010/01/country-of-origin-rules-should-be-expanded-says-which.jsp, accessed 20 May 2010.

Windfuhr, M., and J. Jonsén (2005) 'Food Sovereignty: Towards Democracy in Localized Food Systems' (paper by FIAN [FoodFirst Information and Action Network]; Rugby, UK: ITDG Publishing).

WRAP (Waste and Resources Action Programme) (2010) 'Supply Chain Food Waste'; www.wrap.org.uk/retail/food_waste/supply_chain_food.html, accessed 18 March 2010.

5

Transition towards sustainable consumption and production? The case of organic food in Denmark

Michael Søgaard Jørgensen
DTU Management Engineering, Technical University of Denmark

This chapter discusses mechanisms in the shaping of organic food as a strategy in the Danish food sector since the 1980s as a contribution to the discussion of strategies for the development of more sustainable production and consumption of food. The background to the chapter is not only the major achievements in Denmark within organic food since the 1980s but also the reduction in recent years in area given over to organic agriculture.

Organic agriculture is defined here as agriculture, where use of artificial fertiliser and chemical pesticides is not permitted and where animal husbandry is based on organic fodder and some stricter ethical demands than conventional husbandry.

The development towards around 6% of the agricultural land being organic and around 25% of the milk consumed by Danish consumers being organic is seen as a radical innovation; it is also a system innovation because it involves changes to the food chain, from changes in agriculture to changes in consumption. The analysis in this chapter sees these innovations in terms of the creation of a new technological field—that of organic food—where the process is seen to involve a combination of path creation and path dependence, involving the shaping of new institutions, new structures and new knowledge fields as well as the reshaping of existing institutions, structures and knowledge fields. These processes involve an ongoing interaction between production, consumption or use, knowledge and regulation, with these systems constantly shaping each other (Karnøe and Garud 1998). The technological field itself is shaped as part of the transition. For example, the definition of organic agriculture in Danish governmental regulation focuses on environmental issues rather than on social issues, which were important to the first organic farmers.

The analysis shows how the initial conditions in the shaping of a technological field can be a constraining as well as an enabling factor in a transition. The main Danish export—pork—is produced in specialised farms with indoor sties, thus limiting the possibility of converting easily to organic pork production. The focus of organic production has been more on milk production, since the existing system of dairy farming is more easily converted to organic production. The close links between two large dairy companies and a major cooperative retail chain have been important in the development of retail sales of organic milk products. The retail chain was able to influence the dairies to start buying and processing organic milk and, later on, to convince them to lower their prices as part of a more ambitious retail strategy for organic food.

Governmental policy has played and is still playing a major role in the development of organic food in Denmark. This includes national governmental regulation and international regulation, especially within the European Union. Denmark was the first country to introduce a national support scheme for organic agriculture and for research and development (1988), based on a law on organic agriculture (1987).[1] The idea was to generate consumer confidence in organic food in order to develop a bigger market for such food. This market-based approach, however, has also caused several problems in the interaction between demand and supply because of strong price competition in the Danish retail sector. Furthermore, organic agriculture and food have a limited role in national strategies to reduce the environmental impact of agriculture. Danish governments have seen organic food mostly as a strategic product niche and not as an environmental strategy for agriculture in general, although organic agriculture is given some preference in the granting of applications for environment-related subsides to agriculture in environmentally vulnerable areas.

5.1 Case description

5.1.1 Overview

Since the 1970s in Denmark, organic food has changed from being a social and environmental vision of a few environmentalists and people interested in new ways of ownership in agriculture to an important strategic market niche in a highly competitive food market, but without societal, environmental or social targets. Integration into the conventional food sector has changed the principles of organic food production. Some

1 Lov om Økologisk jordbrugsproduktion. Lov nr. 363 af 10.10 1986 (Law on Organic Agriculture. Law no. 363 of 10 October 2006; available at http://retsinformation.w0.dk/Forms/R0710.aspx?id=79789&exp=1; accessed 24 May 2010).

actors still have visions about the possible societal role of organic food production, as a way of transforming the whole food sector, but this is not the dominant vision driving its development today (Ingemann 2002).

When the development within organic food in Denmark is assessed, the criteria will depend on the stakeholder who is making the assessment. The following are some of the aspects that are brought forward when different stakeholders discuss either the success or the problems related to organic agriculture and organic food:

- The environmental impacts of agriculture are reduced as a result of, for example, the reduced risk of pesticide residues or nitrate in the groundwater
- Food is healthier because of the lower pesticide residues and nitrate content
- Food is healthier because of the higher content of so-called 'secondary metabolites'
- Biodiversity in agricultural fields is improved
- The food does not contain genetically modified (GM) ingredients
- Products taste better
- Food is produced through more ethical animal husbandry
- Income is generated from food exports
- Regional development is strengthened through regional interaction along the product chain, from field to table

This list shows that organic food is supported by a number of different stakeholders and for a number of different reasons.

Table 5.1 presents a division into phases of the development of organic food in Denmark since the 1970s. Each phase is characterised by some milestones and some structural characteristics. Table 5.1 lists also some important stakeholders in this development.

5.1.2 Case context: landscape and regime

It is not possible to talk about a specific landscape or a specific regime when discussing comprehensive changes as the development within organic food in Denmark involves agriculture, food processing, distribution, retailing and consumption. Instead, the development should be characterised as the result of interdependent developments between societal development, the food sector in general and the production and consumption of organic food. Furthermore, one cannot talk about a planned change organised by some specific stakeholders. However, based on an analysis of the development one can point to some of those characteristics of Danish society and the food sector that have turned out to provide important potential and barriers:

TABLE 5.1 Phases in the development of organic food production and consumption in Denmark since the 1970s

Source: developed from Kristensen and Nielsen 1996; Ingemann 2002

Period	Milestone	Characteristics	Description
1970s	Small increase in number of organic farms	Pioneer phase	No integration with conventional agriculture. Different value concepts are evident in the organic movement, with a focus on new forms of ownership and everyday life and the environmental impact of the agriculture
1981–87	Start of a national organisation for organic agriculture	Expansion Formation of a national organic agriculture organisation and school	National organisation and school are dominated by agricultural expertise and less by people with a broader societal perspective on organic agriculture. Some expansion of organic agriculture is seen. Sales are primarily through health-food stores and local buying clubs
1987–92	A national label for organic food is created	Integration Governmental acceptance Expansion	Organic sale companies are established. There is now some integration with conventional food companies and organisations. National law is passed on organic agriculture, with a focus on the closing of nutrient cycles, avoidance of hazardous chemicals and animal welfare. Governmental law and a subsidy scheme are introduced for conversion to organic agriculture
1992–98	Retail chains make organic food a strategic area	Organic food as a strategic niche Consolidation of the industrial processing of organic food	After some market saturation, organic food achieves a more strategic competitive role in the retail sector. Consolidation occurs in the industrial processing of organic food. More professional organic organisations are seen within agricultural politics and in sales and marketing
1998–2007	New forms of production and supply	Direct supply to consumers Small, independent dairies re-emerge	Some box schemes appear for direct supply to the consumers of vegetables and some other types of fresh food are developed New small, independent dairies are started A stagnating, and later shrinking, organic agricultural area A stagnating, and later increasing, market share for organic food

- A strong, export-oriented agriculture and food industry with a number of big farmer-owned cooperatives and many institutions owned by the farmers and their organisations, such as an agricultural advice service, a meat research institute and so on
- Ongoing centralisation in all sectors along the food chain from field to table: agriculture, food industry and food retailing
- Big farmer-owned cooperative food-processing industries within several product areas, especially within the meat industry and dairy industry
- A retail sector highly focused on price and on the constant launching of new products as important competitive parameters
- An increasing societal focus on the negative environmental impacts of agriculture and increasing national and EU regulation of the environmental impact of agriculture and of some nutritional and hygiene aspects of food

As the analysis later in this chapter will show, the relationships between conventional and organic food stakeholders can be characterised by elements of competition and cooperation and have been instrumental in shaping each other.

5.1.3 Case history and development

Denmark was the first country to introduce a national support scheme for organic agriculture and for research and development (1988), starting with a law on organic agriculture (1987).[2] Also a governmental inspection and control system and a national label for organic food were introduced. The idea was to generate consumer confidence in organic food (Ingemann 2002). A private certification system was introduced in 1981 and this triggered the initial growth of organic agriculture (Michelsen and Søgaard 2001). In this earlier system, confidence was based on the independent control of farmers by other farmers, based on an inspection scheme by the national organisation for organic agriculture (Landsforeningen Økologisk Jordbrug [later Økologisk Landsforening]).

As well as acting as a general support for conversion to organic agriculture, the national support scheme has also been used as a more specific instrument, such as the provision of increased subsidies to pig farmers and plant farmers to convert their farms to organic agriculture. The background to this was a lack of organic grain in the mid-1990s and the much smaller conversion rate within pig husbandry compared with the conversion rate for dairy cattle farming and vegetable growing (Michelsen and Søgaard 2001).

5.1.3.1 A market-based approach to organic food

The relation between supply of and demand for organic food has been an ongoing topic of discussion since the 1980s between the organic farmers' organisation, the food indus-

2 See footnote 1 on page 83.

try, retailers and the ministry regulating agriculture and food, because the approach to the development of organic food production has been an approach based on developing supply and demand on a market basis. Although the environmental impacts of conventional agriculture and the animal welfare in conventional agriculture are discussed, organic food production has not been developed into a joint vision for a transition of conventional agriculture into a sustainable system of food production. An explanation for this seems to be the large differences between, on the one hand, highly specialised conventional farms and an ongoing centralisation in agriculture and food industry and, on the other hand, the principles underlying organic agriculture: closed nutrient cycles, the precautionary principle and close relations between farmer and consumer.

The conversion rate to organic agriculture has been many times greater within dairy cattle farming than within pig farming. The difference in the necessary changes for a conversion to organic agriculture between dairy cattle farming and pig farming may be part of the explanation for the different conversion rates within the two types of agriculture. Both types of farm are important in Denmark, but it is much easier to convert a dairy cattle farm into organic agriculture than a pig farm. A number of conventional dairy cattle farms keep their cattle outside for part of the year, so conversion to organic agriculture could be done without large investments in new sheds. The opposite is the case for pig farming, where the demand for pigs to be able to go outside for a part of the year demands a big change in animal husbandry and large investments in changes to the sties. The difference in conversion rate can be seen in the relative number of farms in organic production: 18.7% of all organic farms are dairy farms, whereas 13.7% of all conventional farms are dairy farms; 8.4% of organic farms have pigs, whereas 18.7% of conventional farms have pigs (Plantedirektoratet 2006).

5.1.3.2 Cooperation, competition and interactions in the development of the production and consumption of organic food

The major dairies and pig slaughterhouses have traditionally been farmer-owned cooperatives, where the farmers sign in as partners and suppliers for a number of years and in return receive a guarantee that the company will buy all the milk or the pigs the farmers supply. During the past 20 years relations between these farmer-owned food-processing industries and organic farmers have shifted several times between competition and cooperation, and the transition towards more organic food also shows elements of the creation of new companies that are reshaping existing companies to take on new roles within the organic food industry.

The cooperative retail chain, now 'Coop', has played a significant role in the development and expansion of organic food production in Denmark. The chain has provided and is providing an important market for organic products, but the conditions that the cooperative chain has demanded from farmers and supplier companies in order to be able to supply all shops via a centralised national distribution system have also shaped organic food production towards, for example, fewer suppliers, implying that organic farmers had to cooperate in terms of distribution, packaging and so on. Perhaps the first step towards integration between organic food production and the conventional food industry was taken when, in the mid-1980s, the cooperative chain, as a big customer, asked the then two major conventional dairy companies to start buying milk from organic farmers; and Coop said they were considering starting to buy organic milk directly from a small organic dairy. The cooperative chain and the two big conventional cooperative dairy companies were at that time closely related and had, for example, jointly developed a centralised system for fresh food supply to the retail sector. After some time the two dairies finally accepted the request from the cooperative chain and guaranteed the farmers a premium price of 40% compared with conventional milk. The increase in number of organic farms by the end of the 1980s was probably encouraged by this decision of the two big conventional dairy companies (MAF 1995).

5.1.3.3 The development of organic food into a strategic market niche

Besides being the first retail chain to start selling organic food, as encouraged by a number of organic farmers, the cooperative retail chain was also approached by key stakeholders in the organic sector at the beginning of the 1990s and encouraged to develop a more proactive market strategy on organic food by reducing retail prices and expanding the retail assortment of such products. When this strategy was implemented in 1993 it more or less immediately made organic food a strategic product area in the retail sector and even caused a deficit of organic milk. At that time, a new small organic dairy was able to obtain an agreement with a supermarket chain to stock its products, because this would give the chain a unique profile with regard to its organic food.

Since the mid-1990s the most organic food has been sold via supermarkets, whereas early on in its history health food stores were the main retail channel. In 2002 a consumer survey showed that nearly 69% of consumers had listed the cooperative retail chain as their preferred channel, 42% listed private supermarket chains and only 8% listed food markets and 3% health food stores as their preferred channels (Økologisk Landsforening 2002). In 2006 retail shops (discount shops, supermarkets, etc.) accounted for around 70% of the total sale of organic food; small shops and petrol stations accounted for around 8%; health food stores, box schemes and direct farm sales accounted for around 10%; and public and private catering (canteens, schools, restaurants, etc.) accounted for around 12%. The share of retail sales from discount shops is now almost as high as from traditional supermarkets. Of the 70% of sale through retail outlets, each of these two types of outlet accounts for around 40% of sales.

There are big differences between retail chains with respect to the ratio between their market share of organic food and their total market share for food. One supermarket chain has an organic food market share that is almost twice its total share of the food market as a whole. One discount chain and all the different cooperative retail chains have an organic food share that is slightly above their total share of the food market, and the other chains have organic food market shares that are substantially lower than their total share of the food market (Økologisk Landsforening 2007).

The integration of organic food into the basic offerings of major supermarket chains as a strategic area is now being followed by a phase in which organic food has to comply with general conditions of competition in the retail sector. These conditions have turned out to be strong price competition, a continuous demand for new products and high demands regarding the turnover per metre of shelf in the supermarkets. In some cases, a sharpening of price competition has caused supermarkets to reduce their assortment of organic food. The busy lives of many consumers may mean that they will buy the corresponding conventional product if they do not find the organic product in their usual supermarket, as they do not have time to search for the product in another shop (Økologisk Landsforening 2002).

In recent years, options for the delivery of organic food (especially organic produce) to households, either at their home address or at the work—so-called box schemes—have been developed. This kind of service has probably grown because it makes it less time-consuming for the consumer to get access to these products, delivering them directly to their front door. It is not clear whether this new distribution channel has implied a higher consumption of organic food or whether it has mainly redistributed some of the shopping from supermarkets into a box-scheme system.

5.1.3.4 Competition from other 'green' concepts

The relative success of organic food has also brought competition from other 'green' concepts developed by conventional food companies. In the mid-1990s the vegetable producers' association tried to get the retailers to recognise their concept of so-called 'integrated production' as a 'green' label and pay a premium price for the products, as for the organic products. The cooperative retail chain refused to recognise the label because it did not necessarily imply an environmental advantage and because the chain found that another 'green' label in addition to the organic label would confuse consumers. The label is still there, but it is not being marketed as a type of eco-labelling.

The relatively high share of organic dairy products and the low share of organic meat products is, as mentioned above, a consequence of the ease with which dairy farms can be converted and the difficulties of converting pig farms. It is also a consequence of the difficulties of obtaining premium prices for organic pork, which is a highly price-competitive product area in the Danish retail sector. However, the increased focus on animal ethics since the end of the 1990s has allowed some increase in the market share of organic meat, but a number of private labels on conventional fresh meat with a focus on quality and animal ethics have made the market opportunities for organic meat in this part of the market more difficult. A survey of consumers shows that some consumers actually think they are buying organic meat when in fact they are buying some of these ethical but conventional products (Økologisk Landsforening 2002).

5.1.3.5 The role of the knowledge system

The knowledge system, understood as the advice system and the research system, has played an important role in the development of organic food production and consumption. New institutions have been developed, but they have gradually developed cooperation with existing institutions in terms of the conventional agricultural advice system and 'mainstream' food researchers.

Denmark has a long tradition of agricultural advisers organised by the farmers' associations. When organic agriculture became more widespread in the 1980s it became clear that there was a need for advice on organic agriculture. Based on an emerging dialogue between the national organic agriculture organisation (Landsforeningen Økologisk Jordbrug) and the general farmers' associations it was agreed that the advice could be organised in association with the general farmers' association (Danske Landboforeninger). It was decided to create positions for dedicated organic agricultural advisers, because the existing agricultural advisory system was very much integrated into the conventional agriculture concept and its use of artificial fertiliser, pesticides and so on and cooperated with the suppliers of these products. Later on, cooperation between the two groups of advisers was organised by training some of the conventional advisers in organic agriculture, because it turned out to be a barrier to the organic advice system that the knowledge of the organic advisers was not 'deep' enough compared with the depth of knowledge the more specialised conventional advisers had. By combining advice from the two groups it became possible to offer more specialised advice to the organic farmers. It seems as if there also has been some transfer of experience from organic agriculture to conventional agriculture. For example, some conventional farmers have adopted the organic farmers' use of pig fodder with more fibre because it reduces the risk of infections in the guts of the pigs (Sørensen 2003).

Research into organic food production in Denmark has been carried out on different topics, including:

- The practice on organic farms in relation to nutrient flow, yield and so on
- Experiments with the development of more efficient and less polluting organic agricultural practice
- The environmental impact of organic agriculture
- Good practice in organic food processing
- Action research in the development of standards for processed food , the development of organic food products and the initiation of production of these organic products
- The implementation and use of organic food in food catering
- The attitudes and practice of consumers towards organic food
- Scenarios for further transition of agriculture in general towards organic agriculture

Most of the research has focused on the agricultural side and a smaller part on food processing and strategies for increasing private and professional consumption of organic products. Some of the agricultural research has been carried out by researchers who previously did research in conventional agriculture and some by researchers who have worked only within organic agriculture and organic food. Much of the research has been carried out in cooperation with stakeholders within the organic food sector.

5.1.3.6 The role of government regulation

Government regulation has played and still plays a major role in the shaping of organic food production in Denmark. This includes regulation at different levels: regional, national and international, especially within the European Community (EU). Also, the regulation of conventional agriculture plays a role in shaping organic agriculture. Several types of government regulation have been implemented. Elzen and Wieczorek (2005) describe three general governance paradigms, which can also be recognised as part of the governmental regulation of organic food production in Denmark:

- **Classical steering**, with a central role for government and hierarchical relations; the Danish law on organic agriculture and the rules for organic agriculture and food processing are examples
- **Market-based regulation**, based on financial incentives; the national support scheme for the conversion of farms and companies, the national support scheme for research and development and public 'green' procurement, with a focus on organic food are examples
- **Policy networks**, based on interactions between actors, through which information and resources are exchanged; the national council for organic food, the development of national action plans, working groups on agricultural strategies and demonstration projects are examples

Some regulatory initiatives have been dedicated to organic food production whereas others have been directed at agriculture in general. Whether or not these general schemes have been applicable to organic agriculture has been part of the social shaping of the transition. In the 1990s, when the Danish parliament implemented some European Community subsidies for environmental friendly agriculture, defined as a reduction in the use of pesticides and fertiliser, the national organic agriculture organisation (Landsforeningen Økologisk Jordbrug) had to 'work hard' to get organic agriculture accepted as an eligible strategy, although it does not use pesticides. Later on, organic agriculture became a high priority area within schemes to protect the aquatic environment, but this does not seem to have attracted organic farmers.

Today, in addition to voluntary schemes, conventional agriculture is regulated through a number of restrictive regulations which are administered by the Ministry of Food, Agriculture and Fisheries or by the Ministry of Environment. The regulation focuses on:

- Limits to the annual number of pesticide sprayings at a national level[3] combined with taxes on pesticides
- A request for nutrient accounts for nitrogen from individual farms

The number of annual pesticide sprayings has increased to 2.5 in 2007 from around 2.0 in 2000. The goal is 1.7 sprayings a year (Environmental Protection Agency 2004, 2008). The request for nutrient accounts and control was aimed at reducing nitrogen and nutrient loss and making farmers recognise and utilise the fertiliser value in manure. This regulation seems to have been the basis for a 40% reduction in the use of

3 Calculated as the number of times that it is possible annually to spray the conventionally grown agricultural land with the amount of pesticides sold in a certain year.

artificial fertiliser since the mid-1990s over an almost constant agricultural area (Mikkelsen *et al.* 2005).

The regulation of conventional agriculture may be part of the reason for an emerging focus on further technical optimisation within conventional agriculture. The new emerging concepts focus on the development of more advanced equipment for precision farming based on GPS technology and computing technology in order to reduce further the amount of pesticides and fertiliser (MSTI 2003). Also, the addition of enzymes to pig fodder is considered a strategy to improve the digestion of the fodder and thereby reduce the amount of phosphate in the resulting manure.

The use of organic food in public catering was originally initiated by staff in nurseries, hospitals and other public institutions who wanted to contribute to more sustainable development. Its use increased because governmental and municipal institutions also saw the value of organic food as a way of practising public 'green' procurement, which has been a request made of governmental institutions for a number of years and on which several municipalities have also developed a strategy. The conversion has been enabled partly by local funding and partly by funding from the national council for organic food. Given the fact that the use of organic food often implies a diet with less meat and dairy products and more bread and vegetables (because of the rather high extra costs connected with the use of organic dairy and meat products), the conversion may also have indirect environmental advantages, as bread and vegetables are less polluting and resource consuming than animal products (Kristensen and Nielsen 1994; Kristensen *et al.* 2002).

5.2 Results of the transition towards organic agriculture

5.2.1 Main results: market shares and percentages of converted farms and land

This section gives an overall picture of achievements in the development of organic agriculture and food in Denmark. A first question to be raised could be whether there has been a transition in the Danish food sector towards organic food. The answer could be a combined 'yes' and 'no'. 'Yes' because organic products, especially within the dairy sector and the vegetable sector, play a major role and, for some products, have reached domestic market shares of 20% to 25%; 'no' because the amount of organic products in other product areas has not increased beyond 1% or 2%. This is, for example, the case for pork, which is a dominant sector in the Danish food sector together with the dairy sector. The recent changes in the domestic market shares for some products are shown in Table 5.2, which shows large differences in market shares between the different product groups.

The total market share for organic food stagnated in 1999 at a level of around 5%, but it increased again during recent years to a level of 7–8%, despite a recent reduction in area of organic agricultural land, which is discussed later in this section. Sales of dairy

TABLE 5.2 Market shares of some organic food products in Denmark, 1999, 2002 and 2006

Source: Økologisk Landsforening 2002, 2007

Product	Market share (%)		
	1999	2002	2006
Oat flakes	24.9	27.2	27.0
Milk	21.1	23.5	24.7
Egg	18.2	16.8	17.2
Carrots	14.6	12.8	16.2
Wheat flour	10.7	8.2	10.7
Rye bread	7.8	5.0	3.0
Yoghurt, etc.	7.9	5.4	7.7
Coffee	3.4	3.5	4.1
Potatoes	4.3	3.2	3.2
Beef	1.3	0.9	2.4
Pork	0.5	0.4	0.7
Total market share	**5.5**	**5.5**	**6.2**

products account for more than 40% of all organic food sales. The recent increase in market shares seems to be based on improved economy in many households, more so-called 'food scandals' and a more strategic role for organic food in one of the discount retail chains (Økologisk Landsforening 2007). Increasing imports of new organic food products and probably an increased sale of some agricultural products, such as milk, as organic food and not as conventional food seem to explain the increasing share for organic food, although the area given over to organic agricultural land is decreasing (CASA/IDA 2008).

The biggest recent changes in market shares (seen in Table 5.2) are an 80% increase for beef, a more than 50% reduction for rye bread, a 25% reduction for potatoes and a 15% increase for milk. Although the market share for organic beef has almost doubled since 1999, the market share is still very low compared with the high market share for milk. However, dairy cattle herd are raised for milk production and, when cattle are slaughtered, there seems to be little interest from farmers or slaughterhouses in handling and selling this meat as organic meat, although 90% of Danish cattle is raised as dairy cattle and sold as beef afterwards; only 10% of cattle are raised only for beef and are primarily sold for export.

Table 5.3 shows the development of organic agricultural land and the number of organic farms since 1989, representing almost all the present transition towards organic

TABLE 5.3 Index of change in number of organic farms and area under organic agriculture in Denmark, 1989–2006 (1989 = 100) and the number of organic farms as a percentage of the total number of farms and the total agricultural area 1995–2006. The figures include farms and agricultural land under conversion

Source: own calculations based on Plantedirektoratet 1996–2007

	Year								
	1989	1990	1991	1992	1993	1994	1995	1996	1997
Number of organic farms									
%	–	–	–	–	–	–	1.5	1.7	2.5
Index	100	130	168	168	150	169	262	291	403
Area under organic agriculture									
%	–	–	–	–	–	–	1.5	1.7	2.4
Index	100	121	188	195	210	221	428	483	673

	Year								
	1998	1999	2000	2001	2002	2003	2004	2005	2006
Number of organic farms									
%	3.5	5.2	6.4	6.5	7.3	7.2	6.9	6.1	5.6
Index	556	773	864	879	926	875	789	722	665
Area under organic agriculture									
%	3.7	5.5	6.2	6.5	6.7	6.3	6.1	5.5	5.3
Index	1,037	1,535	1,730	1,816	1,867	1,759	1,703	1,603	1,534

agriculture, which took off in the 1980s. Also, the corresponding figures for conventional agriculture are shown.

In 2006 some 5.3% of Danish agricultural land was under organic agriculture or was in transition to organic agricultural land compared with less than 0.5% at the end of the 1980s. The figure of 5.3% represents, however, a decrease of around 20% since 2002. Around 6% of farms (that is, about 3,000 out of around 47,500) are organic farms, indicating that the average organic farm is a little smaller than the average farm (conventional and organic together). The percentage of organic farms has decreased by more than 25% since 2002. Compared with the development in the total number of farms, the (relative) increase in the number of organic farms since the end of the 1980s is greater, since the total number of all farms in the period decreased. During 1995–2003 the total number of farms decreased by 30%, whereas the number of organic farms increased by more than 300%. The total area under agriculture has been almost constant throughout the period so the increase in organic area relative to all land under agriculture has been between 300% and 400% during 1995–2005.

The table shows a sixfold to sevenfold increase in the number of organic farms and about a fifteenfold increase in organic agricultural area. This covers an above-average number of small organic farms and an above-average number of large organic farms. The high number of larger organic farms can be explained by the fact that, among animal husbandry with dairy cattle, an above-average number of large farms have been converted into organic farms.

5.2.1.1 The role of consumer concern

The question of which consumers buy organic food, and why, has been an issue in the public debate about organic food for many years. It appears as if some sceptical advisers and researchers want to divide consumers into those that buy organic food because they care for their own health and those that do it in order to protect the environment, which some researchers see as a more altruistic concern than a more self-centred concern for health. However, other research seems to show that consumers do not distinguish between health and environment. According to a survey, improved animal welfare and environmental protection are the two most important features of organic food production to citizens. Health attributes are ranked lower, and most citizens who perceive organic food as healthier do so because of the absence of pesticide and medicine residues (Wier and Andersen 2003).

The role of premium prices has also been a major issue in discussions about the possible role of organic food. It appears as if the food industry and (some) retailers think that consumers should be willing to pay more for organic products, whereas environmental organisations tend to think that it should not be the organic products, as the less polluting products, that should be the more expensive products but the more polluting products. According to Wier and Andersen (2003) between 20% and 33% of consumers are willing to pay more for organic products.

Danish consumers seem, in general, to be positive towards organic food. In 2002 93% of Danish households had bought organic food at least once and 87% at least twice during the past year (Økologisk Landsforening 2002). There are, however, significant differences regarding the amount of organic food consumers buy. It is probably less than 1% of households that buy only organic food. An investigation based on the actual practice of consumers divided them into 'heavy users' (who spent more than 10% of the food budget on organic food), medium users (2.5–9.9% of food budget), 'light users' (up to 2.49% spent) and 'non-users' (Økologisk Landsforening 2003). According to that investigation, in 2002, the 13% of the sample that were classified as 'heavy users' accounted for 61% of all organic food purchases (in relation to value; Økologisk Landsforening 2003). The analysis shows that a relative large contingency of the 'heavy users' live in and around the capital, Copenhagen: 39% of the 'heavy users' compared with 25% of the total population. The opposite tendency is seen in more rural areas. However, the local market share of organic food is not only a demand-based tendency but also a question of supply. A region that has shops providing a wide range of organic foods can reach a market share of organic products that is above the average for that type of region (Økologisk Landsforening 2002).

5.2.2 Change in sustainability performance

This section provides a qualitative discussion of the environmental aspects of conversion to organic agriculture. The change in organic agriculture from a pioneer phase into a phase governed by national governmental regulation began dialogue between the organisation for organic agriculture (Landsforeningen Økologisk Jordbrug) and the conventional family farmers' association (Dansk Familielandbrug) and later with the (larger) farmers' association (Landsbrugsrådet) (see Section 5.1.3.5). The cooperation between conventional organisations and the shaping of national law implied a definition of organic agriculture; this definition did not address all the values on which the organisation of organic agriculture focuses. These values can be characterised as (Ingemann 2002):

- The **cyclical principle**, focusing on the interplay between the farm and the surrounding natural systems in terms of nutrients, energy and so on
- The **precautionary principle**, aiming at prevention of problems and concern for the environment, although there might not be scientific evidence for the negative environmental impacts of alternative methods
- The **proximity principle**, focusing on creating geographic proximity by focusing on local cooperation regarding fodder, manure and so on and social proximity, focusing on close and open relations between farmers, the food industry, consumers and so on

According to Ingemann (2002), since governmental regulation was implemented and since the beginning of cooperation with conventional agricultural organisations, the focus in the development of organic agriculture has primarily been on the cyclical principle on the single farm and with farms in the neighbourhood by putting limits on external inputs to the farm. The focus has not been on more widespread closed cycles on a larger scale between agriculture, industry and consumers.

The precautionary principle has been practised in relation to, for example, the ban against the use of genetically modified organisms (GMOs) in organic agriculture and restrictions on the use of medicines in animal husbandry as well as restrictions on the use of additives in fodder and food.

The proximity principle has focused on setting limits on the distance of transportation of some external input to the organic farm but has not addressed the social proximity between farmers, industry, consumers and other actors, probably because this principle is very different from what is required for the specialisation met in conventional agriculture and the physical and economic centralisation in the food industry and the retail sector.

Nor has the cyclical principle been fully met. This is because of the fact that it has been impossible to get 100% organic fodder for organic husbandry; so that certified conversion is not delayed, this requirement was 75% organic fodder until 2000. However, the possibility of using conventional fodder is being reduced gradually over the years as part of the EU regulation of organic agriculture. This shows how the principles behind the transition to sustainable farming has been negotiated and shaped according to the available implementation options.

5.2.2.1 Organic food as local and/or global concept

Most Danish organic products are sold in the domestic market, although 60–80% of the conventional meat and dairy products are sold to foreign markets. This shows that organic food is first of all a domestic market niche. However, there is some export of organic food; in particular, some of the small organic dairies seem to be rather successful in this respect. The large conventional dairy companies have had some problems and their sales to the UK market have met problems because of the increased national demand in UK for organic food (Økologisk Landsforening 2002). This highlights contradictory values in organic agriculture; although it is to be a more local way of producing and supplying food, it is also an export strategy impacting on the shaping of the role of organic food production in a country.

5.2.3 Lessons learned from the Danish experience

The most important lessons for future strategies for sustainable transition from the development of organic agriculture food production in Denmark are as follows:

- The interactions between production, consumption or use, knowledge and regulation and interactions between these need to be kept in focus in the creation and ongoing assessment of strategies for development towards new, more sustainable, paths. Market-based governmental regulation is not enough to ensure a substantial transition but needs to be combined with support for the development of a supply of more sustainable products and restrictions on the conventional production paradigm.
- The creation of more sustainable paths may build on a combination of the creation of new institutions and structures and the use and reshaping of existing institutions and structures, as seen in relation to processing plants, the role of

the supermarkets, and agricultural advice and research within the organic food industry

- The technology area itself is shaped by the transition as well as playing a part in shaping the transition. Success in pursuing a more sustainable path demands an ongoing process of enrolment in other actors' discourses and enrolment of other actors into one's own discourses

The analyses of organic agriculture and food in Denmark confirm Karnøe and Garud's (1998) hypothesis about the role of initial conditions in the shaping of a transition as acting as constraining as well as enabling factors. Among the important conditions have been the significant pork export market and the specialised farms, which have led to a limited focus on organic pork production and a greater focus on that agricultural production which can be converted more easily (such as farms focusing on vegetables and dairy cattle). The existing close links between the large dairy companies and the cooperative retail chain have been important, because the cooperative retail chain has been able to influence the dairies to start buying and processing organic milk.

The analyses also show how the different actors are shaping each other. For example actors in organic agriculture have shaped the strategy of the cooperative retail chain and of the food retail sector as a whole; in turn, organic food processing has been shaped by the demands of the retail sector with regard to packaging, ordering schemes, number of suppliers and so on.

The analyses have also shown how governmental regulation has shaped the concept of organic agriculture. The definition of organic agriculture in national regulations focuses on only some of the key values of organic agriculture as developed by the national organisation for organic agriculture. In terms of proximity, the focus in the regulation and in practice is on the principles of geographic proximity and not on the values of social proximity, probably because social proximity seems to be too far removed from the system of centralisation prevalent within conventional agriculture, the food industry and retail chains.

Organic food is seen as a threat by some actors within conventional food and agriculture. They have tried to question whether organic food is healthier or better than conventional food, seeking to highlight, for example, comparative analyses of the two types of food that do not show higher amounts of vitamins or better taste in organic food. Actors wishing to promote organic food and agriculture have highlighted where organic food comparisons have shown organic food to be better than conventional food and have also argued that the health-related and environmental advantages from not applying pesticides, the better nutrient management and improved biodiversity should be enough to show the advantages of organic agriculture and food.

The reduction in organic farmed area during recent years and the limited market shares for a number of organic food products seems to show the limits of a market-based approach to a transition in Danish agriculture towards more organic agriculture. The role of initial conditions in the planning of a transition towards more sustainable production and consumption shows the need for the development of transition strategies that recognise the national role of agriculture and the food industry and the national environmental discourses in relation to agriculture.

5.3 Conclusions: potential for and barriers to diffusion and scaling up of organic agriculture and food in Denmark

5.3.1 Environmental achievements

The development of the production and consumption of organic food in Denmark can be seen as a success because of the transition to agricultural land that has been achieved and the substantial market shares obtained within some product areas. At the same time organic agriculture and food have been shown to have made positive contributions to:

- Animal welfare
- Human nutrition through the change towards a more healthy diet, sometimes associated with a change to organic food in public catering
- Regional development
- Increased employee satisfaction from the use of better tasting raw materials in cooking and industrial food production

Organic agriculture has also in some cases inspired conventional agriculture such as through the use of other types of fodder, which gives better animal health.

However, the developments in the production and consumption of organic food in Denmark have not reached the point of sustainability, because the present market-based strategy may have reached its maximum for a market-oriented, voluntary conversion from conventional to organic agriculture. At the same there has not been a substantial change in the Danish diet away from animal products and exotic food products towards a more vegetarian, local and season-based diet, although the conversion to organic food in public catering, as mentioned, sometimes seems to imply a change towards a more vegetarian diet.

Among the barriers to further transition is the highly specialised nature of conventional agriculture, with a major focus on meat and dairy production and receiving major export income and employment from these areas; however, the societal benefits of conventional systems have been questioned as there is a major import market for fodder to support such agriculture, and this fodder production, in the case of production of soy protein in South America, causes environmental problems.

5.3.2 Strategic considerations for radical conversion towards organic agriculture and food

If the environmental benefits of organic agriculture were to be developed further in coming years through diffusion and scaling up of organic agriculture in Denmark, it will be necessary to consider whether and how it is possible to make organic agriculture more acceptable as a mainstream strategy.

5.3.2.1 Organic agriculture and food as a multi-stakeholder strategy

Organic agriculture has potential as an environmental strategy in relation to several environmental topics being discussed in Denmark. Cleaner groundwater, reduced eutrophication of streams, lakes and seas, and the safeguarding of biodiversity are some of the environmental problems where organic agriculture may make a positive contribution. Since organic products have also been highlighted, by different stakeholders, as having other advantages—such as being of good quality, being free of genetic modification and as strengthening regional economic development—organic agriculture and food should be seen as having potential for gaining further support from regional politicians, cooks, health professionals and so on.

One of the tools to achieve a more substantial conversion could be the creation of contracts between the government and stakeholders in the food sector, including a strategy for the so-called 'surplus products'. 'Surplus products' is a notion a large dairy company has used regarding its organic products: when they cannot sell the organic products at a price that covers the premium price that the farmers require to compensate them for the lower yield of organic agriculture. Other regulatory measures that could support further conversion are higher taxes on pesticides, taxes on artificial fertiliser and nutrient loss from agriculture and more funding for research and development.

The question is the strategy conventional farmers will choose in the case of strengthened environmental regulation. Precision farming as a technological strategy for reducing the environmental impact of conventional agriculture seems be receiving some attention as an environmental strategy, because the strategy fits into the present specialisation of farms and because they can develop technology with potential for export. In a Danish governmental 'green technology foresight' publication in 2002–2003 the potential to organic agriculture of some of the technologies that are under development for precision farming were considered (MSTI 2003). The focus was especially on pattern-recognition technology and its potential use in the development of automated weed management in organic agriculture. If this development were to support the development of technology in organic agriculture it would be necessary to request that such technological development in precision farming also address organic agriculture and involve organic agriculture stakeholders. Otherwise, the development of precision farming might strengthen the conventional agriculture paradigm and at the same time weaken the organic agriculture paradigm.

5.3.2.2 Potentials and barriers to a conversion of pig farming

As part of a strategy for further conversion, it is necessary to develop a strategy for future Danish pig farming. It seems impossible to produce the present large number of conventional pigs as part of an agriculture that is based on organic agriculture as the dominant strategy. However, many workplaces and export income depends on pig production so strategic considerations are necessary if a more substantial transition to organic agriculture and food production is to be a feasible future development path that can gain broader societal support.

Earlier scenario analyses have shown that it is theoretically possible to convert agriculture 100% into organic agriculture while maintaining the present focus on meat and dairy products (Norholt 1997; Teknologirådet 1997). However, such substantial animal-based production is not environmentally sustainable because of the significant resource losses and environmental impact associated with it.

One possibility for support for a radical conversion and restructuring, including a reduction of pig production, could be the use of stronger pressure on the export markets of other countries with increasing meat and dairy production that may be able to compete with Danish products on price. One example is the growing meat production in Brazil, including for export. However, it appears that there is an emerging middle or upper class in many countries, such as Russia, who are willing to pay more for Danish products and thereby secure the Danish export of pork and dairy products.

A strategy based on high-quality products, for example, inspired by the principles of the Slow Food movement and the so-called 'new Nordic food' concept, with a focus on high-quality products with good taste, could be a way to change food consumption towards less food but of a higher quality. In relation to meat consumption, such a strategy could contribute to better health, reduced resource consumption and reduced emissions of greenhouse gasses, while ensuring the companies' profitability by moving from a price-competitive market towards market niches.

References

CASA/IDA (Centre for Alternative Social Analysis/The Danish Society of Engineers) (2008) 'Grøn indikatorrapport: en kritisk rapport om miljøindikatorer' ('Green Indicator Report: A Critical Report about Environmental Indicators') (Copenhagen: CASA/IDA).

Elzen, B., and A. Wieczorek (2005) 'Transitions towards Sustainability through System Innovation', *Technological Forecasting and Social Change* 72: 551-61.

Environmental Protection Agency (2004) 'Bekæmpelsesmiddelstatistik 2003' ('Pesticide Statistics 2003'), *Orientering fra Miljøstyrelsen* 9: 2004 (Copenhagen: Environmental Protection Agency).

—— (2008) 'Bekæmpelsesmiddelstatistik 2007' ('Pesticide Statistics 2007'), *Orientering fra Miljøstyrelsen* 4: 2008 (Copenhagen: Environmental Protection Agency).

Ingemann, J.H. (ed.) (2002) 'Økologisk landbrug mellem historie og principper' ('Organic Agriculture between History and Principles') (OASE Working Paper; Aalborg: Aalborg University).

Karnøe, P., and R. Garud (1998) 'Path Creation and Dependence in the Danish Wind Turbine Field' (Papers in Organisation 26; Copenhagen: Institute of Organisation and Industrial Sociology, Copenhagen Business School).

Kristensen, N.H., and T. Nielsen (1994) 'Anvendelsen af økologiske fødevarer i Københavns Amt' ('The Use of Organic Food in Copenhagen County') (Report 3; Lyngby: Interdisciplinary Centre, Technical University of Denmark).

—— and T. Nielsen (1996) 'From Alternative Agriculture to the Food Industry: The Need for Changes in Food Policy', *IPTS Report* 20: 20-25.

——, T. Nielsen and B.E. Mikkelsen (2002) 'Anvendelse af økologiske fødevarer i kommuner og amter' ('The Use of Organic Food in Municipalities and Counties') (Copenhagen: National Food Agency, Ministry of Food, Agriculture and Fisheries).

MAF (Ministry for Agriculture and Fisheries) (1995) 'Aktionsplan for fremme af den økologiske fødevareproduktion i Danmark' ('Action Plan for the Promotion of Organic Food Production in Denmark') (Copenhagen: MAF).

Michelsen, J., and V. Søgaard (2001) *Policy Instruments Promoting Conversion to Organic Farming and their Impact in 18 European Countries, 1985–1997* (Political Science Publications 1; Odense: Faculty of Sciences, University of Southern Denmark).

Mikkelsen, M.H., S. Gyldenkærne, H.D. Poulsen. H, J.E. Olesen and S.G. Sommer (2005) 'Opgørelse og beregningsmetode for landbrugets emissioner af ammoniak og drivhusgasser 1985–2002' ('Estimation and Estimation Method for Agricultural Emissions of Ammonia and Greenhouse Gases') (WP-204 Roskilde: National Environmental Research Institute, www2.dmu.dk/1_Viden/2_Publikationer/3_arbrapporter/rapporter/AR204.pdf, accessed 9 July 2009).

MSTI (Ministry of Science, Technology and Innovation) (2003) 'Grønt teknologisk fremsyn' ('Green Technology Foresight') (Copenhagen: MSTI).

Norholt, H. (1997) 'Samfundsøkonomiske konsekvenser af en omlægning til økologisk jordbrug, Delrapport 1: Agronomiske konsekvenser af en omlægning til økologisk jordbrug' ('Socioeconomic Consequences of a Conversion to Organic Agriculture. Report 1: Agronomic Consequences of a Conversion to Organic Agriculture' (Copenhagen: Royal Agricultural and Veterinarian University).

Økologisk Landsforening (2002) 'Forbrugernotat' ('Consumer Note') (Århus, Denmark: Økologisk Landsforening).

—— (2003) 'Forbrugernotat' ('Consumer Note') (Århus, Denmark: Økologisk Landsforening; www.alt-om-okologi.dk, accessed7 July 2009).

—— (2007) 'Markedsnotat' ('Market Note') (Århus, Denmark: Økologisk Landsforening).

Plantedirektoratet (1996–2007) *Økologiske jordbrugsbedrifter: Autorisation, Produktion* (*Organic Farms: Authorisation, Production*) (annual publications covering data for the years 1995 to 2006; Denmark: Plantedirektoratet).

Sørensen, J.T. (ed.) (2003) 'Produktionsstyring med fokus på husdyrsundhed og fødevaresikkerhed i økologiske svinebesætninger' ('Production Management with a Focus on Husbandry Health and Food Safety in Organic Pig Production') (Internal Report 54; Tjele, Denmark: ICROFS, Internationalt Center for Forskning i Økologisk Jordbrug og Fødevaresystemer [International Centre for Research in Organic Food Systems]).

Teknologirådet (1997) 'Visionen om økologisk landbrug: Komparativ økonomisk analyse af fuld omlægning af dansk landbrug til økologisk drift' ('A Vision of Organic Agriculture: Comparative Analysis of Full Conversion of Danish Agriculture to Organic Farming') (Copenhagen: Teknologirådet, www.tekno.dk, accessed7 July 2009).

Wier, M., and L.M. Andersen (2003) 'Consumer Demand for Organic Food: Stated Attitudes and Actual Behaviour' (SØM paper; Copenhagen: AKF).

6
Socioeconomic aspects of farmers' markets in Sweden

Helen Nilsson and Oksana Mont
International Institute for Industrial Environmental Economics, Lund, Sweden

The modern food industry is based on large-scale facilities, long transportation distances and absence of contact between food producers and consumers. It is associated with increasing social and environmental impacts, food security issues and the health and safety of food. Consumer awareness regarding food production and food content is steadily growing. An alternative to the current model of global food chains is local production of food and local sales of food at farmers' markets. Farmers' markets present consumers with the opportunity to access fresh produce and to learn about the conditions in which food has been produced and present producers with the marketing opportunity and possibility to socialise with consumers and learn about their preferences. In addition, farmers' markets provide increased income for people in the local community. Farmers' markets represent a voluntary bottom-up initiative, a so-called social innovation case, and comprise both production and consumption sides, with the main focus on the point-of-sale.

The aim of the current case is to investigate the economic and social aspects of farmers' markets as well as to get a better understanding of the driving forces for the establishment of farmers' markets in Sweden. It is based on interviews with and surveys of a number of stakeholders involved in establishing and running two farmers' markets, in Halmstad and Malmö, Sweden. Interviewed stakeholders include producers, consumers, local authorities, organising committees at the markets and farmers' associations. The collection of primary data was conducted in autumn 2005.

The farmers' market concept in Sweden started in 2000 through the initiative of one person who took the concept from examples in the UK and USA. John Higson, a British national, initiated a series of seminars where the idea of farmers' markets was presented, and support for pilot markets was obtained. Since 2000, 14 farmers' markets have been established in Sweden.

The markets are a relatively new phenomenon, and their potential benefits, difficulties and impacts on the community are not totally clear. In order to ensure this progression towards local markets is a positive one for the community, investigations of the social and economic aspects are important.

6.1 Case description

6.1.1 Overview

The purpose of this study was to assess the social and economic benefits of farmers' markets both for the farmers taking part in the markets and for the consumers purchasing the products on sale. More specifically, for each of the markets a number of questions were posed:

- What are the social and economic benefits perceived by the consumers attending the markets?
- What are the social and economic benefits perceived by the producers attending the markets?
- Are there other values recognised by the producers?

Three basic types of questions were included when designing the questionnaires. These focused on: the behavioural, where the objective was to find out what people do rather than their opinions; on attitudes, where opinions or basic beliefs were gathered about the products that consumers buy; and, finally, on classification, to classify the

TABLE 6.1 Organisations represented by stakeholders interviewed during the study

Stakeholder	Organisation interviewed	No.
Consumer	Consumers interviewed directly at the markets	355
Producer	Individual interviews and questionnaires with producers, including those on the board of the Farmers' Market organisation in Sweden	33
Organiser	The organisers of the two markets and the initiator of the markets in Stockholm	3
Local authority	Representatives from Malmö and Halmstad	3
Farmers' organisation	*Representatives from the Federation of Swedish Farmers (Lantbrukarnas Riksförbund; LRF) in two regions: Halland and Skåne* *Ecological Farmers and Ecological Marketing centre*	43

information once it was been collected, for example, on gender, employment and so on (Hague 1993).

In order to answer the aforementioned questions a large number of interviews and surveys was conducted. Table 6.1 provides an overview of interviewed actors—the stakeholders in the farmers' markets.

6.1.2 Global food systems and local alternatives

The food industry has followed the same trend of technological development as in other industries. Processes have become more elaborate, with food being transported great distances and packaged and processed to fit our modern lifestyles. These developments have brought benefits to the consumer, with greater choice, lower prices in the supermarkets and a level of convenience not encountered before by the consumer. However, this mechanisation of the industry has led to unforeseen consequences, which have come to light in recent years, in the form of health scares (e.g. BSE, salmonella, foot and mouth, avian flu, etc.) and the impact of food production on the social infrastructure of rural communities and the natural environment. These revelations have shaken the confidence of consumers in food producers and have acted as one of the drivers to more active and aware consumerism. Consumers are now becoming aware that our supposed abundance of cheap and healthful food is to some extent illusionary. They are beginning to see that the social, ecological and economic costs of such cheap food are, in fact, great (Berry 1995).

As a result of globalisation there has been an increasing division between the producer and consumer in the food production chain. The international trade in food has led to an increase in the distance food travels. Consumers no longer know where and how food is produced and producers do not know who is buying the food they are producing. Additional processing steps have been added between the farmer and the consumer leading to an increase in transport costs, energy use and a loss of connection between farmers and consumers: consumption has become disconnected from production (Murdoch and Miele 1999). Instead of selling food to their neighbours, farmers sell into a long and complex marketing chain of which they are a tiny part and are paid accordingly (La Trobe 2001; Halweil 2002; Nilsson 2005). Studies show that only 26% of checkout prices go to farmers, compared with approximately 50% only 50 years ago (Smith *et al.* 2005).

Owing to food scares and general disquiet regarding the state of the food industry, there has been the beginning of a renaissance in how some people think about the food they eat and the relationship that its production has to the local community and economy. The growing globalisation of the food industry has changed the relationship between the economic institution of the market and its social context; local alternatives have often developed as a response to this globalisation trend (O'Hara and Stagl 2001). The local food movement has seen a shortening of the supply chain as consumers are taking the opportunity to buy directly from the producer, providing increased income for people in the local community (La Trobe 2001). It is these consumer concerns and the desire to re-establish connections that are seen as the main motivating factor in moving away from the homogenised products of the global agro-food industry in the western world (Winter 2003). Examples of attempts to reconnect the producer and consumer include community-supported agriculture,[1] vegetable box schemes and farmers' markets.[2]

The outlet the farmers' markets create, along with farm shops, vegetable box schemes and community-supported agriculture are also examples of local producers creating their own economic system for marketing and selling their produce. Local systems of production and consumption fit into a different mind-set for regional development. This is where a selective share of production is distributed to regions where a diverse range of activities are organised in the form of small-scale, flexible units that are synergistically connected with each other and prioritise quality in their production (Johansson *et al.* 2005). It is not an abandonment of large-scale production; it is more about finding a balance between the large and small scale, where producers of all sizes, in the case of farmers' markets small-scale farmers, can find a place in the marketplace and fulfil consumers' desire for quality, reasonably priced, products.

There are two approaches to the concept of local food that can be investigated. The first approach places a focus on 'locality as a closed or bounded system', where food is produced, processed and retailed within a certain geographical area. These local systems are often characterised by alternative channels, such as direct sale to consumers and farmgate outlets. The second approach identifies the 'locality as value added for export': for example, Italian organic wine sold in Germany. In either case the attention to localness can be associated with speciality, traditional and quality foods (Morris and Buller 2003; Jones *et al.* 2004). Local food carries with it the trappings of certain expectations, that it provides a quality product and that it brings some loosely specified environmental, animal welfare, employment, fair trade, producer profitability and cultural conditions (Sustain 2002).

Previously, farmers who were unhappy with modern production systems and demands from industry had very little support if they wanted to take an alternative path, but things are changing. The new European rural development policy options designed to support the provision of environmental goods and services,[3] and the marketing and processing of quality and/or speciality produce, are now supporting farmers who wish to shift their business emphasis away from commodity production to preserve the European model of agriculture (Banks and Marsden 2001) and to ensure that sustainable development goals are met. The recent reform of the EU Common Agricultural Policy (CAP)[4] has made it easier for farmers to decide what they want to produce, and this brings flexibility to the market, where farmers and producers can listen to the demands of the consumers and produce food that is demanded. A growing niche market is for quality local products often grown organically or using other more sustainable methods (such as integrated crop management).

1 Community-supported agriculture: where a local community of individuals financially supports a farm so that the grower and consumer share the risks and benefits of food production. For more information, see www.localharvest.org/csa (accessed 28 April 2010).

2 The concept of farmers' markets can be defined as a common facility or area where several farmers gather on a regular, recurring basis to sell a variety of fresh farm products directly to consumers (USDA 2002).

3 For more information see ec.europa.eu/agriculture/envir/cap/index_en.htm (accessed 28 April 2010).

4 See ec.europa.eu/agriculture/capreform/index_en.htm (accessed 28 April 2010).

6.1.3 Actors and their perspectives

The main stakeholders of the farmers' markets were identified during in-depth interviews with the organisers of the markets (Nilsson 2005). Based on this information a stakeholder map was constructed distinguishing between the direct and indirect stakeholders involved in the markets (see Fig. 6.1).

FIGURE 6.1 Direct and indirect stakeholders in farmers' markets

Source Nilsson 2005

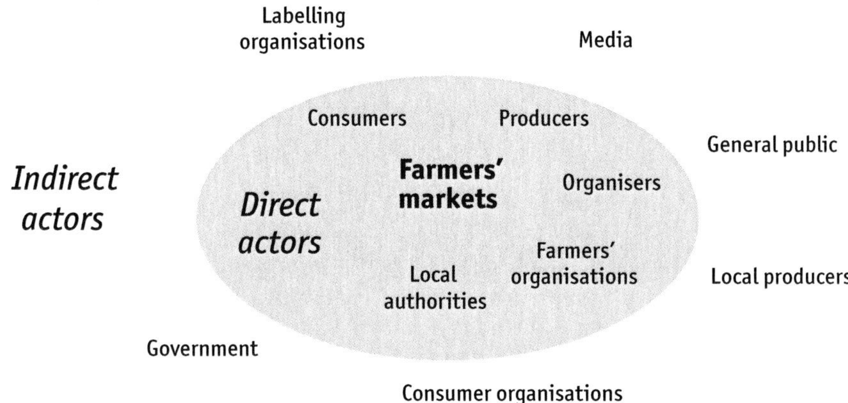

6.1.3.1 Primary actors

The primary actors for farmers' markets are, of course, producers (farmers) and consumers, but also to be included are local authorities, farmers' organisations and the organisers of the farmers' markets.

People attending the markets are loyal consumers and visit frequently during the season. First-time visitors make up 15% of the total number of consumers at each market, which indicates the potential growth rate if these consumers return. The majority of consumers visiting the market in both cities were over 50 years old. This indicated that the attractiveness of the markets could be the result of a nostalgic feeling created by the markets. It also illustrates that those attending the market could well be more financially comfortable as they do not have any dependents and can afford to spend their income on specialised products. The majority of those interviewed were women, reflecting not only that it is mostly women who make food purchases for the family but also that women have a greater willingness to answer questions. When asked to evaluate the market, an overwhelming majority of consumers in Malmö (79%) were happy with the market, with requests for fish and more organic products coming from 13% of consumers. Half of consumers in Halmstad were happy with the market (56%), but a third (30%) would like more products. A wider variety of products would increase the attractiveness of the event and quite possibly the number of visitors and the amount of money made by the producers attending, creating a win–win situation for all.

The majority of producers attending the market were small-scale farmers with less than 100 acres under cultivation. A quarter of producers in Halmstad had no land at all. This means that the markets are attractive outlets for small-scale producers as well as for small farmers. The majority of the producers represent established businesses, with 60% of the producers in Malmö having been in operation for more than 10 years. The majority of Malmö producers did not use conventional growing methods, instead opting for integrated production (IP) and organic production methods. This was reflected in the number of organic stalls, compared with Halmstad, where only 25% of producers use alternative method, compared with 70% in Malmö. Organic produce was very popular at the market in Malmö and many consumers cited it as something they valued at the market.

6.1.3.2 Secondary actors and externally involved parties

A number of different themes were brought up by the stakeholders interviewed in the course of this study. The fact that the markets have invigorated the squares used for the markets is something that all of the stakeholders commented on. The benefits to Drottningtorget in Malmö have included an organic shop, an organic hairstylist, plus a café and a number of delicatessens. This increased activity has made the area more 'alive' and safer for residents. So the community has not only received an outlet for local produce and a chance to meet producers, they have also gained a more urban space, with an increase in the flow of people in and around the square. In fact it was indicated in interviews that the square is at the heart of a regeneration process for that part of Malmö.

FIGURE 6.2 A shopper at a stall in the farmers' market in Halmstad

In addition, it was noted by stakeholders that one of the main benefits of the markets was that they gave power back to the consumers by giving them the opportunity to buy directly from the producers. This means that the power of choice and decision that consumers have lost as a result of the rationalisation of the food industry has begun to return to them.

From the perspective of the Federation of Swedish Farmers (Lantbrukarnas Riksförbund [LRF]) it is a model project. It has supported the creation of a large number of such markets in Sweden. This support has taken the form of both financial and administrative support and has enabled the markets to establish themselves. When the support was taken away at the end of the project period, the markets have had the momentum to continue, a sign of success for LRF. The success has been put down to the fact that consumers feels powerless when dealing with the supermarkets; they do not know what they are getting. At the farmers' markets they get a little of the power back by buying directly from the producers.

The farmers' market in Malmö was a new experience for consumers. Before the market there was a lack of access to locally produced products for those living in Malmö; now, visitors are increasing and the word is spreading throughout the city. There is no problem among those championing organic food products that the market sells a mix of organic and conventionally grown products. The market is seen as being a 'shop window' for organic products and an effective form of marketing. The sale of organic and conventional food together simply gives the consumer the opportunity to choose and compare.

6.1.4 Case history and development

The development of the farmers' market concept—Bondens egen Marknad—in Stockholm in 2000 followed success stories from the UK and USA. The start was motivated by observations from both the UK and USA of a pleasant social atmosphere and the sale of high-quality products. The branding expert John Higson was the brainchild behind the introduction of the markets. After touring around both the UK and USA he constructed guidelines of what should and should not be at a farmers' market (Source 1). It was found among consumer groups that there was support for this idea and, after approaching a number of organisations and authorities in Stockholm (Ekologiska Lantbrukarna, Lantbrukarnas Riksförbund, Stockholm city), a small pilot project was initiated to run a farmers' market in Stockholm. Of the 2,000 farmers that were contacted 45 agreed to take part, and the huge success of the market ensured that there would be more markets, with even the local businesses confirming that their profits had increased during days (Source 1). This was not, however, the first initiative in Sweden. In 1992 a farmers' market was started in the town of Trelleborg by Erika von Boxhoevden. This was not a success, although the reasons for that were not researched at the time; one commentator suggested that it was before its time (Source 2). With the success of the market in Stockholm, there was enthusiasm to continue with similar markets in the rest of Sweden. Key organisations, such as LRF and Hushållningssällskapet, sponsored pilot projects in places such as Göteborg, Halmstad and Malmö, with 13 in total falling under the official trademark for farmers' markets in Sweden, Bondens egen Marknad. All of the markets follow the rules laid out by the organisation[5] and producers must sign an agreement that they will follow the rules.

A management committee is in charge of the running of the market, with an organiser taking the lead motivational role and coordinating the administration of the market. At the beginning of the year the committee has a meeting to start the market process. Permits and permissions are applied for by sending applications to the local authority, which then forwards the applications to the police, who have to approve the permit. Permissions and permits are needed for such things as cleaning and car parking at the market. Each of the stallholders at the market pay a fixed fee for selling their products at the market for the entire season. Malmö Stad assist the administration of the market by offering a lower rent on the market, as well as helping with organising electricity and parking permits.

Farmers' markets are not seen as a source of competition for the other market squares in Malmö as they are not year-round events. Instead they are seen as a positive addition to market life in Malmö that may also encourage an improvement in the quality of other products. The market in Halmstad is located away from the main square, and not seen as competition.

The markets are advertised in the local press and through posters set up around the cities. This represents a large cost for the markets, but the organisers are in agreement that the benefits outweigh the costs as the advertisements attract new consumers to the market and inform regulars when the markets are starting again at the beginning of the season.

6.2 Results

This section presents the main results from the socioeconomic evaluation of farmers' markets and provides discussion on their economic, social and environmental performance.

6.2.1 Main results

This study evaluated the economic and social aspects of farmers' markets recently established in two Swedish cities, Malmö and Halmstad. The evaluation demonstrated that it is mostly social benefits that attract farmers and consumers to the markets. Economically, farmers' markets are not very important to farmers: they are earning relatively little there as a percentage of their yearly income; rather, the markets provide a platform for communication with consumers. Farmers' markets are still only a niche activity within the existing food system. The following sections elaborate in more detail on this main outcome of the study.

5 For more information see www.bondensegen.com.

6.2.2 Change in sustainability performance

6.2.2.1 Economic dimension

6.2.2.1.1 Economic value for consumers

There was a mixed understanding among the consumers as to whether the products at the markets were more expensive or cheaper than conventional food outlets. The economic value of the market for the consumer cannot therefore be proven, with only 17% of consumers in Malmö and 26% of consumers in Halmstad believing that they were paying lower prices at the market, meaning that the other consumers found other values present in the products if a purely economic perspective is taken.

For many consumers price was not an issue; 12% of consumers attending the market did not have an appreciation of the prices of the products on sale, and even if they found them to be higher than the supermarket, they were prepared to pay more for quality and freshness. When one consumer was buying apples at the market she observed that one of the stallholders had two qualities of apples on sale. The best apples were on sale for SEK 5 per kilogramme, and the 'pie apples' were on sale for SEK per kilogramme.[6] This second price was comparable with the Swedish apples in the supermarket, as the consumer noted. When she asked the stallholder why the apples were not sold at full price, the stallholder answered that they came from the centre of the tree and had not received as much sun as the others so were not as sweet. This level of care and attention to the ripeness was appreciated by the consumer who commented 'there is no way of knowing the level of ripeness and quality of the apples on sale in the supermarket'; this is just one example of how the economic benefits cross over into other benefits and values. Even if consumers are not gaining from buying cheaper apples, they can purchase fruit about which they have more knowledge.

6.2.2.1.2 Economic value of the market for the producer

The answers received from the producers indicate that there are more complex reasons and benefits to the producers attending the market than simply a pure economic gain. In fact, for more than 50% of the producers the markets provide less than 5% of the producers' yearly income, and for many it is not deemed an important source of income. It is interesting to note, however, that the producers in Malmö value its income and believe they benefit despite the fact that they only receive a fraction of their yearly income from the market. The producers in Halmstad admit to receiving more of their yearly income from the market, but do not value its input as much.

Producers in Malmö appear to have a more positive attitude to the market; this may be due to the fact that the market in Malmö is well visited by consumers. The average amount of money earned at each market is between SEK4,000 and SEK6,000. This is a not a significant amount for most of the producers but is still appreciated and the producers interviewed admitted that they would not be attending if they were not making money through participating in the market. The producers in Halmstad valued the use of the market as a marketing opportunity first and foremost, followed by the chance to meet consumers.

6 Note: €1 is worth approximately SEK10 (10 Swedish crowns, or krona [SEK]), as of February 2010.

Some of the producers have noted an increase in visitors to their farm shops and in their sales as a result of their attendance. Even a small amount of sales at every market session is an addition to the overall income of the farmer. From the results it can be said that economic profitability is not the major benefit to the producers attending the market.

6.2.2.2 Social dimension

6.2.2.2.1 Social values for consumers

From the question where consumers were asked to identify what they valued about the market, a wide range of answers were given. The availability of organic produce, the environmental benefits and access to local produce were all ranked as things consumers appreciated.

In the case of both Halmstad and Malmö, the freshness of the product was the most popular element they valued, which was clearly associated with quality and with willingness to pay more.

The market is a form of nostalgia for some, remembering how the purchasing experience was in the past, before supermarkets; this is why atmosphere scored high as an appreciated property at both markets. The market experience was a determining factor when buying products for some. This atmosphere was not created purely by the producers, but also by the consumers participating. Groups of friends met and talked in between buying their fresh vegetables and bread.

Evidence of a humanistic relationship with the producers was illustrated by the value placed on the connection between producer and consumer. This is a feeling of interdependence and solidarity with the producers; by trading with them they are supporting

FIGURE 6.3 Shoppers at the farmers' market in Malmö

them and their way of life (Prigent-Simonin and Hérault-Fournier 2005). Consumers at both markets cited the ability to support small farms as a factor they valued in the markets. This correlates with the result from the question regarding what they felt they were contributing to by attending the market. Here, an overwhelming 89% of consumers in Malmö and 77% in Halmstad stated that they felt that they were supporting small farmers through attending the market and purchasing products there.

6.2.2.2.2 Social values for the producer

The contact with the consumers was a very important part of the farmers' market experience for the producers interviewed. As one producer commented, 'farming is a lonely profession'. The chance to meet consumers and other producers is a prime driver for many producers' attendance. One farmer described meeting the consumers as a 'priceless experience', with another farmer commenting that it allows the development of a relation of trust between the producer and consumer. An additional bonus is, naturally enough, a sales point for their goods. While it is not seen as a major revenue generator, it is still a positive boost and allows producers to inform consumers where they can purchase their goods.

The market was also seen as a marketing tool, with one producer commenting that it gave them a chance to show the consumers 'who I am'. Another producer mentioned that the market gathers together like-minded consumers who want to buy local, quality products and who are often willing to pay more for these goods. The marketing opportunities are valued by all the producers interviewed, with it encouraging consumers to purchase and eat more of their products. So what could be called the social benefits of the markets include the contact that the producers gain with the consumers during the process of the market as well as the marketing opportunity. An additional value is the interaction between the different producers. It gives them a chance to meet each other, and a number of those interviewed commented on the camaraderie created at the markets among the different producers.

Many of the producers admitted that it was a lot of work attending the market, with many having a long distance to travel if they live at the edge of the market area. The picking and packing of products, or extra production of, for example, bread, also takes time. For a number of producers, their enterprise is a part-time occupation, fitted in alongside other income-generating activities, such as other food production, tourism and external employment. Their commitment to producing and selling their products indicates that they enjoy the social contact and the actual act of food production and cherish the use of traditional production methods and recipes.

6.2.2.3 Environmental dimension

We have not been able to find environmental evaluations of farmers' markets. Several aspects might contribute to their improved environmental performance and improved sold produce. The most significant is perhaps the potential in reducing transportation by use of regional distribution systems. Of course, the further away from the final user the products are produced, the higher the environmental impact associated with transportation. However, if a more systemic picture is taken into account it might be better from the environmental perspective to transport some products than to grow them locally. A study demonstrated that the environmental impact of tomatoes brought to the UK from Spain is smaller than tomatoes grown in the UK in heated greenhouses (Smith *et al.* 2005). The energy used for their heating is more environmentally burdensome than that associated with transportation.

Another issue is, of course, the availability of organic food; most, but not necessarily all, of the food sold on the market is organic. Studies demonstrate that the environmental impact of organic food is lower than that of conventional food, if the amount of fertiliser applied is used as indicator. However, organic farming in many cases requires much larger areas to produce the same amount of food as traditional farming. For example, organic tomato yields are only 75% of that for non-organic tomatoes. 'Thus, the lowest yielding organic, specialist tomatoes incur about six times the burden of non-organic, loose classic tomatoes' (Williams *et al.* 2006). These are just a few factors that affect the complex issue of environmental impacts of food produce and therefore each case should be studied specifically (Anderson 2007).

At the local level, our survey showed that 50% of the consumers in Malmö and 40% of consumers in Halmstad drove to the market. The number of consumers who cycled, approximately 20%, was the same for both markets. The slightly larger number of car journeys to the farmers' markets in Malmö may be the result of better parking facilities there: there was a local car park that offered free parking at weekends. This might have made it more attractive for consumers to take their car instead of public transport. However, a number of consumers mentioned that, when a lot of purchases were made, it was easier to transport them in a car rather than by bicycle or on foot.

It is clear that our understanding of factors that influence environmental performance of farmers' markets is rather limited and therefore more studies are needed.

6.2.3 Learning experiences

From the presented and evaluated farmers' markets it is clear that several factors affect their success.

6.2.3.1 Barriers for farmers' markets

The number of producers participating in the market is seen by all the stakeholders interviewed as the main barrier to the continued success of the markets in both Malmö and Halmstad. The organisers of the markets each acknowledge that it is difficult to attract new producers as there is seen to be a level of risk in attending the markets and using a new sales point for the transaction of goods. Some of this apprehension may be explained by the traditional system of food sales in Sweden. Instead of having to find a market for their products, farmers have traditionally been a member of a farmers' cooperative through which they can sell their products. They may get a slightly lower price than if they were to sell products on the open market, but they are guaranteed a sales outlet and a price for their products. This has lulled many farmers into a sense of security where they do not have to look for markets for their products. This is the difference compared with many other countries, where farmers are constantly looking for new sales outlets and have embraced the farmers' market concept.

There is also a barrier in the age and attitudes of the producers. The majority of farmers in Sweden today are middle-aged or approaching retirement age. The average age for farmers in Sweden is 53 years, and one in six are over 65 years old. Only 6% of farmers are under 35 years old (Jordbruksverket 2003). It is not the time for trying new production techniques (organic or the IP method) or new crops that might appeal to consumers at the markets such as different varieties of beetroots and carrots instead of just the usual varieties.

Another comment was that Swedes did not go to markets as a rule and were more likely to treat such as visit as an excursion rather than as a normal way of buying the weekly grocery order. The market culture was judged to be more common in other countries where such markets have expanded rapidly in recent years (such as the UK and USA). Even Norway, with half the population of Sweden, has more markets per capita.[7]

6.2.3.2 Opportunities

The markets present a tremendous opportunity to act as an outlet for locally produced food products. Many of those interviewed, stakeholders and consumers, noted that before the markets were established they did not know where they could purchase locally produced food. The market acts as an important shop window for the region and for small-scale producers who attend the market to inform the public of their produce and production methods. The farmers' market concept is a great educational and marketing tool for the small-scale and medium-scale producer who often sell niche products that sometimes have difficulty finding a pathway to the market.

As mentioned above, the market has been praised for regenerating the local area in Malmö. Although a similar effect has not been noted in Halmstad there was an increase in the number of people around the harbour area when the market came to town. Transference of the farmers' market concept to another town or city has the potential to facilitate a similar transformation around the square or street where the market is situated. A project in Stockholm, called 'Street', is focusing on revitalising a neglected area of Stockholm through installing a local food shop, art centre, café and restaurant.[8]

Expanding the number of markets and the way they are marketed would increase sales and contact with consumers. Many stakeholders involved with the concept seemed reluctant to take the step to increase the number of markets in southern Sweden, mainly because of the barriers mentioned in Section 6.2.3.1.

7 For number of markets see www.bondensmarked.no and www.bondensegen.com.

8 See www.streetinstockholm.se, accessed 9 July 2009.

6.3 Potential for diffusion and scaling up

The future development of the concept was discussed, but there was no united voice as to which direction the markets should go. A number of different ideas were brought up, with the stakeholders having diverse suggestions as to how the market can evolve in the future. These ideas focused around expanding the number of market locations, the frequency of the markets and alternative sales points, such as a farm shop in town and a home delivery service. The organiser of the market in Malmö would like to see it become independent and self-financing within three years; she believes this is possible if they ensure sufficient funds to cover marketing and advertising.

Views were divided as to whether more markets would have a positive or negative influence on the concept. Some believed that it was important to get the message to as many people as possible and that this could be done by getting more towns and cities involved, either by having a travelling farmers' market visiting the smaller towns or by establishing new markets in other towns in the region. However, there were some who registered the problem of having too many markets, with the producers involved in the markets being spread too thinly to give the consumers the added value of a wide variety of products that are currently on offer at the markets. It was admitted by the stakeholders that one of the keys to the attractiveness of the market was its exclusiveness. It was deemed important to maintain this, which is why some stakeholders were not keen on having more markets or expanding the current market in size or frequency. It was seen as important to follow the seasons as well and not to have a year-round market. There was, though, some support for the idea of having some sort of farm shop in the centre of Malmö, where products normally available at the farmer's market could be for sale, for example honey, oil and preserves. Attracting a wider audience was suggested, with a more festive atmosphere attracting more of a cross-section of the community. Additional outlets for local produce were suggested, for example the development of a delivery service to restaurants.

6.4 Overall conclusions

This study of social and economic benefits of farmers' markets has taken first steps in assessing the benefits and value of having such markets in towns and cities. While small definite economic benefits were found, it was the social benefits and values that were most apparent for the producers and consumers alike. Farmers' markets although growing in Sweden are still only a small phenomenon and can be seen as a niche activity within the food system. However, some studies suggest that they could gain greater importance and have the possibility to become a real alternative system for food distribution and sales (Carlsson-Kanyama *et al.* 2004).

In order to overcome the main barriers to the continued success of the markets, a number of initiatives can be promoted to ensure the growth of the concept in Sweden.

- Encouraging farmers to grow different varieties of the same crop, such as different-coloured beetroots and carrots, would offer consumers the variety that they want and allow farmers not to have to focus on one or two main crops. This does not imply any new skills that need to be learned or equipment to buy, but provides the consumer with quality value products that are attractive and appealing
- Local authorities could promote educational programmes that inform the local community about food produced in the area and where they can purchase it and could also promote a shift away from the dependence of farmers on farmers' cooperatives in order for farmers to realise that there are alternatives to the current system
- A national support scheme for new and young farmers who want to take over farms or start their own businesses would also assist the transfer of farms to the next generation. There are some European schemes already in existence that could be supported more at the national and regional level.
- The establishment of long-term contracts between local farmers and municipalities to promote sustainable public procurement would benefit locally produced food and ensure the survival of small and local food producers

References

Anderson, M.D. (2007) *The Future of Food Systems: Global to Local* (Amherst, MA: The Environmental Institute, University of Massachusetts).

Banks, J., and T. Marsden (2001) 'The Nature of Rural Development: The Organic Potential', *Journal of Environmental Policy and Planning* 3: 103-21.

Berry, W. (1995) *Another Turn of the Crank* (Washington, DC: Counterpoint).

Carlsson-Kanyama, A., Å. Sundkvist and C. Wallgren (2004) *Lokala livsmedelsmarknader: en fallstudie Miljöaspekter på transporter och funktion för ökat medvetande om miljövänlig matproduktion* (*Local Food Markets: A Case Study. Environmental Aspects on Transportation and their Function for Increasing Awareness about Environmentally Sound Food Production*) (Stockholm: Centrum för miljöstrategisk forskning).

Halweil, B. (2002) *Home Grown: The Case for Local Food in a Global Market* (Danvers, MA: Worldwatch).

Hague, P. (1993) *Questionnaire Design* (London: Kogan Page).

Johansson, A., P. Kisch and M. Mirata (2005) 'Distributed Economies: A New Engine for Innovation', *Journal of Cleaner Production* 13: 971-79.

Jones, P., D. Comfort and D. Hillier (2004) 'A Case Study of Local Food and its Routes to Market in the UK', *British Food Journal* 106.4: 328-35.

Jordbruksverket (2003) 'Jordbruksföretag, företagare och ägoslag 2003', *Statistiska meddelanden* JO 34 SM 0401.

La Trobe, H. (2001) 'Farmers' Markets: Consuming Local Rural Produce', *International Journal of Consumer Studies* 25.3: 181-92.

Morris, C., and H. Buller (2003) 'The Local Food Sector: A Preliminary Assessment of its Form and Impact in Gloucestershire', *British Food Journal* 105.8: 559-66.

Murdoch, J., and M. Miele (1999) ' "Back to Nature": Changing "Worlds of Production" in the Food Sector', *Sociologica Ruralis* 30.4: 465-83.

Nilsson, H. (2005) 'The Economic and Social Aspects of "Bondens egen Marknad" in Sweden: Results of a Pre-study' (Lund, Sweden: IIIEE).

O'Hara, S.U., and S. Stagl (2001) 'Global Food Markets and their Local Alternatives: A Socio-ecological Economic Perspective', *Population and Environment* 22.6: 533-54.

Prigent-Simonin, A.H., and C. Hérault-Fournier (2005) 'The Role of Trust in the Perception of the Quality of Local Food Products: With Particular Reference to Direct Relationships between Producer and Consumer', *Anthropology of Food* 04.

Smith, A., P. Watkiss, G. Tweddle, A. McKinnon, M. Browne, A. Hunt, C. Trevelen, C. Nash and S. Cross (2005) *The Validity of Food Miles as an Indicator of Sustainable Development* (Didcot, UK: AEA Technology).

Sustain (2002) 'Sustainable Food Chains, Briefing Paper 1. Local Food: Benefits, Obstacles and Opportunities' (London: Sustain).

USDA (US Department of Agriculture) (2002) 'US Farmers Markets—2000: A Study of Emerging Trends' (Washington, DC: USDA).

Williams, A.G.E., E. Audsley and D.L. Sandars (2006) *Determining the Environmental Burden and Resource Use in the Production of Agricultural and Horticultural Commodities* (Cranfield, UK: Natural Resource Management Institute, Cranfield University).

Winter, M. (2003) 'Embeddedness, the New Food Economy and Defensive Localism', *Journal of Rural Studies* 19: 23-32.

Sources

Source 1: telephone interview with John Higson, conducted by Helen Nilsson, Stockholm, 2005.

Source 2: personal interview with Charlotte Norrman-Oredsson, conducted by Helen Nilsson, Skea Gård, 2005.

Transcripts of interviews available from authors on request.

7

Open Garden: a local organic producer–consumer network in Hungary, going through various levels of system innovation

Edina Vadovics

Central European University, Department of Environmental Sciences and Policy, and GreenDependent Sustainable Solutions Association

Matthew Hayes

Institute for Environmental Management, Szent István University, and Nyitott Kert Alapitvány (Open Garden Foundation)

The Open Garden Foundation (in Hungarian, *Nyitott Kert Alapítvány*) was registered in 1999 with the aim of promoting sustainable agriculture and community-supported local food systems. It is based in Gödöllő, a town with a population of 31,000 people, about 30 kilometres to the east of Budapest, Hungary.

Starting a year before its official registration in 1999, the Open Garden has successfully been managing an organic market garden, and since 1999 a local food delivery system the organisation of which went through different stages (a detailed description is provided below). The Foundation has also been running education and training programmes for consumers as well as producers. In 2003, in cooperation with other non-governmental organisations (NGOs) in the region, a social gardening programme was initiated to provide employment opportunities for people with special needs.[1]

1 More information on the Open Garden Foundation is available at www.nyitottkert.hu, accessed 27 July 2009.

The Open Garden is a voluntary, bottom-up initiative with the vision and mission of establishing a local food system based on an adaptation of the principles of community-supported agriculture (CSA) in the Central Eastern European (CEE) region in a socio-economic context not supportive of such radical and at the same time creative solutions to the challenges faced by the agricultural sector and rural communities.

The system, established by a group of colleagues at the Institute for Environmental Management at Szent István University (KTI-SZIE) and the Open Garden Foundation, is the first CSA scheme in Hungary and, as far as we are aware, also in CEE. One of the most important aims of the case is to create a local food community by linking and bringing together producers and consumers, thus both the production and consumption of food are influenced.

7.1 Case description

7.1.1 Overview: the Central Eastern European region, Hungary and community-supported agriculture

The idea of CSA was originally conceived of in Japan in the 1960s and 1970s in response to the increased industrialisation of the food system. Japanese women, worried about the quality of the food they gave to their families, joined together their efforts in trying to buy fresh and healthy food from local growers whom they knew. Hence the name *teikei* ('food with the farmer's face on it') was used to refer to this system, which brought together producers and consumers. Farmers could be sure that they would be able to sell their produce at a favourable price, and housewives or consumers could be confident of the quality within the framework of a system beneficial to both parties. (Hayes and Milánkovics 2001)

In the 1970s and 1980s CSA began to be adopted first in Northern Europe and then in the USA, where it became very popular. Following its popularity there, it was reintroduced, growing again in favour in Europe, especially in the Netherlands and Great Britain (Hayes and Milánkovics 2001)

The concept of CSA is believed to have a special relevance for CEE countries, where agriculture is an important production sector. In Hungary, agriculture is an important economic sector because of the characteristics of the country (that is, the climate and geography is appropriate for agricultural production). The agricultural sector provides 3% of the GDP and employs 5% of the workforce (FVM 2006).

In spite of the different political and economic context, the development of agriculture in CEE countries followed a pattern similar to that in Western Europe. This meant that the size of farms grew and production methods became increasingly intensive and industrialised, which meant that fewer and fewer people were employed in agriculture.

The change of regime at the end of the 1980s and the beginning of the 1990s brought recession in the agricultural sector along with the removal of agricultural support systems and the privatisation of state and collective farms (Hayes and Milánkovics 2001;

CEC 2002a). These changes and the restructuring had devastating consequences for people living in the countryside and hoping to make a living in agriculture. In this context, CSA and the opportunities it presented in terms of people-centredness and environmental friendliness meant—and still can mean, as problems still have not been overcome (FVM 2007)—a new way of making a living in rural areas.

7.1.1.1 Community-supported agriculture

The problems presented by industrial agriculture—increasing dependence on energy and fertiliser inputs as well as on transportation, fewer jobs in the countryside, increasing levels of environmental pollution, ever-greater distrust in food by consumers and insecurity of farmers—can be overcome only if consumption and production patterns are changed. The approach taken by CSA offers one particular way of solving this problem through bringing together producers and consumers in a local food network. In such a food network:

- Consumers know the farmer(s) producing their food as well as having access to the field where their food is grown
- Farmers produce for a known group of consumers and can be sure to sell at least a part of their produce to them (in certain cases, all of it)
- Food production and consumption is kept local, and thus food miles are reduced and the local economy is strengthened (Cowell and Parkinson 2003; Pretty *et al.* 2005; Seyfang 2006)
- As usually organic methods of production are used, levels of environmental pollution are greatly reduced
- The risks inherent in agriculture are shared by consumers

7.1.1.2 Different types of community-supported agriculture, referring to different levels of system change

At this point it needs to be emphasised that CSA is an approach rather than a single and fixed method, and it is always important to adapt its principles to the particular local context and circumstances. The most common types of CSA schemes are as follows (Hayes and Milánkovics 2001):

- **Share farms**, where consumers buy 'shares' or, in other words, part of the harvest and pay for it in advance
- **Regular box schemes**, in which a farmer or a group of farmers and consumers enter into an association in which the farmer supplies a regular (usually weekly) box of seasonal farm produce and the consumers commit themselves to be regular customers
- **Delivery schemes**, which are the more flexible version of regular box schemes where consumers order weekly (without commitment to be regular customers) and the farmer(s) deliver to established collection points

- **Farmers' markets**, organised by local producers, where local people can buy fresh, local produce directly from the growers (see Chapter 6)

These different forms of CSA require different levels of commitment both from consumers and from producers, as depicted in Figure 7.1. Furthermore, these different levels of commitment also mean different levels of change in the food system, with share farms representing the most radical system change and farmers' markets leading to more incremental innovation. Hinrichs (2000) argues that although farmers' markets create closer social relationships between producers and consumers the act of supplying and buying food remains a commodity relation. At the same time, share farms are, in a way, an alternative to the existing market and represent a move towards the decommodification of food.

FIGURE 7.1 Scale of commitment and levels of system change or innovation in various types of community-supported agriculture

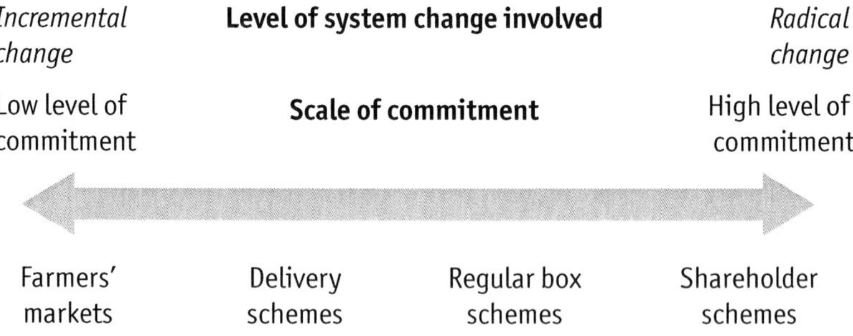

The system established by the Open Garden Foundation originally aimed to operate as a subscription farm with participation in a farmers' market to sell extra produce. However, in response to consumer needs and market demand, it has developed into a more flexible delivery scheme, as discussed in Section 7.1.3.

7.1.2 Case context: organic agriculture, consumption trends, and consumer–producer awareness in Hungary

In this section we focus on the context in which the Open Garden local organic producer–consumer network developed. As a review of all the contributory factors would reach beyond the scope of the present publication we focus on the issues we believe to be most important for the successes and failures of the initiative. These can be divided into three areas, which are all described briefly, with attention to the most relevant factors. The three areas are: the development of organic agriculture in Hungary, trends in food consumption along with existing opportunities for purchasing local organic products, and people's awareness of organic products.

7.1.2.1 Organic agriculture in Hungary

Organic agriculture started in Hungary in the 1980s. The number of organic farms in 1988 was 15, which rose to 108 in 1995, to 471 in 2000 and to 1,420 in 2004 (Kovács and Frühwald 2005). In the meantime, the area under organic cultivation grew from 1,000 ha in 1988 to 128,690 ha in 2004. This might seem like a huge growth but in fact only a little more than 2% of the agricultural land in Hungary is cultivated organically (Kovács and Frühwald 2005).

Apart from the land area under organic cultivation, another important indicator is the percentage of organic produce exported. For Hungary, this is around 90%, with most of the products being exported to 'old member states' of the European Union and Switzerland.

As for policy objectives related to organic agriculture, the objectives set by the New Hungary Rural Development Strategic Plan are not at all ambitious, and the development of organic agriculture is not an aim in its own right (FVM 2007). Thus, in Hungary, policy-makers do not yet see organic agriculture as a sector that could contribute to overcoming the challenges rural communities face today.

7.1.2.2 Trends in food consumption and purchasing

In Hungary, similarly to other countries in the CEE region, people spend a higher percentage of their income on food (30%) than in Western Europe (10–15%; EEA 2005). Because of the recession following the change of regime, during the 1990s household expenditure on food was decreasing (SEI-T 2004) as well as kilogrammes per capita food consumption (Gulyás *et al.* 2006). Since then, expenditure levels have stabilised, and the level of food consumption has remained more or less constant, too; however, some important changes occurred in the content of the annual 'food basket' of the Hungarian population. People today consume more dairy products and at the same time less cereals and potatoes than in the 1970s (EEA 2005; Gulyás *et al.* 2006). Furthermore, people in CEE countries generally eat half the amount of fruit, fish, seafood, cheese and red meat than citizens in the 'old member states' (EEA 2005).

As far as the consumption of organic food is concerned, based on a representative survey carried out recently by the GfK Hungária market research company (GfK 2007a), 76.5% of people never buy organic products, 13.4% only rarely do so and it is only 3.1% that buy them regularly, on a weekly basis.

Somewhat contrary to this result, other surveys show more promising trends. In their report on opportunities for sustainable consumption in Hungary, Gulyás *et al.* (2006) report that, according to Capital Research (2005) and their own survey (ACC 2005), almost half of the population is interested in organic products, and 7–10% buy them regularly.

GfK (2007a) found that those people who buy organic products purchase them in supermarkets and hypermarkets (43%) followed by organic farmers' markets (25%) of which there are only a dozen in Hungary at the moment (Biokultúra 2007). Quite a high percentage of the respondents (17%) reported that they themselves grow the food. This is in line with the estimate of the Ministry of Agriculture and Rural Development, according to which about one-sixth of the Hungarian population grows fruit and vegetables in their own gardens (Cs. Gát 2007). Organic food delivery services do exist; however, based on our own internet search and informal interviews, only three of them (one of them being the Open Garden) place emphasis on the importance of the local as well as the organic feature of products.

The high percentage of people reporting they purchase organic products in supermarkets and hypermarkets is a good indication of how important such products have become in recent years (KSH 2006). At the same time, the number as well as economic importance of markets, especially farmers' markets, has been decreasing continuously (KSH 2003). In addition, people tend to buy more processed food, a fact that is attributed by GfK Hungária (GfK 2002) to the generally faster way of life.

7.1.2.3 Consumer–producer awareness

In the past, the Hungarian public was repeatedly found to be little interested in environmental problems compared with people in other countries. In surveys in the early 1990s it was demonstrated that Hungarians cared the least about environmental problems among the 22 surveyed countries in the 'health of the planet' mammoth poll (Dunlap 1994; Dunlap *et al.* 1992). The situation changed somewhat during the 1990s. In 1996, Hungarians underlined a number of environmental problems as serious (Mészáros 1996), nevertheless, environmental problems without direct immediate impacts on the respondents were not considered to be grave.

News and information about the environment started to attract the attention of the public after 2000. In 2001, a Eurobarometer survey was conducted in the Accession Countries (CEC 2002b) found that almost half of the respondents (48%) in 'new member states' declared they were interested in environmental news. However, many other issues (education, social issues, enlargement, etc.) were ranked to be more important. One year later, another Eurobarometer survey showed that environmental developments were the issues of greatest interest (61%) in Hungary, being ranked even higher than medicine, genetics or astronomy (HGO 2003; see also Kiss 2005; Vadovics and Kiss 2006).

In his study on sustainable consumption, Valkó (2003) reports that quite a high percentage of the Hungarian population (40% of teenagers and 52% of adults) believe that their consumption habits have important environmental impacts. At the same time, both groups think that the government and municipal authorities bear the greatest responsibility for solving environmental problems (57% and 41%, respectively). Both groups put citizens second (27% in both groups), but before companies or NGOs. As for willingness to pay more for environmentally friendly products, which is important in the case of organic products, 27% of teenagers and 26% of adults said that they would be ready to pay a premium price.

For us it is of special significance how the Hungarian population evaluates environmental problems related to agriculture in relation to other issues. In their study called

Eco-barometer, Cognative Ltd and WWF (2004) concluded that Hungarians do not find the environmental impacts of agriculture to be important. In relation to willingness to pay more for environmentally friendly products and services, the majority of the population would be willing to make sacrifices for selective collection of bottles, as opposed to paying more for soft drinks in returnable bottles, environmentally friendly cleaning products or organic food.

In addition to not yet being ready to pay a premium price for organic products, GfK (2007b) found that the Hungarian population is not very health-conscious and people do not know exactly what organic or bio-food products are, or what distinguishes them from non-organic products (GfK 2007c). It is interesting to note here that although the Hungarian population was reported to be not so health-conscious, health is still the most important motivation for purchasing organic products (AMC 2006).

7.1.2.4 Summary

In summary, it can be said that, although the demand for organic products is growing, a large percentage of the population (even some of those who otherwise regularly purchase organic products) cannot define what exactly distinguishes such products from non-organic products. Furthermore, the Hungarian population does not attribute great importance to the environmental and social problems created by industrial agriculture, thus solving them, or contributing personally to their solution, does not have a particular importance for them. Another important factor is that the majority of people do not consume organic food for environmental reasons; rather they do so for health reasons and thus do not pay attention to whether the products were grown locally or not, although they do believe Hungarian organic products do have an added value (AMC 2006). Consumers also think that the origin of the product should be clearly marked on the packaging. At the same time, buyers of organic food express higher levels of well-being, and organic food is closely associated with healthy living, happy families and stronger social ties (AMC 2006).

The area of land under organic cultivation has been growing steadily since the 1980s, but with most of the produce being exported, farmers' markets and potentially also local producer–consumer networks such as the Open Garden initiative can be important channels for farmers to sell their produce at a good price.

7.1.3 Case history and development

7.1.3.1 The beginning (from 1998 to 2002): subscription farm and box scheme

In 1998 a group of people at the Institute of Environmental Management at Szent István University, Gödöllő, decided to put the principles of CSA into practice. At the same time, the Institute wanted to establish a demonstration organic garden on the university estate to provide a model for students, staff as well as other interested people. The aim was to cultivate certain parts of the university's land using organic techniques and to supply local people with fresh organic produce.

As the CSA project in Gödöllő was an entirely new venture, there was no existing knowledge of the local demand for organic produce. There was an indication of some interest through personal contacts, for example, with the local Waldorf (Rudolf Steiner) school; however, the level of this interest needed to be tested.

Thus, CSA was started in 1998 with minimum funds, first with an informal box scheme with the subscription system introduced in 1999. It was based on a weekly vegetable box scheme, with vegetables coming solely from the university's garden. Consumers agreed to 'contract in' to receive a share of the harvest. This worked by people signing up to pay for vegetables boxes for a period of 22–27 weeks, at a price fixed at the beginning of the season. Consumers could either pay on a weekly basis or could support the Open Garden by paying their share in advance if they were able to. Table 7.1 shows that the number of consumers (individuals and families) interested in the system grew from 40 to 100.

TABLE 7.1 Development of the Open Garden community-supported agriculture project

Year	Number of consumers	Weekly price of box (€ equivalent)	Cost of share (25 weeks, in € equivalent)	Number of shareholder families
1998	40	4	–	–
1999	70	4.2	110	15
2000	80	4.5	115	20
2001	100	5	125	40
2002	70[a]	6	150	28

a In 2002 the Real Food Box service started up, with approximately 40 weekly customers

The weekly boxes of vegetables were delivered to a number of collection points from which consumers (individuals or families) were able to pick them up and return the empty box from the week before. Apart from the vegetables and fruit, boxes contained recipes to help with the preparation of less well-known vegetables as well as newsletters to inform people of what was happening at the Open Garden. Additionally, open days, work days and festivals were organised so that people could come to the garden to see where their food was grown, take part in growing and harvesting the vegetables and also to celebrate together.

Apart from the box scheme, the produce of the garden was also sold at a farmers' market in Budapest. Not only was this an excellent way to sell the extra produce but also it provided an opportunity to meet with new customers who might potentially be interested in becoming a member of the box scheme or a shareholder.

While the number of customers in the CSA scheme grew every year, there were two important issues with which the Foundation was struggling. First, in an attempt to provide customers with good value for the shares they agreed to buy and also to overcome the barrier of relatively low environmental awareness, it can now be said that the price of the weekly box was set too low. Thus, the income from the system did not meet production costs, however low they were kept. Second, the system, which provided consumers with a more or less surprise selection of vegetables every week, was not flexible enough for quite a high percentage of consumers. This meant that the turnover of customers was fairly high. To overcome these problems, from 2002 a new system, *Nyitott Kert Futár*, the Real Food Box (RFB), organic delivery service was set up.

7.1.3.2 From 2002 to 2006: Real Food Box organic producer–consumer network

In order to resolve the problems of the CSA system, the Open Garden Foundation decided to re-organise it into a food delivery system that provided more choice and a year-round service. This meant that more producers needed to be involved in the supply of food so that a greater selection of vegetables and fruit could be offered to consumers. However, the producers needed to be coordinated, so the Open Garden had to take on the role of coordination as well as continue growing vegetables in the organic market garden. The schematic of the system is depicted in Figure 7.2.

This system provided more flexibility to consumers as they could either select a weekly box that was offered at a standard price or choose the 'select your own box' option. For both of these options they could either pay weekly when collecting the box or monthly.

FIGURE 7.2 The schematic of the Real Food Box system

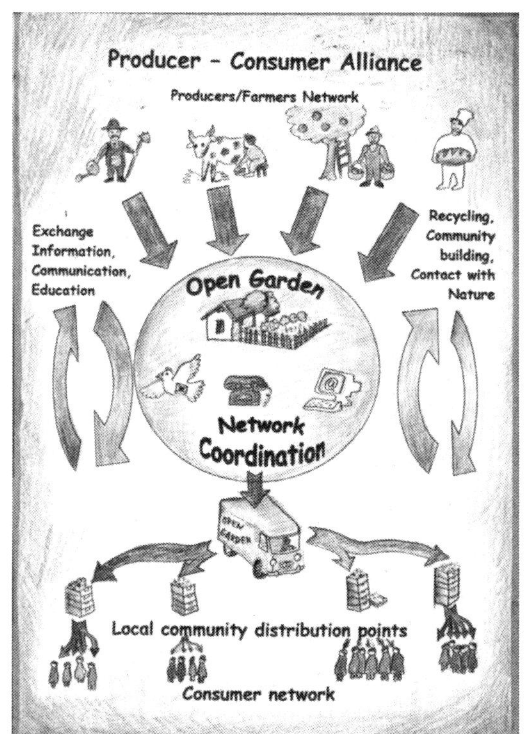

By 2005 the system grew so much that it supplied vegetables basically all year round to an ever-growing number of customers (see Table 7.2).

TABLE 7.2 Development of the Real Food Box network

Year	Number of weeks	Annual number of boxes	Annual turnover from boxes (€)
2002	27	863	12,400
2003	38	2,651	25,600
2004	36	1,834	23,200
2005	50	2,540	31,600

At this point, it is important to mention that NESsT (the Non-profit Enterprise and Self-sustainability Team),[2] a civil organisation established in Latin America and the CEE region to promote social enterprise for non-profit organisations ('profits for non-profits'), had a significant role to play in the development of the Gödöllő producer–consumer network, and in particular the RFB scheme. In 2002 the Open Garden Foundation began to make links with NESsT and in 2003 was included in NESsT's portfolio of organisations in the NESsT Venture Fund.

The Open Garden's efforts as a civil organisation to establish a more businesslike organic food delivery service attracted support from NESsT as a mission-based social venture. NESsT was able to offer the Open Garden Foundation both technical business and management consulting support as well as enterprise development grants over a three-year period (2003 to 2005). NESsT involvement was particularly important in helping to bring business planning and financial management skills into operation within the food box scheme. It is quite possible that the RFB scheme would have collapsed in 2004 had it not been for the consulting support NESsT was able to offer, and the intimate connection between the RFB and the Open Garden as an organisation at the time might have meant a closure of the organisation if the RFB had collapsed.

NESsT has extensively used the Open Garden experience as a case study in social enterprise for non-profit organisations. The relatively high-risk nature of the RFB (relatively high investments for slow returns) has given NESsT the opportunity to explore the methodology it has been applying to non-profit organisations. Not-for-profit social enterprise is seen as a desirable way for civil organisations to develop independent funds for mission causes. The RFB provided both challenging questions and evident successes. The eventual sale of the RFB to the private sector was seen as a creative solution to the daily pressure exerted by the scheme on the life of the Open Garden Foundation. The continuation of the RFB in the private sector is seen as a way of carrying forward the mission goals of the Open Garden Foundation without the ideological baggage that might have been a burden on the commercial success of the organic food box scheme.

2 More information on NESsT is available from www.nesst.hu/default.asp, accessed 27 July 2009.

7.1.3.3 Present: cooperation between various sectors

In January 2006 the management of the Open Garden Foundation decided to try to run the box scheme as a private enterprise by one of the farmers already part of the organic producer network. As it proved to be successful, the system was sold to the farmer (Gódor Biokertészet), who now runs it under the same name (that is, RFB, or *Nyitott Kert Futár*) as a private family enterprise. This family has been practising organic farming since 1996.

At the same time, the market garden of the Open Garden started to be managed again by Szent István University. It is now a university teaching and demonstration garden that still sells its produce through the organic food box scheme and at the weekly farmers' market in Budapest. Thus, the number of activities the Open Garden Foundation performed was reduced and is now perhaps more fitting for a non-profit organisation. It carries out educational activities to promote local, organic, food and in cooperation with the private farmer and the university organises community events at the market garden.

As for the number of consumers ordering the vegetable boxes, it is now between 50 and 60. However, some of these are big orders that need to be packed into two or three boxes. Consumers now have the choice of ordering a set collection of products, a mix of seasonal vegetables, fruit and processed food items (jam, mustard, juices, pasta, etc.) called the 'community box', the price of which is the same every week. Alternatively, they can select any product they want from a list provided, but in this case the minimum price of the order is about a third higher than the price of the community box. To satisfy consumer demand, the farmer imports tropical fruit as well as vegetables that are not seasonal (such as tomatoes in winter). Interestingly, although people ordering the community box can also select additional products from the list, only about a sixth or fifth of the weekly boxes are in this category (that is, about 10). There are still four consumers who pre-order their box, meaning that they order the community box at the beginning of the season or year, and pay in advance. Some consumers require delivery to their homes, but the rest of the boxes are still delivered to collection points (about 12, but growing in number) by a subcontractor.

The box scheme operates throughout the year, though the farmer is considering having a summer break as the number of consumers drops in the holidays. The selection of products is growing almost weekly to satisfy the needs of consumers, who recently started to demand even fair-trade coffee, tea and chocolate.

7.1.4 Actors involved in the initiative and their roles

As described in Section 7.1.3, the number, involvement and role of actors has changed, at times quite dramatically, since the Open Garden producer–consumer network was set up. Thus, for the sake of clarity, primary and secondary actors and their role in the different stages of the project is summarised in Table 7.3.

Table 7.3 Actors and their roles during the different stages of the project

	1998–2002: subscription farm and box scheme	2002–2006: RFB Network	2006–present: cooperation between sectors
Primary actors			
KTI-SZIE	Initiates the RFB project Manages the project (garden and box scheme) in its first year Provides (mainly in-kind) support for the Foundation	Negotiates within the university on how to manage gardens for the future	Owns and manages the organic market garden Sells produce of garden to Gódor Biokertészet and at organic farmers' market Provides office space for the Foundation on its premises
NyKA and its employees	Established in 1999 Manages the market garden and the box scheme Transports the boxes to collection points Organises events for consumers (educational activities, festivals, open days and work days) Communicates with consumers through newsletter and web page	Manages the market garden Organises the Real Food Box scheme Coordinates with producers and consumers Transports the boxes to collection points and homes Organises events for consumers	Organises events for consumers (educational activities, festivals, open days and work days) Communicates with consumers through newsletter and web page
Board of Trustees of the Foundation	Formal role	Role becomes increasingly close to management	Begins to settle into governance role
Volunteers	Work for the Foundation in the market garden	Work for the Foundation in the market garden	Work for the foundation on communication and educational projects
Consumers (subscribers)	Act as 'shareholders' in the CSA and receive weekly vegetable boxes Participate in events organised by the Foundation	Act either as weekly paying subscribers to the food box scheme, or as 'shareholders' paying in advance Participate in events organised by the Foundation	Order their weekly box from Gódor Biokertészet Participate in events organised by the Foundation

continued opposite →

	1998–2002: subscription farm and box scheme	2002–2006: RFB Network	2006–present: cooperation between sectors
Producers (local organic growers)	n.a.	Provide produce to the Foundation to sell through the RFB delivery scheme	One of the producers takes over the management of the RFB delivery scheme (see below) Provide produce to RFB delivery scheme
Gódor Biokertészet (Godor Organic Farm)	n.a.	Provide produce to the Foundation to sell through the RFB delivery scheme	Buys the RFB Manages the RFB delivery service (with the transportation of the boxes now outsourced)
Supporting organisations, such as the local Waldorf School	Assist in finding interested consumers to participate in the subscription farm and the box scheme Act as a collection point for boxes	Act as a collection point for boxes delivered by RFB	Act as a collection point for boxes delivered by RFB
Secondary actors			
Consumers at farmers' markets	Regularly purchase small percentage of garden produce	Regularly purchase about 25% of garden produce	Regularly purchase about 60% of garden produce
ÖISz Foundation (for people with special needs)	n.a.	Small sheltered gardening work programme set up	Sheltered gardening programme continues, with plans for expansion
NESsT	n.a.	NyKA joins NESsT social enterprise venture portfolio Representative on the Board	Board representative and supporter role Uses the Open Garden as case study in social enterprise
Funding organisations: PHARE, REC, etc.	REC-funded project to promote CSA	PHARE-funded project to promote environmental benefits of local organic food systems	n.a.

n.a. = not applicable

CSA = community-supported agriculture NESsT = Non-profit Enterprise and Self-sustainability Team

RFB = Real Food Box (*Nyitott Kert Futár*) NyKA = Nyitott Kert Alapitvány (the Open Garden Foundation)

ÖISz = Foundation for Equal Rights PHARE = Poland–Hungary Assistance for Restructuring the Economy

REC = Regional Environmental Center KTI-SZIE = Szent István University, Institute of Environmental Management

7.2 Results

7.2.1 Main results

The most important result is that an alternative food system, the Open Garden producer–consumer network, was established in 1998 and has been running ever since. Although it has gone through two major restructurings, the network still exists and has been developing constantly through cooperation between different sectors as well as between producers and consumers. The number of consumers interested in ordering weekly boxes of produce is growing continuously, and Gódor Biokertészet regularly receives calls from different parts of Hungary from interested consumers. This indicates that similar systems could be established successfully elsewhere.

A second important result is the establishment of the organic market garden on the estates of Szent István University in Gödöllő. The garden is now a demonstration garden where students studying at the university are taught the principles of organic farming. However, the produce of the market garden is also sold through the RFB scheme as well as a farmers' market. The area given to the garden is being expanded at the moment. In addition, the garden is used as a site for consumer involvement events such as festivals and work days, which are organised by the Open Garden Foundation.

Finally, the non-profit organisation, the Open Garden Foundation, established to promote sustainable agriculture, still operates and is active in educating and involving not only local but also foreign consumer-citizens.

7.2.2 Change in sustainability performance

From the point of view of sustainability, the primary aim of the initiative was to reduce distances in two ways:

- To reduce the food miles associated with products in an increasingly globalised food system (environmental sustainability)
- To reduce the growing distance between producers and consumers (social sustainability)

With regard to reducing food miles, although there was no specific study performed to quantify the change in food miles associated with the products, there are indications that for the consumers involved in the system the change has been marked. Based on consumers' responses to end-of-season questionnaires administered by the Open Garden Foundation in 1999 and 2000, it can be established that the vegetables delivered weekly satisfy 75–100% of the needs of those consumers. Thus, for this group of consumers, food miles are reduced as the vegetables were grown within 50 km of their homes, as at that time imported foods were not offered within the framework of the system. The environmentally positive effect of reducing food miles is confirmed by studies conducted in other countries (Cowell and Parkinson 2003; Pretty *et al.* 2005; Seyfang 2006).

The distance between producers and consumers has also been reduced. Members of the system have information on where their food is grown or, in case of imported food, on where exactly it is from. In the case of locally grown food, they have the opportunity to visit the farms, talk to the growers and also, thanks to the Open Garden Foundation, engage in activities on work days and festivals. If they choose the 'community box', they also make sure that they eat seasonal products and thus share the risk associated with producing food.

Measuring the sustainability performance of the Open Garden local organic producer–consumer food network, on the one hand, has not yet been completed. On the other hand, such as project would be rather challenging as it is difficult to select a system that could be used as a baseline against which to compare it. Nevertheless, based on the literature available (for example, Cowell and Parkinson 2003; Pretty *et al.* 2005; Seyfang 2006), surveys done in Hungary (AMC 2006), our own experience, an interview with Gódor Biokertészet (Source 1) and feedback from consumers involved in the system in 1999 and 2000 (the end-of-season questionnaires),[3] the following is a realistic indication of the improvement in overall sustainability performance as compared with a system where vegetables are grown applying the methods of industrial agriculture, often imported and are bought by consumers in hypermarkets and supermarkets owned by multinational companies.

7.2.2.1 Environmental sustainability

- Fewer food miles equate to less need for the consumption of fossil fuels and lower greenhouse gas emissions (Cowell and Parkinson 2003; Pretty *et al.* 2005)
- More diverse and small-scale production equates to more biodiversity as opposed to the monocultures of industrial agriculture
- There is a greater opportunity for closed resource cycling (for example, recovery of organic waste)
- Positive indirect impacts are improved education and awareness-raising of consumers who, as a result, may exhibit more environmentally friendly behaviour in other areas of their lives
- Organic production methods are used, which are proven to be more environmentally friendly than those applied by industrial agriculture
- There are lower packaging demands and lower associated quantities of packaging waste. As can be seen from Figure 8.3, the weekly produce is delivered in returnable cardboard boxes (which are made from recycled paper) and in as little packaging as possible. Consumers subscribing to the system pay a deposit for the cardboard box to encourage them to return it to the collection points

3 Further information on the questionnaires can be obtained from the authors.

Figure 7.3 Typical weekly box (community box) in spring with a surprise loaf of sweetbread for Easter. Note the reduced level of packaging compared with supermarket food

7.2.2.2 Social sustainability

- The scheme provides an opportunity for consumers and farmers to get to know each other and understand one another better. This way the needs and demands of both groups can be better met
- It is an opportunity to share the risks of food production. This was especially true at the beginning of the history of the Open Garden, when consumers were part of the CSA scheme, or for those consumers who order the community box regularly
- There is increased opportunity for community-building through local people learning about and getting involved in agricultural activities through work days and festivals. Festivals can then become regular annual events to look forward to, as was the case with the Harvest Festival organised by the Open Garden Foundation
- Local jobs are created. The increasing and continuous demand for local organic produce means that growers can extend the area of production and can also invest in growing crops requiring more specialised production methods that are often more work-intensive. The case of the Open Garden is also an exam-

ple of how local food networks can lead to the creation of jobs for people with special needs

- A link is created between urban and rural communities. By participating in food networks, urban people are reconnected to nature and the land, and contact is re-established between urban and rural areas
- This in turn creates new life and prospects for rural communities
- It is an opportunity for people to get closer to the land and relearn the natural cycles of food production, which
  - Encourages seasonal eating and helps people to feel part of nature
  - Gives a higher level of confidence in food quality and sources (food security): consumers know where their food comes from and have the opportunity to communicate with the farmer growing it
- Better food equates to better health. Consumers receive weekly boxes of fresher food that is organic and less processed. Organic food is proven to have a better nutritive content than food produced by industrial agriculture.[4] Most of the people purchasing organic food do so because they believe it to be healthier than conventionally produced food (AMC 2006)

7.2.2.3 Economic sustainability

- Maximum returns to local farmers are guaranteed, thus their livelihood is better ensured than by selling produce through (other) intermediaries. For farmers a regular delivery scheme ensures a more certain income then selling their produce at the farmers' market or to a large (most often multinational) processing company (Source 1)
- Financial resources remain in the local community, which was proven elsewhere to have a
  - 'Multiplier effect': for example, the New Economics Foundation in the UK found that for every euro spent on a local organic food box, €2.40 was contributed to local economy, whereas only €1.20 was generated from every euro spent in supermarkets (see also Sacks 2002)[5]
- Subscribing to the delivery service is economically advantageous for consumers, too. They reported that they believed they were paying a reasonable price for the box, and several of them also mentioned that the regular weekly delivery freed them from the chore of having to go to the market and carrying produce home from there

4 Go to www.ota.com/organic/benefits/nutrition.html, accessed 27 July 2009.

5 See www.neweconomics.org/tex/m6_i231_news.aspx, accessed 27 July 2009.

7.2.3 Learning experiences

Although there are many different learning experiences that could be discussed here, we concentrate on three lines of thought that we consider the most important for the Open Garden story and also for considering the potential for multiplication or scaling up of the case.

7.2.3.1 Adopting the community-supported agriculture concept

Originally, the initiators of the Open Garden aimed at creating a CSA in Hungary. However, as the Open Garden started operating and interacting with consumers, it began to be apparent that the purist CSA was too prescriptive and inflexible for the Hungarian market. At the time when the Open Garden started, Hungarian consumers were just beginning to learn to have a choice after the more restricted economy of the socialist regime, so it was essential that they should also have some choice within the CSA. The Foundation running the scheme needed to be flexible enough to change to suit consumer needs better, who otherwise were eager to participate as organic food was not easily available in Hungary.

The box system offers a particular type of service that is convenient and ideal for some people but will not suit everyone. For the successful running of the box scheme it is very important to establish trust between the producers and consumers, as consumers will not have the opportunity to check the produce before deciding to put it in their basket.

In order to establish and maintain trust, and also because there has been a strong interest from consumers to (re)connect with the land, the farm visits, festivals, volunteer days and other events are essential and have been successful and have become important parts of the system. They form an important part of the communication between producers and consumers, and supplement the newsletters, recipes and phone calls during the year.

7.2.3.2 For-profit business within a not-for-profit organisation

In the case of the Open Garden, it has proved to be very difficult, though not impossible, to run a demanding for-profit business within a not-for-profit organisation, especially in a very market-based sector. Operating as a not-for-profit organisation may have certain advantages; for example, it is perhaps easier to attract funding, there may be more opportunities to test new ideas and consumers often are more trusting of not-for-profit enterprises, a fact that was confirmed during the interview with Gódor Biokertészet, now operating the delivery service (Source 1). However, it appears that for this particular case the best solution, at least for the time being, is to operate the delivery service as a for-profit family enterprise and cooperate with both the university and the Open Garden Foundation in educational and promotional issues.

It also needs to be emphasised that setting the price of the weekly vegetable box right from the beginning is very important for the economic sustainability of the system, and also because studies indicate (such as AMC 2006; Gulyás *et al.* 2006; GfK 2007a) that the price of organic products is an important barrier to increased consumption. Even though for people participating in different forms of CSA social ties and community building are important factors, they still want to feel that they are getting a good deal (Hinrichs 2000). The fact that in the case of the Open Garden the price was initially set too low created difficulties for the Foundation. Thus, in the case of multiplication and adaptation to other areas, this should definitely be an issue for consideration.

7.2.3.3 Producers and consumers

Forming direct relationships between producers and consumers is beneficial for both sides, but many of the benefits are not directly financial (such as satisfaction, sense of food security, mutual understanding of problems, opportunity for feedback, etc.). Because of this, it is important to communicate them to provide continuous positive feedback and information for existing and potential consumers.

As the case of the Open Garden has shown, forming partnerships with other organic producers can strengthen the position of the case in the market. That is, producers can offer a wider range of produce over a longer season, can achieve some economies of scale and can also benefit from specialisation of skills and an increase in capacity. It helped that the Open Garden Foundation took on the role of the coordinator during the RFB period and then could pass on the knowledge, skills and existing network to the private enterprise.

7.3 Potential for diffusion and scaling up

The Open Garden is a case of ongoing success with constant failures that have required the adaptation of the pure and prescribed CSA concept into a more flexible and loose RFB scheme that is sensitive to consumer needs and market demand. The case has proved to be working in Hungary in a context where the urban population was interested in participating in a system that provided regular supplies of fresh, local, organic produce from the surrounding rural areas. It could potentially be diffused to other areas in Hungary (there definitely seems to be some interest; on Source 1) and the CEE region. On the one hand, people are becoming more concerned about their health and the origin of their food and, on the other, in the CEE region local production and products have always been favoured, and seasonal eating still has deep roots and tradition (Vadovics 2006).

At the same time, it needs to be recognised and appreciated that the case is to a large extent embedded in and defined by the local context. It is based on the needs of the local people negotiated through communication with the producers through the involvement of the Open Garden Foundation and the collection-point managers. Local institutions, such as the Waldorf School and Szent István University, played an important role in helping to start up the system and thereafter through acting as collection points and providing a steady group of consumers. Thus, local circumstances, stakeholders, needs, capacities and the potential for cooperation need to be carefully researched and mapped before a system is initiated.

The most important, more global barriers to further diffusion and multiplication are:

- People's lack of awareness of the important environmental impacts of industrial agriculture and the relevance of personal responsibility in relation to selecting 'alternative' products and influencing production in this way
- Infrastructural barriers, such as the lack of policy support and non-existent financial incentives; the fact that the external costs of agriculture are not included in the price of products and that conventional or industrial agriculture receives subsidies is an important issue since the price of products is a significant factor when consumers decide which product they want to purchase

7.4 Overall conclusions

The Open Garden is a case where the management moved from the not-for-profit to the for-profit sector, even though the main actors remained the same throughout its history. This movement also meant that on the scale of commitment shown in Figure 7.2, the case moved from more commitment needed from consumers (share farms) to less (regular box schemes and delivery schemes). As Figure 7.2 indicates, such a movement entails a lower level of system change or innovation. However, the change is still quite radical compared with the mainstream model of food consumption. Although less commitment is required from consumers, the system is still to a large extent built on trust. On the one hand, consumers need to trust that the producer is going to deliver good quality produce to them every week, and, on the other, producers need to trust that consumers will order the boxes. Furthermore, collection-point managers, although they receive some incentives from the producer, to a large extent do the work involved in distributing the boxes voluntarily. In fact, their contribution to the successful operation of the system is essential as they regularly communicate with consumers. Thus, the delivery scheme still represents **system-level change**, though it is more incremental than in the case of subscription farms.

The Open Garden is also a successful case of moving towards the localisation of the food chain by connecting producers and consumers as well as urban and rural areas. Its impact on food consumption as well as production patterns is considerable: it educates consumers about their responsibility for influencing food production patterns and how, by sharing the risks of agriculture, they can contribute to allowing the farmer to practise more environmentally friendly cultivation methods. In this way consumers can also contribute to creating viable rural livelihoods. A member of the Open Garden subscription farm reported that CSA for her and her family is 'a kind of course on how to change our consumer attitudes' (Hayes and Milánkovics 2001: 66) and about learning new consumption patterns that require, for example, the regular ordering and picking up of boxes and the planning of meals on the basis of seasonally available vegetables.

References

ACC (Association of Conscious Consumers) (2005) 'Ethical Consumption Attitudes in Hungary: Research Report' (Budapest: ACC).

AMC (Agricultural Marketing Centre) (2006) 'A magyar öko-élelmiszerek iránti kereslet az ökopiacokon és szupermarketekben vásárlók körében' ('Demand for Hungarian Eco Food Products among Those Shopping in Eco-markets and Supermarkets') (Budapest: Magyar Biokultúra Szövetség).

Biokultúra (2007) 'Biopiacok' ('Organic Markets'); www.biokultura.org, accessed 28 April 2007.

Capital Research (2005) *A Fair Trade rendszer magyarországi meghonositásának lehetőségei* (*The Possibilities for Introducing Fair Trade in Hungary*) (Budapest: Ökotárs Alapitvány).

CEC (Commission of the European Communities) (2002a) 'Agricultural Situation and Perspectives in the Central and Eastern European Countries (Country Report for Hungary)'; ec.europa.eu/agriculture/publi/peco/index_en.htm, accessed 9 July 2009.

—— (2002b) 'Candidate Countries Eurobarometer 2001' (Brussels: CEC, DG Press and Communication).

Cognative Ltd and WWF (2004) 'Milyen környezettudatossággal lépett be a magyar lakosság az Európai Unióba?' ('What was the Level of Environmental Awareness of the Hungarian Population at the Time of the Country's Accession to the EU?'); www.wwf.hu/esemeny.php?id=265, accessed 9 July 2009.

Cowell, S.J., and S. Parkinson (2003) 'Localisation of UK Food Production: An Analysis Using Land Area and Energy as Indicators', *Agriculture, Ecosystems and Environment* 94: 221-36.

Cs. Gát, L. (2007) 'A kiskert most is divat; fenntartása idöigényes, de meghálálja' ('Gardening is Still Popular; It Requires Time, But It's Worth It'); www.delmagyar.hu/cikk.php?id=70&cid=, accessed 9 July 2009.

Dunlap, R.E. (1994) ‚International Attitudes towards Environment and Development', in H.E. Bergesen and G. Parmann (eds.), *Green Globe Yearbook of International Cooperation on Environment and Development* (Oxford: Oxford University Press).

——, G.H. Gallup and A.M. Gallup (1992. 'The Health of the Planet Survey: A Preliminary Report on Attitudes toward the Environment and Economic Growth Measured by Surveys of Citizens in 22 Nations to Date' (Princeton, NJ: George H. Gallup International Institute).

EEA (European Environment Agency) (2005) *Household Consumption and the Environment* (Copenhagen: EEA).

FVM (Földművelésügyi és Vidékfejlesztési Minisztérium [Ministry of Agriculture and Rural Development]) (2006) 'A magyar mezó´gazdaság és élelmiszeripar számokban' ('Hungarian Agriculture and Food Industry in Numbers') (Budapest: FVM).

—— (2007) 'New Hungary Rural Development Strategic Plan for 2007–2013' (Budapest: FVM).

GfK (GfK Hungária) (2002) 'Terjednek a hideg, hamar elkészithetö ételek: Gyorsuló életünk befolyásolja az élelmiszerek forgalmát' ('The Spread of Ready-made Food: Accelerating Lifestyle Influences Food Choice'); www.gfk.hu/sajtokoz/december2002/topoet.html=, accessed 9 July 2009.

—— (2007a) 'A magyarok többsége egyáltalán nem vásárol bioélelmiszert' ('The Majority of the Hungarian Population Never Purchase Organic Food Products'); www.gfk.hu/sajtokoz/articles/200701311200.htm, accessed 13 July 2009.

—— (2007b) 'Egyelőre kevesen táplálkoznak tudatosan' ('The Awareness of the Population about Food is Low'); www.gfk.hu/sajtokoz/articles/200701101200.htm, accessed 13 July 2009.

—— (2007c) 'Sokan nem tudják mitöl "bio" a bioélelmiszer' ('Most of the Population does not Know the Meaning of "Organic" in Relation to Organic Food'); www.gfk.hu/sajtokoz/fr2.htm, accessed 13 July 2009.

Gulyás, E., E. Ujhelyi, A. Farsang and Z. Boda (2006) 'Opportunities and Challenges of Sustainable Consumption in Central and Eastern Europe: Attitudes, Behaviour and Infrastructure: The Case of Hungary', in *Proceedings of the 'Sustainable Consumption and Production (SCP): Opportunities and Challenges', SCORE! Launch Conference, Wuppertal, Germany*.

Hayes, M., and K. Milánkovics (2001) 'Community Supported Agriculture (CSA): A Farmers' Manual. How to Start Up and Run a CSA' (Gödöllő, Hungary: Nyitott Kert Alapitvány).

HGO (Hungarian Gallup Organisation) (2003) 'Public Opinion in the Countries Applying for European Union Membership' (CC-EB 2002.3 on Science and Technology; Brussels: Eurobarometer).

Hinrichs, C.C. (2000) 'Embeddedness and Local Food Systems: Notes on Two Types of Direct Agricultural Market', *Journal of Rural Studies* 16: 295-303.

Kiss, B. (2005) 'Climate Change Literacy in Old Member States and New Member States of the EU' (Master's dissertation; Budapest: Department of Environmental Sciences and Policy, Central European University).

Kovács, A., and F. Frühwald (2005) 'Organic Farming in Hungary'; www.organic-europe.net/country_reports/hungary/default.asp, accessed 13 July 2009.

KSH (Hungarian Central Statistical Office) (2003) 'Piacok, piaci kereskedelem' ('Markets, Retail Trade on Markets') (Budapest: KSH).

KSH (Hungarian Central Statistical Office) (2006) 'Bevásárlóközpontok, Hipermarketek, 2005' ('Shopping Centres, Hypermarkets, 2005') (Budapest: KSH).

Meszáros, J. (1996) *Attitudes about environmental Protection and the State of Environment* (Budapest: TÁRKI).

Pretty, J.N., A.S. Ball, T. Lang and J.I.L. Morison (2005) 'Farm Costs and Food Miles: An Assessment of the Full Cost of the UK Weekly Food Basket', *Food Policy* 30: 1-19.

Sacks, J. (2002) *The Money Trail: Measuring your Impact on the Local Economy using LM3* (London: New Economics Foundation and Countryside Agency)

SEI-T (Estonian Institute for Sustainable Development) (2004) 'Household Consumption Trends in New Member State' (Background Study for European Environment Agency; Tallinn: SEI-T).

Seyfang, G. (2006) 'Ecological Citizenship and Sustainable Consumption: Examining Local Organic Food Networks', *Journal of Rural Studies* 22.4 (October 2006): 383-95.

Vadovics, E. (2006) 'Emerging Sustainable Consumption Patterns in Central Eastern Europe, with Specific Focus on Hungary', in *Proceedings of the Workshop on Perspectives on Radical Changes to Sustainable Consumption and Production, Copenhagen, Denmark.*

—— and B. Kiss (2006) 'Are Sustainable Energy Production and Use Possible? The Hungarian Case', in *Proceedings of the 'Sustainable Consumption and Production (SCP): Opportunities and Challenges', SCORE! Launch Conference, Wuppertal, Germany.*

Valkó, L. (2003) *Fenntartható/környezetbarát fogyasztás és a Magyar lakosság környezeti tudata* (*Sustainable/Environmentally Friendly Consumption and the Environmental Awareness of the Hungarian Population*) (Budapest: Aula Kiadó and Budapesti Közgazdaságtudományi és Államigazgatási Egyetem).

Source

Source 1: interview with the owners of Gódor Biokertészet, 23 April 2007, conducted by E. Vadovics, Galgahévíz, Hungary.

8

Slow Food: counteracting fast food and fast living

Ingrid Kjørstad*

National Institute for Consumer Research (SIFO), Norway

Slow Food is a social movement that started in Italy in the 1980s. It is included here as a case because it addresses issues that may be characterised as 'sustainable food consumption'. It is interesting first of all because of its collective, network-based character. From its Italian origins, Slow Food initiatives have spread out to many other countries. We describe one such case in Poland, an example that clearly demonstrates the context-dependent character of the Slow Food movement.

Section 8.1 presents details of the case study, including a presentation of the context in which the Slow Food movement emerged, the relevant actors involved and the historical development and strategies of Slow Food. Section 8.2 comments briefly on the accomplishments and results of Slow Food regarding change in sustainability performance. In Section 8.3 the potential for diffusion and scaling up of Slow Food is the focus, giving an example of the establishment of Slow Food in Poland. Finally, Section 8.4 gives an overall conclusion and sums up the main points of the chapter.

* A warm thank you to Bruno Scaltriti, lecturer at the University of Gastronomic Science, for valuable comments on the final draft of this chapter.

8.1 Case description

8.1.1 Introduction

The scope of the Slow Food movement is to counteract fast food and fast life—to counter the disappearance of local food traditions and interest in food; it appreciates the origin and taste of food and how food choices affect the rest of the world. Slow Food works to defend biodiversity in food supply, spread taste education and connect producers with co-producers through events and initiatives.[1]

The Slow Food philosophy is based on a belief that everyone has a 'fundamental right to pleasure and consequently the responsibility to protect the heritage of food, tradition and culture that make this pleasure possible'.[2] The movement is founded on this concept of eco-gastronomy: the recognition of the strong connections between plate and planet. Slow food should be good, clean and fair food. The food people eat should taste good, be produced in a clean way that does not harm the environment, animal welfare or human health, and food producers should receive fair compensation for their work. Members of the Slow Food movement consider themselves co-producers, not consumers, because by being informed about how food is produced, and actively supporting those who produce it, they attempt to become a part of (and a partner in) the production process (paraphrased from the website; see footnote 2).

The Slow Food movement is a voluntary bottom-up initiative, and, according to its founder Petrini (2005), it is a *culture*, based on the intrinsic value of local production. Food and taste are considered embedded within tradition and social relations. It offers a critique of 'foods without identity', at once globalised and delocalised and part of large corporations that operate the modern food system (Roos *et al.* 2007). Miele and Murdoch (2003: 26) describe the organisation as

> a structured form of resistance . . . a network which serves to condense cultural norms and in so doing facilitate the spread of particular culinary cultures: . . . Slow Food embodies a European philosophy that 'all food, feeds not only our bodies but also our minds; it is more than just a meal, it is part or our way of life' (Fiddes 1991: 6).

The rapid spread of fast foods, seen as an indicator of the unwelcome advancement of 'fast living' centred on materialism, was the trigger for what Petrini defines as the Italian Slow Food Manifesto: an article written by founding member Folco Portinari, published in the *Gambero Rosso* magazine in 1987. Two years later, the Manifesto was approved at a meeting in Paris where 400 members from 18 nations gathered (Roos *et al.* 2007: 8).

Some main points in the Slow Food Manifesto are:

> We are enslaved by speed and have all succumbed to the same insidious virus: Fast Life, which disrupts our habits, pervades the privacy of our homes and forces us to eat Fast Foods . . . Our defense should begin at the table with Slow Food. Let us rediscover the flavors and savors of regional cooking and banish

1 Go to www.slowfood.com, accessed 9 July 2009.

2 Quote from www.slowfood.com/about_us/eng/philosophy.lasso, accessed 9 July 2009.

> the degrading effects of Fast Food . . . In the name of productivity, Fast Life has changed our way of being and threatens our environment and our landscapes. So Slow Food is now the only truly progressive answer.[3]

8.1.2 Overview

> Slow Food was established in Bra, a small town in the Piedemont region in the North of Italy, in 1986, by a small group of food writers and chefs. The immediate motivation was growing concern about the potential impact of McDonald's on food cultures in Italy. The first Italian McDonald's had opened the previous year in Trentino Alto Adige, a region in the North East of Italy. It was quickly followed by a second in Rome. This latter restaurant, because of its location in the famous Piazza di Spagna, gave rise to a series of protests. These protests provided the spur for the founding of Slow Food (Miele and Murdoch 2003: 32).

Since its constitution Slow Food's focus has developed and expanded; an indication of this was the launch of a new concept in 2006. According to Slow Food they now take a holistic approach to the world of food, as an appreciation of a combination of people, environments and societies.

8.1.3 Case context: landscape and regime

> In the Italian context, traditional eateries retain a close connection to local food production systems; Slow Food argued that their protection required the general promotion of local cultures . . . In short, the movement sought to develop new forms of 'gastronomic associationalism' that link the cultural life of food to biodiverse production spaces (Miele and Murdoch 2003: 32-33).

8.1.3.1 Landscape factors

Slow Food was established as a counteraction to the spread of fast food and its connotations or values. Petrini saw the industrialisation, homogenisation and globalisation of farming as a threat to the biodiversity of vegetables and animals, traditional, local food production and food heritage. Dislocation from nature in food products and consumption were considered negative consequences of modern food production, both in relation to the use of natural resources and in relation to eating practices or experiences. According to Albert Sonnenfeld (quoted in Petrini 2003: xii) Slow Food soon became a

> defence team for vegetable, animal and cultural diversity, taking the stand for animal and human well-being in its periodical *Slow* . . . Its public face has become that of a primarily educational organisation, which realises that food

3 www.slowfood.com/about_us/eng/manifesto.lasso?-session=slowfoodstore_it:C39FB4C2054ce20D36Pmi286CEC9&-session=slowsitestore_it:C39FB4C2054ce20D36lLV286CECA, accessed 17 February 2010.

> consumption cannot be divorced from issues of food production and distribution.

Italians have a very low level of trust in the food they buy compared with people from other European countries (Poppe and Kjærnes 2003). According to Ferretti and Magaudda (2006) they also lack trust in many basic institutions, such as political parties, unions, big industrial companies and courts. Furthermore, when purchasing food they give considerable relevance to personal relationships and the place of origin of food items, they strongly prefer locally and nationally produced food and they have significant concerns regarding taste and safety. In contrast, Italians, unlike many other Europeans, are not very concerned with prices or the dietetic aspects of foods. These findings give some colour to the picture in which Slow Food emerged and provides input to the understanding of the movement's relatively strong position in Italy today.

Ferretti and Magaudda (2006) explain these findings in terms of the types of changes and developments in the Italian national food system over the past 15 years. As a result of institutional transformations in the early 1990s and new international food policy, important shifts in the distribution of responsibility occurred, especially regarding the division of tasks between the private and public sector. Decentralisation of power was one major institutional change that heavily influenced agricultural policies, which as a result became a matter of regional policy. As a consequence, international regulations were difficult to implement. However, the food market sector grew rapidly with respect to innovation, retailing and marketing and it reacted promptly to the first food crises (the BSE scare in 1996) and their negative consequences for consumer trust. The message from the food sector was to choose traditional, typical, local products for quality, which, according to Ferretti and Magaudda, was endorsed by the government. As a result, Italy, compared with other European countries, has the largest number of products recognised at a European level as having Protected Designation of Origin (PDO) or Protected Geographical Indication (PGI) (21% in 2003). Italians prefer quality products because of their link to what is 'typical' and related to cultural identity. According to Ferretti and Magaudda (2006: 164) this connection is 'being widely advertised through food fears and movements such as Slow Food':

> Quality rests in the small producers, into those realities that contribute to make quality and to forming a network. If agro-food products remain connected to their territory, they are not only agro-food products, but they become the cultural identity of our population' (Petrini 2002, quoted in Ferretti and Magaudda 2006: 164).

According to Ferretti and Magaudda (2006), the Italians' distrust in the food they buy and eat is explained by public authorities' abandonment of the Italian consumer. The safety of industrial food relies on public supervision, which consumers in Italy do not trust. The distrust in institutions is thereby transformed into a lack of confidence in food products.

These historical glimpses from Italy provide the landscape in which Slow Food was established, and gives us an idea of how this movement came to be embraced by so many, so soon.

8.1.3.2 Sociotechnical regime factors

Slow Food intended to reinforce the still existing, but weakened, traditional sociotechnical system and to fight the kind of standardisation of food and culture that they argue McDonald's represents. The movement's primary goal is to protect small-scale and local production in order to preserve tradition, heritage, history and diversity of taste. Slow Food works to convince consumers and bureaucrats that 'fresh produce is not dirty, though it comes without cellophane wrapping and looks unsymmetrical, and so unlike the waxen sterility of the processed and prepackaged products of the agri-business industrial revolution' (Sonnenfeld, quoted in Petrini 2003: xiii).

8.1.4 Actors, their roles and perspectives

8.1.4.1 Primary actors

According to the website of Slow Food (www.slowfood.com), the movement is a non-profit association that continually reinvests all profits into activities intended to achieve the aims defined in the association's Statute. The Slow Food International Association deals with strategies for the movement's development worldwide and coordinates the various national offices. Throughout the world Slow Food International, in collaboration with its national offices, has an educational commitment in schools (the 'edible schoolyard') and carries out activities for the preservation of biodiversity through the *Presidia* projects.[4] Slow Food Italy deals with the various Italian *convivium* (see below) and operates in the field of taste education through the Master of Food programme (www.slowfood.com).

The movement was founded by a few food enthusiasts and is based on a local structure. Coordinated from a central headquarters in Bra the local branches engage in a variety of activities aimed at strengthening local cuisine in Italy. The movement rapidly spread into other countries and at the time of writing has 100,000 members in 132 countries.

Germany, Switzerland, the USA, France, Japan, the UK, the Netherlands and Australia have national branches governed by national executive committees.[5] According to Slow Food, the national branches coordinate Slow Food events and projects with a deep knowledge of the needs of their members and their own countries. All members are a part of a *convivium*, which is the local expression of the Slow Food philosophy. They build relationships and networks with producers, campaign to protect traditional foods, organise tastings and seminars, encourage chefs to source locally, nominate producers to participate in international events and work to bring taste education into schools. Most importantly, convivium's aim is to cultivate the appreciation of pleasure and quality in daily life.

The organisation is led by the International Executive Committee, which is elected every four years at the Slow Food International Congress and consists of the President's Committee and the International Council. The Council is made up of representatives

4 www.slowfoodfoundation.org/eng/presidi/lista.lasso, accessed 28 April 2010.

5 See www.slowfood.com/about_us/eng/where.lasso, accessed 9 July 2009.

from those countries that have at least 500 Slow Food members. Carlo Petrini, the founder and president of Slow Food, was recently named 'a great innovator' in *Time* magazine's list of 'European heroes' (Ducasse 2004).

8.1.4.2 Secondary actors and externally involved parties

The Slow Food movement has widened its horizons enormously since its start, and Murdoch and Miele (2003) note that as it seeks to set the local in the context of the global, Slow Food has become cosmopolitan. According to Roos *et al.* (2007: 8-9),

> this transition implies, not only the enlargement of the Slow Food networks across continents, but also collaboration with other alternative food networks. For example, as a result of the meeting between Carlo Petrini, the founder of Slow Food, and Alice Walkers, active in the organic movement in the USA, farmers' markets and school gardens joined the Slow Food project. A meeting with Miguel Altieri, promoter of agroecology, the science of sustainable agriculture, also produced collaboration in the form of The Slow Food Foundation for Biodiversity, founded in 2003.

The Foundation for Biodiversity operates 'around the world with projects to defend local food traditions, protect local biodiversity and promote small-scale quality products, with an increasing focus on investments in countries of the Global South'.[6] This foundation's most important project is the Presidia, which now include more than 300 products (or Presidia) from all over the world. The Presidia involve producers and offer both technical and marketing assistance in order to 'sustain quality production at risk of extinction, protect unique regions and ecosystems, recover traditional processing methods, safeguard native breeds and local plant varieties'.[7] Another large project organised by the foundation is the Ark of Taste, which aims to 'rediscover, catalog, describe and publicize forgotten flavors'[8] and has now registered more than 700 products in 50 countries.

Slow Food has also encouraged the growth of the Slow Cities (*Città Slow*) movement, an autonomous group of towns and cities committed to improving the quality of life of their citizens, especially with regard to food. Slow Cities adhere to a series of guidelines intended to make them more pleasant places to live, such as closing the town centre to traffic one day a week and adopting infrastructure policies that maintain the characteristics of the town. Slow Cities seek to safeguard traditional foods, creating spaces and occasions for direct contact between consumers and quality producers. Slow Cities have sprung up everywhere, from Norway to Brazil, with several dozen in Italy alone (www.slowfood.com).

The United Nations Food and Agriculture Organisation officially recognised Slow Food as a non-profit organisation in 2004, and the two established a collaborative agreement. The non-profit Slow Food Foundation is officially constituted, thanks to the support of the Tuscany Regional Authority, which became a member.

The first 'Terra Madre: World Meeting of Food Communities' was held at the Palazzo

6 www.slowfoodfoundation.com/eng/cosa_facciamo.lasso, accessed 28 April 2010.

7 www.slowfoodfoundation.com/eng/presidi/lista.lasso, accessed 28 April 2010.

8 www.slowfoodfoundation.com/eng/arca/lista.lasso, accessed 28 April 2010.

del Lavoro in 2004, attracting about 5,000 delegates from all over the world. Some 6,000 farmers, 1,000 chefs and 400 academics took part in the second Terra Madre meeting in 2006, where the world of production met culinary *savoir faire* and academic science. The third Terra Madre meeting in 2008 attracted 4,000 producers, 800 cooks, 300 academics, 1,000 young people and 200 musicians. The meeting was held jointly with the 7th Salone del Gusto conference, which attracted 180,000 visitors the same year.[9] In 2009 Slow Food celebrated its 20th anniversary and held a Terra Madre Day on 10 December with a focus at the very local level. More than 1,000 communities and 100,000 participants attended this global event, guided by seven pillars of values and hopes.[10]

A campaign to safeguard outstanding artisan products at risk of disappearing was launched in 2005. The launch gathered 400 producers, university researchers, managers from Italian institutions and the ministries of health and agriculture together with technical experts from the Sicilian Regional Ministry of Agriculture.

The Italian Ministry of Universities and Scientific Research has officially recognised the University of Gastronomic Sciences in 2004, and since then more than 600 students have studied there. The goal of this university is 'to create an international research and education center for those working on renewing farming methods, protecting biodiversity, and building an organic relationship between gastronomy and agricultural science'.[11] The university now offers a three-year undergraduate degree, a two-year graduate degree and two masters' programmes.

8.1.5 Case history and development

8.1.5.1 From Arcigola to Slow Food: historical development

The following outlines the historical development of Slow Food (Slow Food, undated):

1986 The Arcigola association was formed in Piedmont, Italy.

1989 The Slow Food movement is founded in Paris and the Slow Food Manifesto signed.

1990 The first Slow Food international congress is held, in Venice.

1996 The first issue of Slow Food's international magazine *Slow: Herald of Taste and Culture*, was published and distributed to members in Italian, English and German. Slow Food organises an experimental *Salone del Gusto* in Turin, presenting the Ark of Taste project.

1997 The 'Dire fare gustare' (Saying, Doing, Tasting) conference marks the beginning of the Slow Food taste and food education project.

9 www.slowfood.com/about_us/eng/history.lasso?id=3E6E345B0eeb42AAD1wpN2FC40D5, accessed 28 April 2010.

10 www.terramadre.info/pagine/incontri/welcome.lasso?id=4E98738E1d6591914BOJQ2093289&tp=3, accessed 28 April 2010.

11 www.unisg.it/pagine/eng/about/history_and_mission.lasso, accessed 28 April 2010.

2000 The Presidium project is launched, consisting of local initiatives aiming to safeguard or revive small-scale artisan production at risk of extinction. Slow Food USA is founded.

2003 The Slow Food Foundation for Biodiversity is officially set up to support the Slow Food Award, the Presidia and the Ark of Taste.

2004 The non-profit Slow Food Foundation is officially constituted. The first Slow Fish fair is held. The University of Gastronomic Sciences is officially recognised and opens to the first 75 students. The first 'Terra Madre: World Meeting of Food Communities' is held.

2005 A campaign to safeguard artisan products at risk of disappearing is launched.

2006 Slow Food celebrates its 20th birthday and a new concept is transmitted to delegates: an holistic approach to the world of food is defined, consisting of an appreciation of people, environments and societies—Good, Clean and Fair Food. Slow Food USA sets up the Terra Madre Relief Fund to help Louisiana food communities affected by Hurricane Katrina in the summer of 2005.

2007 Slow Food joins the 'GMO-free Europe' coalition. Slow Bier is held in Munchberg, and Slow Food Fair, a German *Salone del Gusto*, is held in Stuttgart. 'Slowfish' is held for the third time in Genoa and 'Cheese' held for the sixth time in Bra, Italy. The first of the *Salone del Gusto*-inspired Algusto food fairs is held.

2008 The 3rd Terra Madre world meeting was held joint with the 7th *Salone del Gusto* conference in Turin, Italy. The first Slow Food Nation is held in San Francisco, USA. Two new national associations are formed: Slow Food Australia and Slow Food Netherlands. Carlo Petrini is named one of the '50 people who could save the planet' by the British newspaper *The Guardian*. Slow Food's international magazine *Slow: Herald of Taste and Culture* is replaced by national newsletters, which are given country-specific names (e.g *L'Escargot* in France, *Snail Pace* in Australia/New Zealand and *Slakkengand* in The Netherlands). These and Terra Madre's newsletter now seem to have joined forces and appear as the monthly *Slow Food & Terra Madre Newsletter*, which are distributed in eight languages to more than 100,000 members as well as thousands of producers, cooks, researchers and youth.[12]

2009 Slow Food celebrated its 20th anniversary (again) and held a Terra Madre Day on 10 December. The first EuroGusto was held on 27–30 November in the City of Tours with the intention of creating a new space for projects and exchanges.[13]

12 www.slowfood.com/about_us/eng/publications.lasso, accessed 28 April 2010.

13 At the time of writing, Slow Food's website has not been updated on this last event.

8.1.5.2 Primary instruments and strategies

Slow Food's strategy to counteract fast food was to disseminate information about local food cultures and the challenges they face. Through this activity Slow Food has become one of the primary sources of knowledge regarding local foods, first in Italy and later on a global basis. Slow Food's own publishing company (established in 1990) is its primary information channel and produces a range of guides to local and 'slow' foods for consumers, books, as well as the newsletters described above.

The Slow Food philosophy includes teaching *gastronomo*—the science of all knowledge revolving around food—in schools. Slow Food argues that everyone should become co-producers and active participants in the production of food rather than consumers (www.slowfood.com). The University of Gastronomic Sciences was founded in 2003 in conjunction with the regional authorities of Emili-Romagna[14] and Piedmont.[15] The objective is to create an international research and training centre, working to renew farming methods, protect biodiversity and maintain a relationship between gastronomy and agricultural science. Its teaching activities are supported by an independent non-profit association (the Association of Friends of the University of Gastronomic Sciences) that oversees organisational and financial matters.[16] In addition, Slow Food has educational programmes at a lower level for everyone: children and adults, members and non-members. Slow Food's school programmes range from training teachers and collaborating on curricula to improving school lunches and organising after-school programmes. The programmes focus on rediscovering the joys of eating, the importance of where food comes from, who makes it and how it is made. Each *convivium* is asked to create a school garden in their town or city in order to teach students to grow plants, understand the seasonal cycle, taste what they have grown and study ways of using the ingredients in the kitchen (www.slowfood.com).

The members, organised in local *convivia*, are an important part of Slow Food's plan to disseminate knowledge and interest in local foods. In addition, the *convivia* organise theme dinners, food and wine tours and tasting courses to underpin national and international initiatives. Some of the seminars and food fairs gather thousands of visitors. The large food fairs offer food and wine tasting, seminars, talks and even collections for charity (www.slowfood.com).

Slow Food has two commercial bodies, which it fully controls, controlling finance, membership policies, organisation and the management of events attended by the public. Slow Food Promozione srl deals mainly with the organisation of large-scale events (Salone del Gusto, Cheese, Slow Fish), fundraising, publicity and sponsorship activities. Slow Food Editore srl is responsible for the association's publishing activities (the internet site, members' magazines and newsletters, as well as guides, essays and cookbooks [about 70 titles]). The quarterly national newsletter, *Slow*, is the most direct means of communication for countries where Slow Food has a strong network of *convivia*. Members contribute stories about regional foods, *presidia* projects, educational initiatives, *convivium* events and all other 'slow' happenings. Slow was born as a vehicle to exchange

14 See www.regione.emilia-romagna.it/uk, accessed 9 July 2009.

15 See www.regione.piemonte.it/lingue/english/index.htm, accessed 9 July 2009.

16 See www.unisg.it/eng/chisiamo.php, accessed 9 July 2009.

experience and knowledge among different countries and traditions and is founded on the acknowledgement of diversity of language, culture and flavour (www.slowfood.com).

8.2 Results

8.2.1 Change in sustainability performance

Kjærnes (2007) argues that it is problematic to decide what issues and aspects should be included in the concept of 'sustainability'. Issues relevant to sustainability may range from the use of natural resources (such as land and energy) to social welfare and solidarity. Food and eating are a fundamental part of people's social lives and of political and regulatory issues regarding social equality and welfare. This can make sustainability issues difficult to delimit from all other issues that are not directly commercial (Kjærnes 2007).

Slow Food has spurred many local initiatives, which often have had positive effects on the environment as well as giving rise to social and economic improvements. Owing to the local framing of these initiatives, each Slow Food project to some degree influences all three areas—environmental, social and economic—at a local level. It might, however, pose more of a challenge to spread the knowledge accumulated and widen the networks on a global basis in a way that causes change leading to sustainable food consumption more generally. Slow Food today depends on private initiatives taken by people in their capacity as consumers (through political and/or ethical consumerism).

The philosophy of Slow Food is that people should slow down their hurried lives, starting by eating locally produced food slowly, taking the time to taste and appreciate the good quality. By asking this, Slow Food is in fact asking consumers to turn to food items provided via alternative types of distribution channels. This may imply organising purchasing activities differently, thus asking consumers to go beyond simple substitution of their conventional food consumption. According to Kjærnes (2007) substitution (based on selecting certain labels) is associated with distinct forms of production and distribution as well as consumption. This means that relevant alternatives must be available on the supply side. But switching to other forms of provisioning may mean organising consumption activities differently, perhaps even influencing a consumer's level of welfare, positively or negatively. Because food consumption is regarded as a highly routine practice, it will take time, effort and convincing arguments to change consumers' food conventions.

But sustainable consumption, Kjærnes (2007) argues, is not only about making people 'do the right thing' when they 'consume', but also related to people themselves making use of their capacity as consumers to influence suppliers and regulators to make their actions more sustainable. These activities cannot be studied in isolation, though—consumption needs to be studied as a social activity, in interaction with an institutional and discursive environment. This argument underlines the problems of evaluating the possible sustainable improvements arising from Slow Food.

According to Kjærnes (2007), Swidler (1986) argues that, in certain situations, previously routine practices may suddenly become explicit and contested, and routines can intermittently break up. New and alternative, often ideologically justified, sets of practices may then be established. These will, however, tend gradually to become routine, eventually also becoming tacit and taken for granted. Kjærnes argues that organic food consumption (in various forms) in this way will become routine and institutionalised, rather than forever remaining new and 'reflexive'. A parallel can be seen in terms of adherence to the advice given by Slow Food.

8.2.1.1 Environmental improvement

Slow food is often linked to organic produce and traditional, small-scale, farming. This underlines Slow Food's holistic and environmental approach to food. Efficiency and mass production is not a goal; rather, the opposite. Since slow production and first-rate taste are the focus, the use of fertilisers and pesticides are not seen to be beneficial, which benefits the environment. Local initiatives throughout the world, inspired by the Slow Food movement, clearly have environmental benefits such as preserving local natural varieties and preventing deforestation locally. It is difficult to assess to what degree Slow Food contributes to environmental improvements from a more global perspective.

8.2.1.2 Social improvements

Slow Food promotes a lifestyle of healthy and traditional food. A higher degree of consumer awareness and concern for food quality and production methods are one of Slow Food's main goals in order to secure the livelihoods of local farmers and to support small-scale production, which again will prevent foodstuffs that embody quality, identity, variety and taste from vanishing.

8.2.1.3 Economic improvement

Slow Food's intention of bringing back taste, tradition and experience to consumers' food consumption will at the same time support small, local food producers and perhaps strengthen their position in the market, as opposed to international and industrial food producers. Slow Food might secure local employment and keep rural communities *alive*. Slow Food wishes to 'offer opportunities for development even to poor and depressed regions through a new model of agriculture' (Petrini 2003: xviii). To accomplish this, Slow Food depends on consumers' willingness to pay more for their products as such products are more expensive to produce and distribute. Sonnenfeld (quoted in Petrini 2003) argues that Slow Food must be aware of the perils of elitism, as support from the 'multitudes' is vital in order to save endangered, traditional, foodstuffs.

In several areas of food-related policies, consumers are attributed agency and are seen as responsible, through their choices, for a number of societal issues: health, food quality, animal welfare in the agri-business and environmental sustainability (Kjærnes 2007). Policy papers refer explicitly to consumer choice and consumers' own responsibility through 'informed choice' and labelling strategies (EU 2000; Reisch 2004). Kjærnes argues that this way of seeing the consumer is insufficient and even misleading if we are to understand the different roles and responsibilities of consumers. Food consumption must be understood as a socially and institutionally created set of practices within national and cultural contexts. This implies that Slow Food, in order to scale up and have an impact on food consumption in settings outside of Italy, must adapt to the distinct character of the national, social, institutional, traditional and historical settings or realities in which consumers are situated.

8.3 Potential for diffusion and scaling up

Kjærnes (2007) argues that institutional conditions, experiences and expectations are, even within a European setting, highly diverse,- in spite of the globalisation of markets and technologies and regulatory harmonisation. This fact clearly has negative consequences regarding the potential for the scaling up of Slow Food in such a way that that the movement can be said to have an impact on global food consumption. Also, scaling up from Slow Food's idealised small-scale and alternative production methods may imply problems for producers if they are to meet retailer requirements and, in addition, for consumers, by reducing transparency and predictability.

According to Kjærnes (2007), food consumption in different countries is not the same nor is 'the consumer' the same figure, and, maybe even more importantly, food consumption and consumers cannot be understood in isolation from interrelations with and between particular food provisioning systems, public policies and public discourses. Slow Food might therefore face quite different obstacles in different countries or regions, depending on how these actors and processes are interwoven and institutionalised. For the consumer, food items and issues are often bundled together, implying that links to sustainability may be difficult to single out and to ascertain. Kjærnes (2007) claims that this 'bundling' is also reflected in marketing strategies where references are often made not to particular sources or production methods but instead to wider conceptualisations and symbols such as a country of origin, organic food or 'naturalness'.

So far, Slow Food appears to have had the greatest impact in Italy and France, where consumers have a somewhat similar relationship to the food industry, distribution chains and production. Results from the Trust in Food project (Poppe and Kjærnes 2003) reveals that Italy has the lowest levels of trust in food in Europe, irrespective of kind of food. Furthermore, Italians give considerable relevance to aspects of food such as local or national production. Such factors seem to provide important space for initiatives such as Slow Food. In other countries, such as Norway, consumers generally have a profound trust in authorities and their control of food production methods and they are correspondingly less active as political consumers (Poppe and Kjærnes 2003).

The encompassing and demanding concept of Slow Food requires considerable effort from consumers, both in terms of changing to a different place to shop and to a different type of shopping (at farmers' markets or on-farm sales) but also in terms of changing their expectations and levels of dedication not only in shopping but also in food preparation and eating. Kjærnes (2007) suggests that the different product varieties, perhaps in smaller quantities, and the effects of seasonality will influence the types of produce on offer, and the use of fresh food and lower degrees of processing may require the consumer to put more effort into cooking and perhaps even storage. These may be necessary adjustments that consumers face if they want to embrace the concept of the Slow Food movement; that is, *if* they live in a community where Slow Food represents a real option.

Most of the alternative food production and distribution initiatives occupy niches, but they highlight the importance of sustainability. Such niches contribute to the definition of new social issues and bring them into public discourse; for instance, Slow Food points at the deterioration of the 'social culture of food' as a consequence of a 'fast-food culture'. The niche market initiatives also influence the 'greening' of the mainstream market. But they are dependent on support through political instruments, media, consumer awareness and consumer demand in order to diffuse or up-scale (Mont *et al.* 2007). Other important elements regarding diffusion or up-scaling relate to how the market is organised, in general, and, in particular, to how the market for sustainable products is organised in terms of production, distribution and availability of products (labels). Furthermore, it will be of importance to consider how the consumers' role is framed; do people feel empowered and take responsibility or do they consider other forums, actors or actions as more relevant and powerful? One must acknowledge the diversity of consumers—they are not a homogenous group, and their consumption practices must be contextualised along with their different roles as parents, citizens, students and so on. With respect to diffusion and up-scaling, the division of responsibility for food ethics and sustainability among relevant stakeholders (such as public authorities, market actors, non-governmental organisations [NGOs], consumers, etc.) must also be taken into account: who is 'in charge' and do they cooperate or not, how do consumers perceive the division of responsibility and do consumers trust and agree with the responsible actors (Mont *et al.* 2007)?

8.3.1 Possibilities and opportunities for repetition in alternative contexts

In this section we present a case study of a Polish Slow Food initiative, as it is described in the EMUDE project,[17] and discuss differences from the Italian case that has been the focus of this chapter.

8.3.1.1 The case of Poland

In Poland (Krakow and Warszawa), the Visana Consulting Group[18] has integrated Slow Food into its network of logistics and distribution of organic produce. Box 8.1 is quoted

17 EMUDE (Emerging User Demands for Sustainable Solutions) is a programme of activities funded by the European Commission, the aim of which is to explore the potential of social innovation as a driver for technological and production innovation, in view of sustainability. See sustainable-everyday.net/cases/?p=50#more-50, accessed 9 July 2009.

18 See vcg.pl, accessed 9 July 2009.

directly from the EMUDE website (see footnote 17 on previous page), written by Grzegorz Cholewiak of the Academy of Fine Arts in Krakow, Poland.

Box 8.1 Comparison case study: Slow Food in Poland

Source: sustainable-everyday.net/cases/?p=50#more-50 (posted 7 April 2006), accessed 9 July 2009

Solution

Working with a group of producers without their own wide-scale distribution network on a wider scale, Visana came up with its own brand 'Soplicowo i okolice', simultaneously creating a logistics and distribution network. Now promoting the idea of 'slow food' Visana trades in organic produce, traditionally farmed. It cooperates with local producers, offering better sales opportunities for their products in exchange for production under the Soplicowo i okolice label. This gives producers a better opportunity of selling their products, and city consumers access to high-quality products from a trustworthy source.

Background

On [the] one hand, consumer awareness of quality is growing. Visana's target group in Warsaw attaches importance to traditional taste, and with little time for cooking instead eats out in restaurants offering regional cuisine or buys ready-made product with 'home-made' associations. On the other hand concern about biodiversity in agriculture is increasing.

Current state of development

The scheme, run by Visana Consulting Group, has been going since August 2003. Warsaw is the main market, with Krakow included recently as well, but the target is to cover the whole of Poland. To increase sales, the company may take over cooperative production companies, or [invest] in new product lines. These would have significant impacts on the profit margin, the prices of raw material purchases and production quality control. When the company started, [its] shouldered [its] own investment, which allowed for the basic needs of the company to be paid for. At present, the company is surviving off dividends. Costs include office rental, storehouses, accountancy, staff salaries and transportation (rental costs and petrol). With the opening of the European Union borders and the large interest in the West in Polish food, the company is planning to enter the European market in the space of the next few years.

The benefits

Society

A new lifestyle is created of healthy and traditional food, and consumer awareness and concern for the source and quality of products. In this way Poland is slowly starting to follow the Western [example] of supporting small regional food producers, particularly original, traditional and healthy [foods] that are unfortunately threatened with obsolescence.

continued opposite →

Environment

Visana's products are made from natural ingredients and grown in environmentally friendly conditions. The soil is fertilised with natural composts, and the fruit and vegetables contain no pesticides. This encourages traditional farming methods with fertilisers that do no harm to the environment.

Economy

There are currently 13 producers, working under the common Soplicowo i okolice brand, and around 40 distributors in Warsaw alone, with all the participants profiting from the cooperative venture. Producers and distributors split the risks and profits evenly. The company is employing more staff, the production firms are expanding, by investing in [machines], and, as confidence grows, new ideas emerge all the time.

Future development possibilities

- Creating services, events and places to present and taste local traditional food
- Conceiving a platform of technical services to support local small farmers and food producers
- Designing a visual identity for the brand which still allows the individual producers their own identity

8.3.1.2 Discussion

This case of Slow Food in Poland is a good example of how the Slow Food philosophy is adapted outside the Italian context to fit another country's reality. Polish and Italian consumers share scepticism and mistrust towards the food industry and systems, but Italy has to a larger extent than Poland maintained a tradition of small-scale farming. According to Eurostat (50/2008, 38/2009), 66% of Italian agricultural holdings made use of less than one AWU[19] compared to 29% in Poland. This said, the distribution of the utilised agricultural area in Italy is generally even over different area size classes, one exception being the class 50–<200 ha, which comprises twice as much area as do all the other classes. Looking at Poland, the picture is different, as smaller classes (5–<10 and 10–<20 ha) comprise the largest utilised agricultural areas. Regarding the economy, the statistics also reveal that 82% of Italian holdings had an economic size of at least 1 ESU,[20] compared to 47% in the case of Poland. In addition, these figures show that in Italy organic farming counted for 5.6% of the UAA in 2007 (a rise from 4.9% in 2005), while in Poland organic farming counted for 1% of the UAA[21] (a rise from 0.6% in 2005). Polish consumers' mistrust, combined with farmers' low income, has given an alternative food provisioning system a head start, compared to, for example, Norway where consumers tend to trust authorities' control of the food industry. The Polish Slow Food

19 annual work unit

20 European size unit

21 utilised agricultural area

initiative focuses on creating a market for small-scale farmers, establishing a distribution chain whereby producers can improve their sales. Quite different from the Italian Slow Food tradition the Polish initiative has targeted two major cities in order to maximise sales and has set its goals on distributing products throughout Poland and, it hopes, further afield to Western Europe. This indicates that the Polish Slow Food initiative's primary focus is to support and secure small-scale producers in mass markets, contrary to the Italian founders' focus on strengthening consumers' ability to choose, taste and enjoy local food in local settings.

8.3.2 Possibilities to fit the project into a portfolio (cluster) of different initiatives in relation to the same sustainability vision

Over the years the Slow Food movement has agreed to collaborate with various food and welfare organisations. Most significant are perhaps the organic movement in the USA, farmers' markets, school gardens and The Slow Food Foundation for Biodiversity, which was founded after a meeting between Petrini and Altier (promoter of agro-ecology, the science of sustainable agriculture). The many 'good causes' that are fitted into Slow Food's manifesto makes the movement suitable for collaboration with many other food-related initiatives related to environmental, social and health-related issues as well as to the rural economy, ethnography and history.

One could probably most easily place Slow Food as an alternative food production and distribution system alongside community-supported agriculture (CSA) initiatives such as farmers' markets, various types of delivery schemes, regular box schemes and shareholders schemes (see also, for example Chapters 6 and 7). They all represent alternative production and distribution channels of food and are likely to help reduce the distance 'from field to fork' as well as the distance between the consumer and the producer. The different schemes demand different levels of system change and consumer commitment and as such might fit different situations (Mont *et al.* 2007).

8.4 Overall conclusions

Roos *et al.* (2007) consider the potential that the 'local' has for creating closer and more direct relations between persons, products and places. They conclude that creating relocalised food systems such as Fair Trade, Slow Food and CSA may endow consumers with an awareness of the conditions of food production and of the origins of food (Lyson and Green 1999). They see the 'localising' of a product, by adding a history and identity, as a possible strategy for shortening the distance within an otherwise complex and fragmented food chain.

The histories differ and, in the case of Slow Food, 'preserving the uniqueness of recipes and flavour combinations, including those arising from specific production methods, held in folklore are among the main elements' (Roos *et al.* 2007: 11). The use of the local dimension to express consumer ethics and food authenticity may be the result of different processes, and three possible elements are pointed out:

> The attention towards post-materialist values typical in affluent societies (Inglehart 1977) may contribute to explaining some forms of ethical consumption. Food scares have also played a part and together with a growing awareness of positive health have directed attention towards the quality and origin of food. Moreover, as suggested by Zward (2000), there has been a shift within food ethics from concerns related to the product itself, to the way it is produced. [. . .] Thus tracing the local dimension of food gives consumers an opportunity to make ethical judgements of food production practices (Roos *et al.* 2007: 12).

'Local' has often been framed discursively as 'nostalgia', or 'going back to nature', but it has now been transformed into a 'local-in-the-global'. As more meanings, including new experiences and new uses of technology and information systems, attach to the term 'local', Roos *et al.* (2007) claim that the boundaries between meanings have become blurred and that the way is now open to conflicting interests. The 'local' is especially valuable as a site of resistance to the 'global' (Dirlik 1996), but there is a risk that 'local-in-the-global' becomes yet another 'market segment' with a premium based on 'added value' (Coff 2005). As suggested by Murray and Raynolds (2000: 67): 'the space that appears to exist for creating a truly alternative trade will be captured by agrofood corporations able to transform this progressive initiative into a niche marketing scheme for products repackaged under "green" and/or "ethical" symbols'.

Roos *et al.* (2007) nevertheless claim that transnational corporations are vulnerable to new forms of consumer and citizen contestation and to the relocation of local purchasing. They argue that the process of globalisation can create space for local food initiatives (discussed by Busch 2004; Hendrickson and Heffernan 2002): 'This clearly indicates that the local–global hybrid is potentially both a site for resistance and a potential transformation consistent with the broad values of the "local movement"' (Roos *et al.* 2007: 13). Miele and Murdoch (2003) argue that it is possible that the two opposites of the food spectrum (slow food and fast food) might co-exist and flourish simultaneously because of the varied contexts in which food is consumed and the respective differentiation in consumption patterns. Slow Food might represent a supplement or alternative to other eating practices, fast food standing at the other end of the spectrum, rather than becoming the only way, as consumers have different needs in different contexts.

Political consumerism has grown stronger in recent years, even in countries such as Norway, but it is difficult to evaluate the impact of Slow Food. While the Italian Slow Food movement explicitly presents itself as a response from 'eaters' (especially of the expert kind), the Polish case appears to lack such a link to the consumer side. As representatives of the movement themselves underline, a turn towards slow food and slow living requires learning and changes to routine practises concerning eating, which is a time-consuming process and probably difficult to measure. Perhaps Slow Food should be judged by how it presents itself—as a form of social mobilisation. The results should then be framed more widely, as the influence on the public and political agenda, on the general development of the food provisioning system and so on.

The case of Slow Food in Poland demonstrates the diversity of this phenomenon. Rather than emerging from the mobilisation of (perhaps professional) consumers and local consumer–producer networks, it is in Poland linked to the efforts of small-scale farmers to increase their income and to reach urban mass markets. This framing makes the Polish Slow Food initiative more similar in nature to the numerous marketing initiatives linking provenance to food quality that we see across Europe, especially in the south.

References

Busch, L. (2004) 'The Changing Food System: From Markets to Networks', paper presented at the *9th World Congress of the International Rural Sociological Association, Trondheim*, Norway, July 2004.

Coff, C. (2005) *Smag for etik: På sporet efter fødevareetikken* (*A Taste of Ethics: Tracing Food Ethics*) (Copenhagen: Musem Tusculanums Forlag).

Dirlik, A. (1996) 'The Global in the Local', in R. Wilson and W. Dissanayake (eds.), *Global/Local: Cultural Production and the Transnational Imaginary* (Durham, NC: Duke University Press): 21-45.

Ducasse, A. (2004) 'The Slow Revolutionary', *Time International (Europe Edition)*, 11 October 2004: 64.

EU (2000) 'White Paper on Food Safety' (COM[1999]719 final; Brussels: Commission of the European Communities).

Eurostat (2008) 'Statistics in Focus. Agriculture and Fisheries. Farm Structure in Poland—2007' (Eurostat 50/2008; European Commission).

—— (2009) 'Statistics in Focus. Agriculture and Fisheries. Farm Structure Survey in Italy—2007' (Eurostat 38/2009; European Commission).

Ferretti, M.P., and P. Magaudda (2006) 'The Slow Pace of Institutional Change in the Italian Food System', *Appetite* 47: 161-69.

Fiddes, N. (1991) *Meat: A Natural Symbol* (London: Routledge).

Folco Portinari (1987) first published in the *Gambero Rosso* magazine, later endorsed and named the 'Slow Food Manifesto' by the founders of Slow Food in 1989.

Hendrickson, M.K., and W.D. Heffernan (2002) 'Opening Spaces through Relocalisation: Locating Potential Resistance in the Weaknesses of the Global Food System', *Sociologia Ruralis* 42: 347-69.

Inglehart, R. (1977) *The Silent Revolution: Changing Values and Political Styles among Western Publics* (Princeton, NJ: Princeton University Press).

Kjærnes, U. (2007) 'Work in Progress: Sustainable Food Consumption, Some Contemporary Issues', paper produced for Sustainable Consumption Research Exchange (SCORE!), National Institute for Consumer Research (SIFO), Norway.

Lyson, T.A., and J. Green (1999) 'The Agricultural Marketscape: A Framework for Sustaining Agriculture and Communities in the North', *Journal of Sustainable Agriculture* 15: 133-50.

Miele, M., and J. Murdoch (2003) 'Fast Food/Slow Food: Standardising and Differentiating Cultures of Food', in R. Almås and G. Lawrence (eds.), *Globalisation, Localisation and Sustainable Livelihoods* (Farnham, UK: Ashgate): 25-41.

Mont, O., E. Vadovics and I. Kjørstad (2007) 'Sustainable Consumption Initiatives: From Niches to Mainstream; Possibilities and Challenges', paper presented at the SCORE! Workshop 3, Milan, 29 November 2007.

Murray, D.L., and L. Raynolds (2000) 'Alternative Trade in Bananas: Obstacles and Opportunities for Progressive Social Change in the Global Economy', *Agriculture and Human Values* 17: 65-74.

Petrini, C. (2002) Presentation at the *4th Salone del Gusto*, Turin, 7 November 2002 (quoted in Ferretti and Magaudda 2006).

—— (2003) *Slow Food: The Case for Taste* (New York: Columbia University Press).

—— (2005) *Buono, Pulito e Giusto* (Turin: Einaudi).

—— and G. Padovani (2005) *Slow Food Revolution* (Milan: Rizzoli).

Poppe, C., and U. Kjærnes (2003) *Trust in Food in Europe: A Comparative Analysis* (Oslo: National Institute for Consumer Research [SIFO]).

Reisch, L.A. (2004) 'Principles and Visions of a New Consumer Policy', *Journal of Consumer Policy* 27.1: 1-42.

Roos, G., L. Teraagni and H. Torjusen (2007) 'The Local in the Global: Creating Ethical Relations between Producers and Consumers', *Anthropology of Food* (Special Issue S2), March 2007.

Slow Food (undated) 'Slow Food: From Arcigola to Slow Food. Chronology of the Birth and Development of an International Association', content.slowfood.it/upload/3E6E345C0791e2A378Oxs408230D/files/02_Chronology.pdf, accessed 9 July 2009.

Swidler, A. (1986) 'Culture in Action: Symbols and Strategies', *American Sociological Review* 51.2: 273-86.

Zward, H. (2000) 'A Short History of Food Ethics', *Journal of Agriculture and Environmental Ethics* 12: 11.

9

Sambazon: creating environmental and social value through marketing the açaí berry; sustainable agro-forestry practices in the Brazilian Amazon

Burcu Tunçer and Patrick Schroeder
UNEP/Wuppertal Institute Centre for Sustainable Consumption and Production (CSCP), Wuppertal, Germany

The 'Sambazon: Marketing the Açaí Berry' business model is guided by triple-bottom-line principles and is in accord with the 'human development through the market' (HDtM) approach, which covers market-based environmentally and socially responsible activities that empower the poor to meet their basic needs. Sambazon Inc., a registered fair-trade company and founded by three young Californians who 'discovered' açaí on a holiday in Brazil, harvest the açaí berry with the help of around 1,500 local families living inside 66,000 acres of the Varzea Flooded Forest Eco-Region 147 of the Brazilian Amazon region.[1] The low-income family producers receive a fixed minimum price for the açaí, which is two to three times higher than the average wage in Brazil, creating stability for the producers against market fluctuations. The açaí palm grows naturally in the Amazonian rainforest and is often the target of 'palm poachers', who kill the entire tree to harvest the edible inner portion of the stem of the palm. Shifting the focus on harvesting the açaí berries, which does not harm the tree, protects the palms as well as the surrounding forest and promotes sustainable agro-forestry.

1 Go to www.sambazon.com, accessed 13 July 2009.

Being the main source of many important nutrients for around two million people in Brazil, the açaí berry was virtually unknown to US consumers before Sambazon began its grass-roots business in 2000. The açaí products are marketed and consumed in the USA, and now also in the UK, in the form of fruit drinks, energy products, health food and, most recently, in cosmetics. Sambazon açaí products have seen a 300% increase in sales each year since 2000 and have moved from being a niche market product to account for 10% of total frozen-fruit sales in natural food channels in the year 2006 (Brewster 2006). This is because of the high nutritional qualities of the açaí and the innovative marketing campaigns of Sambazon focusing on live demonstrations of how to prepare and consume açaí berry products.

Sambazon has become the model of an innovative small business in the USA and won the Secretary of State's 2006 Award for Corporate Excellence (ACE) in November 2006. Following this event Sambazon received nationwide media coverage, and the status of the açaí berry is changing from an exotic trendy fruit drink to become a mainstream product with a wide range of applications. In recent years around half a dozen competitors have emerged in the açaí business; however, Sambazon still accounts for more than 80% of the imported açaí products to the USA. To ensure that the business remains sustainable and benefits local communities, Sambazon has initiated the Sustainable Açaí Project, working together with local non-governmental organisations (NGOs), research organisations and local government. The entire project is centred on institutional capacity building to create a consistent and replicable model that can be used through the entire region. The Nature Conservancy Eco Enterprise Fund and the Eco-Logic Enterprise Venture are involved in the financing of the project.

9.1 Case description

9.1.1 Overview

Sambazon (which stands for 'Save and Manage the Brazilian Amazon') is a for-profit company located in San Clemente, CA, and operates in the USA and Brazil. Sambazon Inc. buys açaí berries, which grow naturally in the Amazon rainforest, from local landowners and farmers of the riverine population, the *ribeirinhos*. The processing of Sambazon açaí products also takes place locally.

Sambazon Inc. is a fair-trade company certified through Fair Trade Federation (FTF) and has the goal to benefit local communities through price premiums paid to local workers. There are two mechanisms used:

- Minimum price: the minimum price for açaí fruit guarantees that the real costs of goods for the growers are covered and provides a necessary profit margin that allows them to live with the basic necessities of health and satisfaction
- Price premium: when the local market prices for conventional fruit are equal to or above the minimum price, Sambazon pays an additional premium on top

of the local market prices for conventional fruit for every basket of fruit purchased

In the Amazon, 90% of *ribeirinho* families survive on less than US$4 (£1.90) a day, which does not allow families to live above the poverty line (Purvis 2007). A large majority of the low-income family producers that supply açaí fruit to Sambazon are able to earn triple this amount on an annual basis during the harvest season. Traditionally, the local *ribeirinhos* were dependent on middlemen who controlled the fruit market in the Amazon estuary region. Owing to the increased demand from overseas the middlemen now have to pay more to the *ribeirinhos* than before Sambazon entered the market. Açaí used by Sambazon typically fetches 44 reais (£12) a sack compared with 15 reais five years ago. In addition to the benefits and security-of-purchase guarantees, Sambazon offers a fixed premium of 5% over the market price (Purvis 2007).

Sambazon also offers training programmes on agricultural management and technological aspects benefiting the local family cooperatives. These initiatives focus on programmes with the local communities, involving demonstration courses aimed at increasing the logistics of the harvest system, agricultural management of the açaí fruit and technological qualification for increased production. This system has been shown to increase exponentially the operational and financial yield of the local producer associations involved.

Sambazon claims the harvesting process to be environmentally friendly and that it contributes to Amazon rainforest conservation efforts. The Sambazon business model includes the Sustainable Açaí Project. The aim of the project is to prove that sustainable agro-forestry in the Amazon estuary can improve living conditions, enhance educational opportunities and promote forest conservation. Landowners who harvest and sell açaí to Sambazon designate a piece of their land as an ecological reserve and carefully manage the rest of their terrain to protect the biodiversity of the rainforest. The Sustainable Açaí Project is a public–private partnership involving and funded by corporations, charitable foundations and other public and private donors. To this goal Sambazon is working together with the Peabiru Institute, Goeldi Museum, Amapa and Para State Federal Universities as well as with local NGOs.[2]

The project consists of three phases:

- First, creating bio-socio indicators to assess the impact of the açaí trade on the local region, through standard metrics
- Second, develop and implement sustainable agro-forestry management outreach projects to positively influence the opportunities and threats to açaí cultivation
- Third, establish an Açaí Technical Centre aimed at developing and disseminating economically feasible technologies to recycle açaí by-products such as the bead-like açaí stones. In one of the projects Sambazon gives the stones, churned out by the factory at a rate of seven million an hour, to the Brazilian Women's Group, a cooperative in downtown Belém, to be used for jewellery-making (Purvis 2007)

2 See the Sambazon website: www.sambazon.com, accessed 13 July 2009.

The entire project is centred on institutional capacity building to create a consistent and replicable model that can be used through the entire region. The goal is to create socioeconomic change and simultaneously protect Amazonian rainforest and biodiversity (Sambazon website, 2007).

Sambazon has been very successful in the marketing of açaí products in the USA. The fruit has a very complex flavour profile, is high in nutritional value and vitamins and, for these reasons, has a competitive edge over other fruit used to make beverages (fruit smoothies), health food, ice cream and energy products. More recently açaí has been used as an ingredient in anti-ageing cosmetics because of its high levels of antioxidants and good anti-cancer and leukaemia activity (University of Florida 2006).

Before Sambazon started its marketing campaign açaí was virtually unknown to US consumers. Now, following the success of Sambazon açaí, a number of other businesses have started using açaí as an ingredient in food products. According to Spins Inc., which tracks natural foods, sales of açaí products in the USA were at US$14 million for the 12-month period ending in April, up from US$2.7 million for the same period two years before (Astor 2007).

Sambazon is still the main importer of frozen açaí (80% of total imports) to the USA and now offers 26 different açaí products. The company expects to export around 4,000 tonnes in 2007. Sambazon açaí products accounted for 10% of total frozen-fruit sales through natural-food channels in the year ended January 2007. Sambazon's revenue grew from $3.1 million in 2004 to $25 million in 2008 (Brasileiro 2009).

Sambazon has become the model of an innovative small business in the USA. On 6 November 2006 Sambazon won the Secretary of State's 2006 Award for Corporate Excellence (ACE). Following this event Sambazon received nationwide media coverage and in March 2007 Sambazon was featured in the *Wall Street Journal* Small Business Report (Bounds 2007).

9.1.2 Case context: landscape and regime

The Brazilian Amazon region, especially the rainforest frontier, is a dramatically changing 'sociocultural' landscape, with a relative absence of formal social institutions (including those of government) often characterised by violent conflict between interest groups. The interests of traditional dwellers of Amazonia—the Amerindian tribes, the riverine population (*ribeirinhos*) and the rubber tappers—stand in stark contrast to those of newly arriving settlers, often poor displaced peasants from other areas of Brazil looking to improve precarious livelihoods (in Brazil half of the agricultural workforce is landless). Other groups with a dramatic impact on the sociocultural, economic and ecological conditions are large landowners, international mining corporations and cattle ranchers (Diegues 1992).

Indigenous communities in Brazil's Amazon regions continuously have to defend their traditional lands against the advance of unsustainable logging, mining and large-scale farming practices, especially cattle ranching which is responsible for 95% of deforestation (Purvis 2007). These economic initiatives are often supported by government policies to integrate the Amazon region into the national economy (the Programme of National Integration, launched in 1970 (Bessa *et al.* 2005). Chief responsibility for most of the massive deforestation lies principally with the government and its development strategy of the past three decades that, through credit, tax and other incentives, has enabled large-scale agricultural and cattle-raising schemes to be established. This development often is detrimental to local indigenous people. Even the support agencies and some NGOs tend to follow the government line.

However, under the pressure of local grass-roots movements (such as the rubber tappers and other forest-based people) that have the goal of protecting the rainforest and the natural resources from which local people derive their livelihoods, 'extractive reserves' (areas designated specifically for the long-term sustainable harvest of forest products) are being created to ensure sustainable use and conservation of renewable natural resources while protecting the ways of life and culture of the traditional inhabitants of these lands. In this context the concept of 'neo-extractivism' has evolved, describing alternative forms of extensive forest use with little environmental disturbance. Extractive reserves are legal entities under Brazilian law and tacitly legitimise the traditional rights of communities to the land and its natural resources.

With land-use rights protected the question of how indigenous communities can achieve sustainable economic development remains. Challenges are how to develop constructive alliances with outsiders to achieve this end and how to ensure that external perspectives do not dominate local initiatives and undermine traditional practices (Frost *et al.* 2006).

The Sambazon business case offers an example of a sustainable alternative to collaborate with indigenous communities within extractive reserves, protecting traditional lifestyles and ecosystems. The development of non-timber forest products (NTFP) such as açaí offers great potential for conservation and support for rural livelihoods. With the increased demand for the fruit, and through the promotion of sustainable agro-forestry systems, the value is being shifted to renewable resources with the incentive on sustainability. This change of paradigm gives great ammunition not only to the environmental movement but also to socially responsible business worldwide.

9.1.3 Actors and their roles and perspectives

The primary actors are the two brothers Ryan and Jeremy Black plus Ed Nichols, who raised the initial business capital of US$100,000 through private contacts to set up Sambazon Inc. The motivation for setting up the business was not solely driven by pure commercial interest. Having spent time in the Amazon region as travellers and living with local people, the business idea is intended to support local communities. According to Ryan Black his personal motivation to start the business was to 'do positive things in the world and to do good in the Amazon, not necessarily to get a big payout' (Bounds 2007).

Secondary actors and externally involved parties include Sambazon do Brasil Representação Comercial Ltda, a wholly owned subsidiary of Sambazon Inc. and located in Rio de Janeiro. Through this subsidiary the organic açaí berries are bought directly from around 1,500 native local families of the *ribeirinhos*, which are organised in four cooperatives. During the initial years the fruits were processed by local companies; in 2006 Sambazon set up its own fruit-processing facility.

For the initial distribution and marketing of the products Sambazon associated with the Juice It Up! Franchise Corporation, which has around 130 stores and juice bars in California and other states. After the initial period Sambazon has moved 'further up' and its açaí products are now distributed at stores such as Whole Foods, Wild Oats and Jamba Juice, as well as many conventional grocery chains. Sambazon açaí is also increasingly found in hundreds of other companies' products, including those of Stonyfield Farm and Haagen-Dazs. Sambazon has recently expanded its business overseas into the European market supplying companies such as Happy Monkey drinks (available at Waitrose, Holland & Barrett, Fresh & Wild, Planet Organic and Harrod's) and The Berry Company (most major supermarkets, food halls and health food stores). UK-based Innocent Drinks now buys 300 tonnes of frozen açaí pulp every year from Sambazon (Purvis 2007).

Sambazon products are certified organic, kosher and fair trade through US Department of Agriculture (USDA), Orthodox Union and US Fair Trade Federation. In addition, Sambazon is working with the USDA to produce guidelines for other companies interested in using açaí ingredients.

To bring the company to its present scale, the founders of Sambazon had to raise the capital of US$9 million. The investors in the USA now own around a quarter of the company (Bounds 2007).

The Sustainable Açaí Project, aimed at benefiting local communities and rainforest protection, is a public–private partnership involving and funded by corporations, charitable foundations and other public and private donors. On the ground in Brazil, Samazon is working together with the Peabiru Institute, Goeldi Museum, Amapa and Para State Federal Universities as well as with local NGOs such as the Federation of Organisations for Social and Education Assistance (FASE-PA) and the Fairtrade Federation (FTF). Scientific support is provided by the Para Federal University Food Engineering Department (UFPA). The Nature Conservancy Eco Enterprise Fund and EcoLogic Enterprise Venture were involved in the financing of the initial stages of the project.

9.1.4 Case history, development and main results

The Sambazon business model is in accord with the principles of the 'human development through the market' (HDtM) approach, which covers market-based environmentally and socially responsible activities that empower the poor to meet their basic needs. Having been introduced by the United Nations Environment Programme (UNEP) at the Costa Rica Meeting in September 2005 (Barbut 2005), HDtM approaches seek to harvest the economic and social benefits of sustainability. HDtM is building its foundations on the strengths and learning experiences of similar business approaches targeting 'the base of the pyramid'.

HDtM stresses that commercial activities involving environmental programmes such as sustainable production, sustainable procurement or sustainable product design are not only good for the environment, but also might serve wider goals. First, they are good for the economy (saving costs and developing domestic and export markets) and, second, they can support social progress such as helping to spread good labour conditions and creating good jobs (see also Tunçer *et al.* 2008).

The activities that can be considered HDtM are manifold, but three key principles define the concept:

- They aim to create social and environmental value
- They function collectively through the market
- They aim to create opportunities for the poor

The Sambazon case is an example of successful realisation of the HDtM principles in the agro-forestry sector by providing equitable employment and income generation aimed at poverty alleviation while promoting natural resource management and environmental protection.

FIGURE 9.1 The 'human development through the market' (HDtM) business model

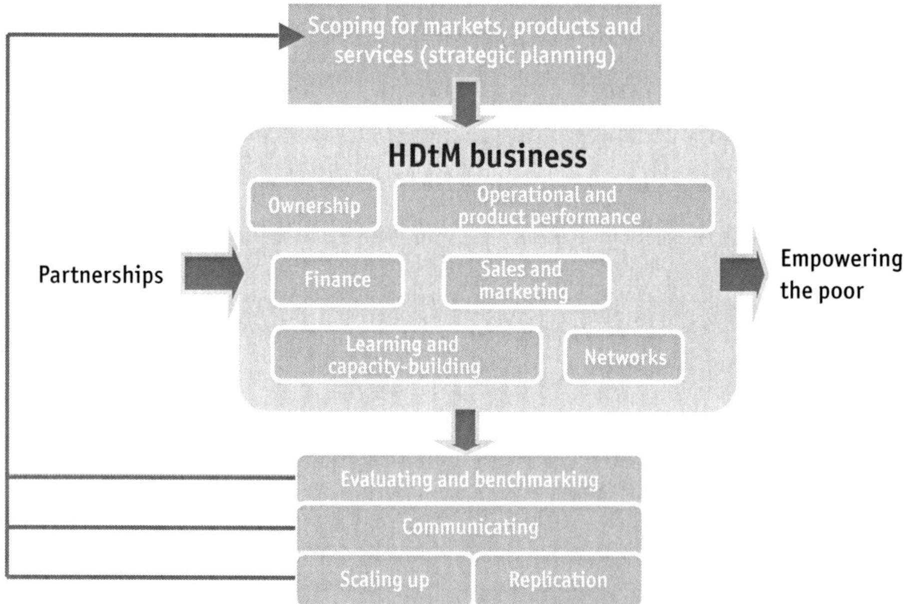

9.1.4.1 Ownership

The founders of Sambazon, the two brothers Ryan and Jeremy Black plus Ed Nichols, hold the main ownership of the company. The investors who provided capital to expand the capacity of the business in the USA now own around a quarter of the company.

The forest from which açaí fruits are harvested continues to be owned by local small-scale landowners and farmers, the *ribeirinhos*, who live along the Amazon estuary. To be able to sell their açaí to Sambazon the *ribeirinhos* are required to designate a piece of their land as an ecological reserve and carefully manage the rest of their terrain to protect the biodiversity of the rainforest.

Initially, the fruit processing was undertaken in local facilities in the Para district. As of 2006 Sambazon has opened a new processing facility that is vertically integrated into the business.

9.1.4.2 Finance

After initially purchasing, importing and successfully marketing a first batch of açaí berries, the Black brothers raised US$100,000 from friends and family to set up the infrastructure in Brazil to harvest açaí organically and in a way that protects the rainforest and promotes fair trade. To bring the company to its present scale, the founders of Sambazon had to raise capital of US$9 million. There is potential conflict of business interest between the investors and the sustainability goals of the Black brothers.

Since 2003 Sambazon has a recurring debt-financing arrangement with the Nature Conservancy Eco Enterprise Fund to purchase the açaí fruit. The credit granted is repaid during the year. Its profits rely on an increasing demand for açaí products for the US market. In addition, EcoLogic Enterprise Ventures, a microcredit fund based in Cambridge, MA, provides low-interest loans of US$25,000–US$500,000 in environmentally sensitive rural areas in Latin America and has been supporting the açaí cooperatives in the Amazon since November 2003, when it lent Sambazon US$175,000 for the previous season's harvest. In 2004 the amount disbursed to Sambazon rose to US$400,000 (*New York Times* 2004).

9.1.4.3 Sales and marketing

Initially, Sambazon emerged as a movement, not simply the purveyor of a healthy beverage choice. It was promoted through grass-roots marketing at music and athletic events, letting word of mouth about Sambazon spread among surfers, skateboarders, world class athletes and health conscious consumers, making sustainability not far-off and 'nerdy', but approachable, desirable and cool. They canvassed smoothie vendors, such as the Californian Juice It Up! Stores and juice bars to build their first customer base. In the words of Mr Sidoti of Juice It Up! 'they'd put on quite a show, going from store to store and putting on this Barnum and Bailey act' (Bounds 2007).

Customer education was crucial, because açaí was virtually unknown to US consumers. Sambazon also used an 'educate-the-customer-by-giving-it-away' approach. Business really picked up after Dr Nicholar Perricone named açaí the Number 1 'superfood' in his 2004 book *The Perricone Promise*. 'All of a sudden, the floodgates opened.', says Jeremy Black; 'it really helped accelerate the knowledge of what açaí is' (*Entrepreneur Magazine* 2007)

Despite the recent success and media coverage Sambazon still has a strong focus on active marketing through promotion, not so much focusing on traditional advertisements, but live in-store demonstrations of how to use the fruit pulp in making smoothies and other drinks. Sambazon spent US$500,000 in 2006 on in-store demonstrations. The Sustainable Açaí Project, even though a high priority for Sambazon's success, is now only secondary in the more mainstream marketing process. Many consumers are concerned mainly with consuming a healthy and tasty product. To that end, Sambazon has boiled its main message down to simple slogans such as: 'Superhealthy. Supertasty. Superfood', and 'Get with the purple berry' (Bounds 2007).

9.1.4.4 Operational and product performance

Growing consumer awareness demands social and environmental accountability and transparency in the supply chains of products sourced from developing countries overseas. Sambazon, by sourcing straight from organic local growers with products that are fully traceable to the person who grew it, fulfils these demands, which are increasingly necessary to establish brand equity and reputation (CSCP 2007).

In terms of transport and logistics, the recent discussion about food miles that emerged in the UK, which implies that fuel use and carbon dioxide (CO_2) emissions will be lower for food that is transported shorter rather than longer distances, would be an issue to be considered for Sambazon in their expansion into the European market. However, it could be argued that the socioeconomic benefit for the *ribeirinhos*, who only have a minuscule carbon footprint, outweighs the emissions produced through shipping transport of frozen açaí products (Muller 2007).

An important step that improved Sambazon's product performance was the opening of a new processing facility to provide products that could be marketed in the USA. To enter bigger retail stores Sambazon initially developed products that had a 30-day shelf-life, but realised that that was not a long enough window to get bottles shipped, stocked and sold. Looking for a packer that could bottle and flash-pasteurise the açaí so it would be stable for 90 days was necessary. Finding that partner delayed their entry into bigger stores and required a US$1 million investment

To meet growing demand in the USA the company is currently vertically integrating its supply chain and in 2006 inaugurated its own industrial unit to process the sustainable açaí fruits in Macapá, near the mouth of the Amazon. Sambazon has partly reduced its reliance on industrial units outside its system, which has advanced the company's ability to be flexible with its production and to adapt to US market requirements and demand.

9.1.4.5 Learning and capacity building

Teaching the US consumer about the açaí fruit and its applications was an important learning and teaching exercise and crucial for building a following. Having set up the business Sambazon is increasingly learning how to deal with competition, especially now that large corporations are entering the market.

Ongoing capacity building in Brazil to keep the harvesting practices and production of frozen Sambazon products sustainable is necessary. As the business expands the pressure on ecosystems increases. The Sustainable Açaí Project is crucial in dealing with newly emerging environmental issues, such as waste products resulting from the food processing. As the leading foreign company in the region Sambazon considers itself to be responsible in developing a set of standard metrics that can be used as a replicable model by other companies entering the market.

9.1.4.6 Networks

Sambazon had to establish networks in two main areas: first, on the sustainable production side in Brazil, working together with local *ribeirinhos*, NGOs, local government and research institutes and, second, on the consumption side in the USA setting up marketing and distribution networks. Over the past seven years Sambazon has been constantly expanding its networks in both areas. Networking through personal contacts was crucial in the initial stages of the business.

Sambazon works together with retail stores (organic, gourmet) in the USA. It promotes the distribution of free samples at festivals and cultural events. Sambazon also established links with beauty magazines that publish articles about the positive social and environmental effects related to the purchase of the açaí fruit. Now, through national media coverage and extended distribution networks, the product is becoming more mainstream.

9.1.5 Change in sustainability performance

9.1.5.1 Economic and profit improvements

Generally speaking, the economic profit of one hectare (2.471 acres) used as cattle pasture will produce just US$148 per year. Clear-cutting that same land for timber will produce eight times more, or around US$1,000, but this is a one-time gain, not a sustainable harvest. Sustainable harvesting that land for fruits such as açaí, latex and timber will produce almost seven times more, or US$6,820 per year, continuously.[3] Açaí berry production and harvesting are—together with sustainable eco-tourism, rainforest honey and sustainable 'heart-of-palm' production—part of the sustainable 'new economy' in the region that is based on non-timber forest products and already accounts for 9% of the total Amazonian economy (Purvis 2007).

Income for families harvesting açaí for Sambazon and workers in the processing factories is up to three times as high as the average income in Brazil. Sustainable long-term income harvesting açaí protects rainforest and the resource base, which contributes to the ability to sustain value creation in the future

Benefits on the consumption side include increased health benefits through consumption of organic açaí products. Because of the way açaí products are marketed, the fruit sets new standards of what is 'cool' among young people in the USA and could help combat the 'obesity epidemic'.

9.1.5.2 Environmental improvements

Harvesting the açaí fruit provides locals with an economic alternative to destructive economic practices, such as logging, mining, cultivating cash crops or, in the worst case, cattle ranching. Açaí berry harvesting protects açaí palms and forests that are otherwise cut down or slashed and burnt. The harvesting process is environmentally friendly, because the fruit is collected from trees naturally occurring in nature where no pesticides have been used. This increases incentives for the long-term preservation of the forest and provides locals with long-term, sustainable, access to the fruit.

3 See the webpage of Zola Açaí, at zolaacai.com/why-zola/sustainability, accessed 29 April 2010.

Apart from the açaí fruit the other commercially valuable product of the palm is the soft inner growing tip, particularly from *Euterpe edulis* but also from *Euterpe oleracea,* which is called heart of palm (*palmito*), and can be consumed, often in salads. Although the extraction of the palm's heart does not necessarily involve the inevitable death of the entire palm tree, over-harvesting and illegal trade of *palmito* has resulted in the demise of many palm populations in the Amazon estuary (Galetti and Fernandez 1998). Palm heart dishes are regarded as a delicacy more than as a staple diet; heart of palm is sometimes called 'millionaire's salad' because of to its high price. *Palmiteiros*, or palm poachers, are illegally cutting down 5,000 to 10,000 palm trees per week for these hearts of palm. Important markets for heart-of-palm are France and the USA.[4] Sustainable harvesting of heart-of-palm is possible and could be promoted alongside the açaí berry trade.

9.1.5.3 Social and cultural improvements

Currently 1,500 low-income rural producer families (with an average of 15 members per family) benefit from selling their harvested açaí to Sambazon. They receive a fixed minimum price creating stability for the producers against market fluctuations. Higher prices paid by Sambazon and increased demand from abroad have been an incentive for local middlemen to pay more to local *ribeirinhos* for açaí berries.

Regarding the *ribeirinho* communities, at this point in time local NGOs ensure that fair trade premiums and human rights are guaranteed, as well as improving the community by initiating health and educational programmes. Therefore, access to health and education has also improved. The açaí boom has also contributed to a revival of traditional ways of life. Young people who otherwise would have moved to the cities and be exposed to the violence and crime of the *favelas* are encouraged to keep their traditional way of life and instead simultaneously become guardians of the rainforest (Purvis 2007).

9.1.6 Learning experiences

9.1.6.1 Strengths and opportunities

The strengths and benefits of Sambazon's business model have been described in detail above. Regarding opportunities, Sambazon has much growth potential. As the product is becoming widely renowned, not only in specific society circles in the USA but also among mainstream consumers, the positive impacts of the business outlined above have the potential to benefit a larger number of people. In addition, the mission and ideology of Sambazon have the potential to be replicated in other countries with other organic produce, such as papayas, maté tea, hemp and so on.

The partnerships with local NGOs, fair trade associations and research institutes have proven effective and greatly contribute to the success of the company and improvements

4 See the webpage of El Servicio de Información Agropecuaria del Ministerio de Agricultura y Ganadería del Ecuador (SICA): www.sica.gov.ec/ingles/agro%20oportunidades/docs/business_op/palmito/palmitog.html, accessed 13 July 2009.

to social standards in the area. The potential for innovation is possible, if input from Sambazon's collaborating institutions is taken seriously.

9.1.6.2 Weaknesses and threats

There are a range of economic, social and environmental weaknesses and threats that Sambazon needs to deal with. First of all, the company relies on a single product and is therefore not sufficiently diversified to handle unforeseen market shifts or environmental disasters related to the açaí fruit. For example, if the fruit suddenly becomes 'out of fashion' the company's survival would be difficult.

The supply of açaí fruits is not always stable. A higher demand for the açaí fruit from North American markets might put pressure on the supply side of the company, especially since the amount of fruit growing wild is limited and must be harvested by hand. In addition, other companies have entered the market via other subtropical regions where the açaí fruit grows, which has created competition for Sambazon and might affect Amazon rainforest conservation efforts.

Around half a dozen serious players with names such as Zola and Bossa Nova have emerged as competitors. In 2003 Sambazon sued the rival company Zola, claiming the company used similar marketing verbiage (Bounds 2007). Zola juice products are in some 3,000 accounts nationwide competing directly with Sambazon's new juices, and the rivalry is intense. Positive is that Zola and Bossa Nova (certified to the Forest Stewardship Council) also run sustainable harvesting projects in Brazil and operate under similar principles as Sambazon. However, billion-dollar beverage giants, including Coca-Cola Co., PepsiCo Inc. and Anheuser-Busch Co., are adding the fruit to their beverage lineups. The involvement of these transnational corporations might threaten conservation efforts and lead to the establishment of monoculture açaí plantations. The trend towards intensification of palm tree management in some cases has resulted in the cutting of other trees and vegetation from around açaí palms. This facilitates the conversion of local flood plain forests into açaí-dominated forests closely resembling plantations, which has already been observed in the regions around Belem (Weinstein and Moegenburg 2004).

Another issue relates to the consumption of açaí by locals. While the açaí export industry has the potential to uphold a sustainable economy in the Amazon region, it also has the potential to rob locals of a nutritious part of their diet. For example, in the regional capital of Belem, the largest city in the Amazon region, which has 1.2 million inhabitants, around 400,000 litres of açaí are consumed daily. Many middlemen and vendors are threatened to be squeezed out of the market through Sambazon's involvement in the local economy. One negative side-effect of rising prices is that it now puts açaí out of reach of many local consumers. A few years ago, a litre of açaí juice cost US$0.52, in 2007 it was US$2.60, and in 2009 US$4.70—too expensive for many in an area where the monthly income is often less than US$200 (Astor 2007; Brasileiro 2009).

If foreign demand and the prices for açaí continue to increase, many locals will no longer be able to afford açaí, potentially depriving locals of the protein-rich nutrient they have relied on for generations. Local businesses and communities might then begin to oppose the growing influence of Sambazon on local economic and social conditions. For example, there have been reports from local sources that disagreements with local fruit processors in Para over work standards have motivated Sambazon to open its own processing facilities.

The unprecedented wealth through açaí has solved many problems but also brought unprecedented social problems among the *ribeirinhos*. According to João Filho of the Peabiru Institute 'the money they have now, they have never seen in their lives, so one thing they need is financial education. We see people buying a fridge when they have no electricity, just using it as furniture' (Purvis 2007). Part of the Sustainable Açaí Project run by Peabiru Institute will therefore include monitoring the social impacts before and after Sambazon.

While the impacts of the waste associated with food processing are being addressed through the Sustainable Açaí Project, the energy and emission issues resulting from freezing of the pulp and transport to the USA and EU are currently only marginal to the sustainability considerations of Sambazon operations but need to be accounted for in a sustainability assessment relating to food miles debates. However, through the preservation of forest sinks, the transport emissions are easily offset.

9.1.6.3 Main lessons learned

- Innovative and successful business models can be socially responsible and reduce poverty while being environmentally sustainable. However, for Sambazon to be truly sustainable not only the local environmental ecosystem impacts in Brazil but also further aspects of the supply chain such as energy consumption, transport and waste need to be considered
- Increased demand for a resource does not necessarily lead to unsustainable agro-forestry harvesting practices. Problems will arise when larger companies who do not follow the same sustainability principles enter the market. The development of monoculture plantations might be the result
- Initially successful and sustainable businesses encounter problems when reaching a certain scale that threatens sustainability as well as socially responsible business practices. In the case of Sambazon this includes the growing demand for the product in the USA and Europe, resulting in possible pressure on local prices and affordability for local consumers

9.2 Potential for diffusion and scaling up

Scaling up and replicating business models based on the HDtM approach that have proven their viability and capacity to deliver tangible benefits in specific circumstances present a great opportunity to make an impact on an even larger scale. Successful scaling up and replication of HDtM businesses has often been accomplished through the setting up of decentralised organisational structures, franchise models or the involvement of larger, already established and recognised, organisations. In the case of Sambazon the company is already at the scaling-up stage. In the eighth year of activities, the Sambazon Sustainable Açaí initiative achieved considerable impact through the company's sustainable business model based on market-driven conservation and a focus on the triple bottom line.

According to Sambazon the business has the potential to lift around 50,000 local families out of poverty, if the potential demand for certified sustainable açaí palm berry fruit products continues to grow at the current fast rate. According to Peabiru Institute the potential is even larger; a 20-year commitment by Sambazon and other companies to sourcing açaí organically, sustainably and responsibly, could result in revolutionary positive changes in the region.[5]

This increase in demand has a direct impact on the low-income family producers who harvest the fruit because it provides them with increased wages and an economic alternative to destructive economic practices such as logging or palm harvesting.

The Sambazon Sustainable Açaí initiative might be a model that can be replicated and implemented throughout the Amazon Basin as the demand for açaí fruits and other value added agro-forest products continues to grow. The cases of Sambazon competitors Zola and Bossa Nova show that there is a spillover effect and investments from other competitor companies does not necessarily threaten sustainability but can assist and support HDtM approaches. The demand for sustainable açaí throughout international markets and investment in the sustainability of the supply chain (through increased technical assistance), in order to guarantee that açaí harvesting will remain sustainable and benefit low-income family producers, can improve socioeconomic and environmental conditions.

In addition to increased industrial capacity, it is necessary that foundations, local government and local and international NGOs also invest in technical assistance aimed at providing these low-income families (who will benefit from the increased international demand for açaí-based products) with forest management programmes and business education so that the supply chain will remain sustainable for future generations.

Sambazon's long-term vision is to cooperate with like-minded companies and consumers against a higher triple-bottom-line standard, merging good business practices with global responsibilities. Cluster building under the HDtM concept with other companies operating under similar principles is being considered, these include Guayakí (maté drinks), Manitoba Harvest (hemp foods), Adina World Beat Beverages (fruit drinks) and Jungle Products (oils from tropical plants).

9.3 Overall conclusions

The Sambazon business model combines the marketing of organic açaí products in the USA and Europe with the Sustainable Açaí Project in Brazil. It has proven that profitable

5 See the website of the Peabiru Institute: www.peabiru.org.br/index3.htm, accessed 13 July 2009.

business models can have a positive socioeconomic impact on low-income rural communities in the Amazon region while simultaneously promoting sustainable agro-forestry. So far Sambazon has successfully built its innovative business model around a traditional method that uses the forest ecosystem resources in a sustainable way. This economic activity is preferable to the prevalent economic paradigm in the area, which includes poaching of the Açaí palms and promotes logging, mining and large-scale agriculture, especially raising cattle. If Sambazon continues to be successful in balancing the alleviation of rural poverty and long-term conservation of biodiversity, the business could become an important factor in promoting radical changes to the existing paradigm.

While the growing demand for açaí in the USA has the potential to provide incentives to involve more communities and actors in Sambazon's sustainable agro-forestry projects, there are also new challenges and threats that might impact on the existing sustainable business model. The threat to sustainability and biodiversity through an enrichment of forests with açaí palms to improve the harvest output is to be taken seriously. Also worthy of mention are possible social conflicts of interest with local actors, such as middlemen and businesses, who might feel threatened by another too powerful US company. In addition, other larger US companies that have not the same sustainability criteria have begun getting involved in the açaí trade. Growing demand from the USA (and most likely Europe within the next few years if the fruit continues being in fashion) might possibly result in a steep rise of prices and make the fruit unaffordable for locals for whom açaí has traditionally been an important dietary component.

Sambazon, and all other actors involved in açaí harvesting, will need to tackle these issues to retain the balance between rural development and conservation.

References

Astor, M. (2007) 'Acai Boom Puts the Purple Berry out of Reach for Those Closest to it in Brazil', Associated Press, 16 June 2007, www.pr-inside.com/acai-boom-puts-the-purple-berry-r155119.htm, accessed 13 July 2009.

Barbut, M. (2005) 'The Implementation Challenge of the Marrakech Process: It is Time to Focus', Costa Rica Meeting, Opening Session, 5 September 2005, San Jose, Costa Rica.

Bessa, L.F.M, M.M. de Oliveira and R. Abers (2005) 'The State of Forest Management Resources in Brazil. Green Governance, Green Peace: A Program of International Exchange in Environmental Governance, Community Resource Management, and Conflict Resolution' (University of Berkeley, California; globetrotter.berkeley.edu/EnvirPol/LucePapers/Brazil2005.pdf).

Bounds, G. (2007) 'The Perils of Being First', *Wall Street Journal*, 19 March 2007, online.wsj.com/article/SB117390281601437167.html?mod=THEJOURNALREPORTSMALLBUSINESS_2_1281.htm_1, accessed 13 July 2009.

Brasileiro, A. (2009) ' "Superfood" promoted on Oprah's site robs Amazon poor of staple', Bloomberg, 14 May 2009; www.bloomberg.com/apps/news?pid=20601109&sid=ai8WCgSJrhmY, accessed 29 April 2010.

Brewster, E. (2006) 'Healthy Beverages Ride the Wave', *Nutritional Outlook*, www.nutritionaloutlook.com/article.php?ArticleID=2084, accessed 13 July 2009.

CSCP (UNEP/Wuppertal Institute Collaborating Centre on Sustainable Consumption and Production) (2007) 'Retailers' Calendar: Exploring New Horizons in 12 Steps towards Long-term Market Success', CSCP; www.scp-centre.org/projects/ongoing-projects/retailers-role-towards-scp.html, accessed 29 April 2010.

Diegues, A.C. (1992) *The Social Dynamics of Deforestation in the Brazilian Amazon: An Overview* (Geneva: United Nations Research Institute for Social Development).

Entrepreneur Magazine (2007) 'Young Millionaires: Sambazon', *Entrepreneur Magazine*, October 2007; www.entrepreneur.com/magazine/entrepreneur/2007/october/184394-6.html, accessed 13 July 2009.

Frost, P., B. Campbell, G. Medina and L. Usongo (2006) 'Landscape-Scale Approaches for Integrated Natural Resource Management in Tropical Forest Landscapes', *Ecology and Society* 11.2 (Article 30); www.ecologyandsociety.org/vol11/iss2/art30, accessed 13 July 2009.

Galetti, M., and J.C. Fernandez (1998) 'Palm Heart Harvesting in the Brazilian Atlantic Forest: Changes in Industry Structure and the Illegal Trade', *The Journal of Applied Ecology* 35.2 (April 1998): 294-301.

Müller, B. (2007) 'Food Miles or Poverty Eradication? The Moral Duty to Eat African Strawberries at Christmas', Oxford Institute for Energy Studies; www.oxfordenergy.org/pdfs/comment_1007-1.pdf, accessed 13 July 2009.

New York Times (2004) 'Acai berry sales to US brings security to Amazon farmers', *New York Times*, 4 August 2004; www.fao.org/forestry/50059/en/#P10_378, accessed 29 April 2010.

Perricone, N. (2004) *The Perricone Promise: Look Younger, Live Longer in Three Easy Steps* (New York: Warner Books).

Purvis, A. (2007) 'Can a Smoothie Save the Rainforest?', *The Observer*, 18 November 2007; observer.guardian.co.uk/foodmonthly/story/0,,2210774,00.html, accessed 13 July 2009.

Tunçer, B., F. Tessema, M. Herrndorf and N. Pratt (2008) 'Towards "Human Development through the Market": A Comparative Review of Business Approaches Benefiting Low-Income Markets from a Sustainable Consumption and Production Perspective', in P. Kandachar and M. Halme (eds.), *Sustainability Challenges and Solutions at the Base of the Pyramid: Business, Technology and the Poor* (Sheffield, UK: Greenleaf Publishing; www.greenleaf-publishing.com/bop): 475-500.

University of Florida (2006) 'Brazilian Berry Destroys Cancer Cells in Lab, UF Study Shows', 12 January 2006; news.ufl.edu/2006/01/12/berries, accessed 13 July 2009.

Weinstein, S., and S. Moegenburg (2004) 'Acai Palm Management in the Amazon Estuary: Course for Conservation or Passage to Plantations?', *Conservation and Society*, 2 February 2004: 315-46.

10

Fairtrade Max Havelaar Norway: the Norwegian labelling organisation for fair trade

Ingri Osmundsvåg
National Institute for Consumer Research (SIFO), Norway

Fair trade started in the 1950s as a partnership between non-profit importers and retailers in the North and small-scale producers in developing countries. Producers were at the time struggling against low market prices and high dependence on intermediaries.[1] The development of increasing globalisation with internationalisation of trade and finance has led to the concentration of great resources and power in multinational corporations. Outsourcing production to wherever it is cheapest creates what is often referred to as a **race to the bottom**. Globalisation weakens national governments' authority to enforce legislation in industries or introduce new labour laws. The unbalanced trading relationships between poor, small-scale producers in the South and big multinational corporate buyers from the North has created unpredictable and low prices for commodities such as coffee. Commodity prices have in some cases been sinking to beneath the level of production costs (FTAO 2005). The farmers' payment is shortened by middlemen in the form of buyers, brokers, wholesalers and retailers who receive most of the profit from end-sales to consumers.

The producers in the South are small-scale and unorganised, with little access to information and markets, financial services, new technologies or innovations; this creates dependent producers with a weak bargaining position. Large corporations have economic resources, power of scale and access to information. Fairtrade is seen as a form of innovation within the capitalist system, a form of trading partnership that 'seeks

1 See www.fairtrade.net, accessed 13 July 2009.

greater equity in international trade'. Fairtrade organisations therefore support producers in 'awareness-raising and in campaigning for changes in the rules and practices of conventional international trade'.

The goal of fair trade is to contribute 'to sustainable development by offering better trading conditions to, and securing the rights of, disadvantaged producers and workers' in the South. The motivation behind fair trade is to work with marginalised producers and workers in order to help them move from a position of vulnerability to security and economic self-sufficiency.

The definition of fair trade by FINE[2] is as follows:

> Fair trade is a trading partnership, based on dialogue, transparency and respect, that seeks greater equity in international trade. It contributes to sustainable development by offering better trading conditions to, and securing their rights of, disadvantaged producers and workers—especially in the South. Fair-trade organisations (backed by consumers) are actively engaged in supporting producers in awareness-raising and in campaigning for changes in the rules and practices of conventional international trade.[3]

10.1 Case description

In 1988 the first Fairtrade **labelling** was created in the Netherlands for coffee sourced from Mexico.[4] Creating a label was a way to bring fair-trade products to mainstream retailers. In 1997 an association of national labels was established, the Fairtrade Labelling Organisation International (FLO),[5] and with it a set of common international standards for fair trade. FLO is the largest fair-trade standard-setting and certification body in the world, an association of 20 labelling initiatives that promote and market the Fairtrade Certification Mark in 15 European countries plus Australia, New Zealand, Canada, Japan and the USA. The Fairtrade label was developed for coffee but today is available for 20 product categories and for hundreds of individual products.[6]

2 The acronym FINE stands for the cooperation between the Fairtrade Labelling Organisations International (FLO), the International Fair Trade Association (IFAT), the Network of European Worldshops (NEWS!) and the European Fair Trade Association (EFTA), an international network of fair-trade organisations.

3 See 'Fair Trade Rules', www.fairtrade.net/fileadmin/user_upload/content/FairTradeRules_EN.pdf, accessed 4 February 2010.

4 See www.fairtrade.org.uk/what_is_fairtrade/fairtrade_certification_and_the_fairtrade_mark/the_fairtrade_mark.aspx, 4 February 2010.

5 FLO International is divided into two organisations: FLO eV is responsible for setting international fair-trade standards, facilitating and developing fair-trade business and working for trade justice. FLO-CERT GmbH ensures that producers, traders and retailers comply with Fairtrade standards; www.fairtrade.net, accessed 13 July 2009.

6 See the website of the Fair Trade Advocacy Office: www.Fairtrade-advocacy.org, accessed 13 July 2009.

The Norwegian fair-trade labelling organisation, Max Havelaar, was started in 1997 by 12 non-governmental organisations (NGOs), most of which where labour unions, environmental groups and Christian organisations.[7] Today the organisation is called Fairtrade Max Havelaar Norge (FMHN). FMHN is a member of Fairtrade Labelling Organisations International (FLO) and uses the International Fairtrade Certification Mark (CM), which guarantees that products comply with international fair-trade standards set by FLO.[8]

The pioneers of fair trade were the NGOs working with alternative trade organisations (ATOs). From this alternative trade movement the first labelling initiatives appeared. However, this was not the case in Norway, where the alternative trade or 'world shops' never caught on in the same way as they did in many other European countries. When FMHN was established in Norway there was no tradition for world shops. Today there is only one fair-trade shop in Norway. In late 2006 the Friends Fair-trade shop opened in Oslo; it has since become a member of the International Fair Trade Association (IFAT).

FMHN works with importers for product development and to recruit new licence holders. FMHN also helps importers to find fair-trade labelled products already on the market in other countries. Another important task is to improve distribution for fair-trade-labelled goods in supermarkets and shops. An important task for FMHN is to promote the label to consumers and businesses. Large target groups are the private catering industry and public catering for institutions.

10.1.1 From aid to trade

> World market prices for coffee, rice and other commodities are highly volatile and often below the costs of production. A stable price, that covers at least production and living costs, is an essential requirement for farmers to escape from poverty and provide themselves and their families with a decent standard of living.[9]

In Norway, as elsewhere, there has been an ongoing debate about the World Trade Organisation (WTO) and whether the trade system is contributing to relieve or preserve poverty among people in the South. NGOs have intensified the mobilisation of members and others to act as political consumers. There has also been great media attention on corporate social responsibility (CSR).

Fair trade is driven by many of the same motivations found behind more traditional charity, but in fair trade the importance of producer empowerment is emphasised. The slogan is 'Trade not aid', and the idea is that trade will help poor countries develop better and faster than if they were only to receive aid.

Terragni and Kjærnes point to some cultural factors judged as favourable for the fair-trade market in Norway:

7 See www.maxhavelaar.no/Internett/Om_Fairtrade/Fairtrade_Max_Havelaar_Norge/?_to=94, accessed 13 July 2009.

8 www.fairtrade.net, accessed 13 July 2009.

9 *Ibid.*

the fact that Norway is a wealthy country with a high degree of highly educated people; Norwegians engage themselves in politics and are often active citizens in local communities and members of ideal organisations; environmental issues and environmental organisations have a high visibility in the Norwegian society; solidarity with less developed countries is an issue promoted both by NGOs and by initiatives taken at government level (Terragni and Kjærnes 2005: 7).

10.1.2 Actors and their roles

10.1.2.1 Non-governmental organisations

Each of the national fair-trade labelling organisations were initially bottom-up, voluntary, initiatives taken by NGOs. In Norway as well, the strongest advocacy for fair trade has come from civil society. Many NGOs have been involved in both the starting up of FMHN and in actively promoting it since.

One of the most important early contributions by NGOs was to put focus on the politics behind the products. NGOs worked for increased awareness of these issues by gathering, analysing and spreading information to the media and to consumers. In the absence of international legislative authorities, many international NGOs have taken on the role of watchdogs. Traditionally, NGOs in Norway have had a tendency to work towards influencing the government to pass laws rather than to mobilise consumers.

Slowly but increasingly NGOs have started to call for people to become more politically aware consumers. They are appealing to people as consumers to make their views known to companies directly by what they are buying in the market. The NGOs also use their network of members as campaigners. The FMHN ambassadors and volunteers are often also members of supporting NGOs. The NGOs have played an important role both as sources of information as well as inspirational leaders to empower people to engage in political consumption.

10.1.2.2 Consumers

Fair trade can be seen as a form of political consumption. The political consumer represents a shift in the way the market, consumers and politics have previously been understood. In classical economic theory the consumer is seen as a rational agent acting in his or her own interest. In fair trade, politics and ethics are integrated in the economic arena (Sørensen 2004).

The buying of fair-trade goods can be seen as the practising of consumer power and a form of political consumption (Micheletti 2003). In Norway the consumer role has traditionally not been particularly politicised. Consumers have been protected by government institutions, and this has led to what Terragni and Kjærnes (2005) refers to as 'the institutionalisation of the consumer'. When it comes to issues related to the politics behind products, it seems that consumers want the government to make sure that all products sold on the Norwegian market are made in accordance with ethical standards.

For fair trade to go mainstream it is important to make ordinary consumers feel empowered. Market-driven fair trade depends on the will of consumers to pay more for a product. To motivate this, consumers should feel that they are making a difference. Therefore it is important to report back to the market and tell the good stories about improved quality of life for real people.

10.1.2.3 Retailers

The retailers are the gatekeepers for Fairtrade-labelled goods to enter the market. The structure of the Norwegian food supply sector is characterised by concentration of power, with a small number of big actors dominating the market. It has been difficult to gain access to retail space and as a result it has been complicated for Norwegian consumers to find fair-trade goods in supermarkets.

10.1.2.4 The government

Until recently NGOs and private businesses have been the most active in promoting fair trade in Norway (Nordic Council of Ministers 2001). However it appears as if the government plans to take a more active role in the future. The government has stated it wants to 'contribute to the promotion of social responsible consumption patterns through information, transfer of knowledge and "sway of attitude" '. It sees voluntary labelling initiatives such as fair trade as one way of meeting the need for easily available, standardised and quality checked information (Statsbudsjettet 2007). The government has granted funds to an environmental NGO to further develop an internet site for consumers about ethical consumption. FMHN receives funding to help pay administration and marketing costs from the Foreign Secretary through the Norwegian Agency for Development Cooperation (Norad).[10] FMHN is also indirectly supported by local authorities in fair-trade towns, where local councils act as supporters in buying and promoting the Fairtrade label locally.

10.1.2.5 Fair-trade towns

FMHN hopes the concept of fair-trade certified towns could be a way of boosting sales of Fairtrade-labelled goods in Norway. After strong pressure and much criticism the Norwegian government decided in 2004 to establish ethical standards for investments by the *Statens pensjonsfond avdeling utlandet* (a big retirement fund created by income from the petroleum sector). The government wants to use the fund to actively promote CSR.[11] In the wake of this debate the question of ethical standards for public acquisitions was raised. Ministers and local politicians have taken on challenges from NGOs to work for the introduction of ethical standards for public acquisitions. This coincides with the start of FMHN's certification of fair-trade towns. In the UK this has been a great success and many see the 150 fair-trade towns as a key factor for understanding the boost to fair trade in the UK. As of today there are three fair-trade councils in Norway, but a list of 21 towns and councils are working to become certified.[12]

10 See the webpage of Norad, at www.norad.no/default.asp?V_ITEM_ID=1169, accessed 13 July 2009.

11 See www.regjeringen.no/nb/dep/fin/tema/Statens_pensjonsfond.html?id=1441, accessed 13 July 2009.

12 See www.maxhavelaar.no/Internett/Fairtrade-kommune, accessed 29 April 2010.

10.1.3 The development of Fairtrade labelling in Norway

10.1.3.1 Sales development for Fairtrade-labelled goods in Norway

Table 10.1 lists the most important Fairtrade-labelled products available in the Norwegian market today in terms of estimated retail value for volumes sold in 2006.

TABLE 10.1 The most important Fairtrade-labelled products available in the Norwegian market today in terms of estimated retail value for volumes sold in 2006

Source: Max Havelaar, Norway, Market Research, 2006

Product	Retail value (NKr)	Year of introduction
Coffee	34,691,886	1997
Tea	2,243,585	2000
Bananas	15,607,893	2001
Cocoa and chocolate	1,622,192	2002
Orange juice	4,122,934	2002
Rice	45,660	2004
Roses	10,732,198	2005
Sugar	1,397,264	2006
Wine	422,936	2006

According to a survey by the National Institute for Consumer Research (SIFO) in Norway, 16% of Norwegian consumers in 2005 and 18% in 2006 said they had bought goods marked by the fair-trade label Max Havelaar (Berg and Terragni 2006). This is a small increase, but it is possible that each person is buying more. In FMHN's own market research, 20% of consumers said they buy products with the guarantee mark for fair trade once a week or more (FMHN 2006).

In the following sections the development of sales volumes for the different product categories are presented with some brief comments.

10.1.3.1.1 Coffee

Coffee had a market share of approximately 1% of the total market in 2006. It is the oldest labelled product and shows a steady growth (Fig. 10.1). Tea and coffee are the largest fair-trade product groups, bought by private and public catering institutions. Almost all the wholesalers selling directly to businesses in Norway stocked Fairtrade-labelled coffee in 2007.

10.1.3.1.2 Bananas

Fair-trade bananas had an estimated market share of 1.5% in Norway in 2005, up from 0.5% in 2004 (this is similar to the market for organic bananas). Sales of fair-trade

FIGURE 10.1 Sales volumes of Fairtrade-labelled coffee in Norway

Source: Max Havelaar, Norway, Market Survey, 2006

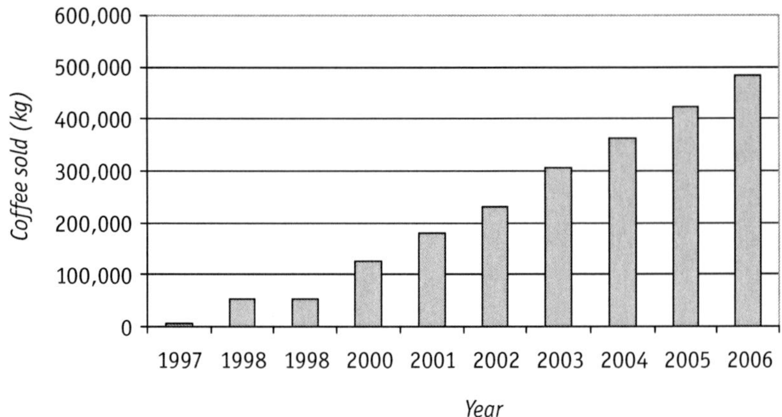

bananas increased by 147% from 2004 to 2005. One possible explanation for the remarkable rise in sales volume could be the effect of favourable coverage in the media. In April 2005 the popular weekly programme focused on consumer issues, 'Forbrukerinspektørene', on the national TV network, aired a special report on the living and working conditions on banana plantations in Latin America. After this programme the sales of Fairtrade-labelled bananas increased dramatically (Aftenposten 2006; see also Fig. 10.2). FMHN's market research from 2006 shows that coverage in newspapers, periodicals and such like are consumers' most important sources of knowledge about the label. Shops and TV programmes are the second most important sources (FMHN 2006).

FIGURE 10.2 Sales volumes of Fairtrade-labelled bananas in Norway

Source: Max Havelaar, Norway, Market Survey, 2006

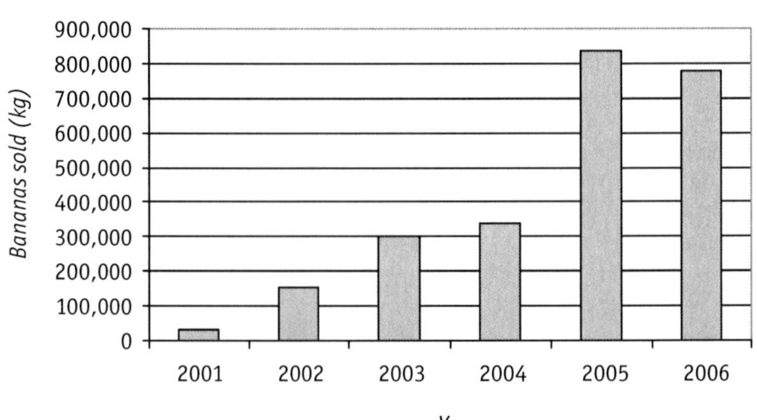

10.1.3.1.3 Orange juice

Fairtrade-labelled orange juice has a very small market share (Fig. 10.3); one of the problems for this product has been distribution. Another is packaging; the juice is sold only in a large carton (1.5 litres) and for one of the brands the design might appear a bit outdated.

FIGURE 10.3 Sales volumes of Fairtrade-labelled orange juice in Norway

Source: Max Havelaar, Norway, Market Survey, 2006

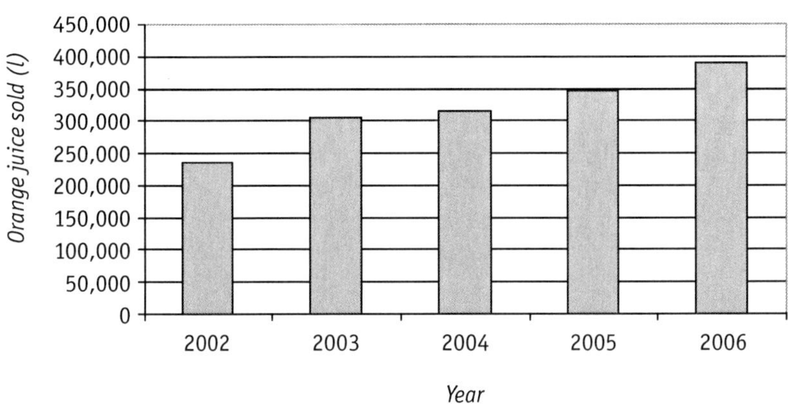

10.1.3.1.4 Tea

Fairtrade-labelled tea has reached relatively small sales volumes and market shares (0.3%) in Norway (Krier 2006: 57; see also Fig. 10.4). One explanation may be that the

FIGURE 10.4 Sales volumes of Fairtrade-labelled tea in Norway

Source: Max Havelaar, Norway, Market Survey, 2006

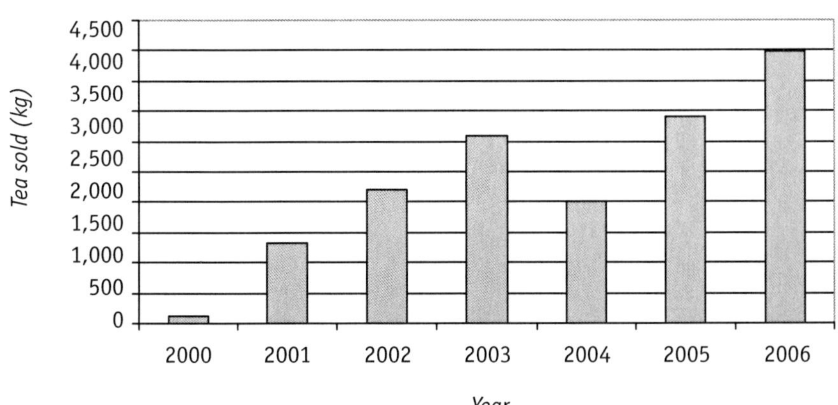

price difference between Fairtrade-labelled tea and conventional tea has been too high. The relative price difference has been much higher for tea than for coffee and bananas. Although surveys show that as many as 70% of consumers are willing to pay more for Fairtrade-labelled products, to reach the mainstream market the price difference should not be more than 15%. There have also been relatively many licence holders for tea (five in 2005) but no big suppliers. For distribution reasons it could be an advantaged to have lager licence holders. The big actors in the food market have better distribution systems and easier access to more attractive retail space.

10.1.3.1.5 Cocoa and chocolate

Sales of Fairtrade-labelled chocolate is low in Norway compared with other countries (see also Fig. 10.5). One of the reasons could be that there have been many but very small license holders and that they have struggled with distribution. The chocolate lines in all the supermarket chains are dominated by two big actors, both of which have resisted pressure from NGOs to launch fair-trade alternatives (Sætre 2004). One can only speculate, but one possible problem for suppliers that want to launch fair-trade goods is that these might put the other products in the brand line in a bad light. From the consumers' perspective they may seem unethical. This might be one reason why The Cooperative retailer in the UK changed all its chocolate to fair trade.[13] Still, there are many examples of brands that offer one or two fair-trade options in their product lines.

FIGURE 10.5 Sales volumes of Fairtrade-labelled cocoa and chocolate in Norway

Source: Max Havelaar, Norway, Market Survey, 2006

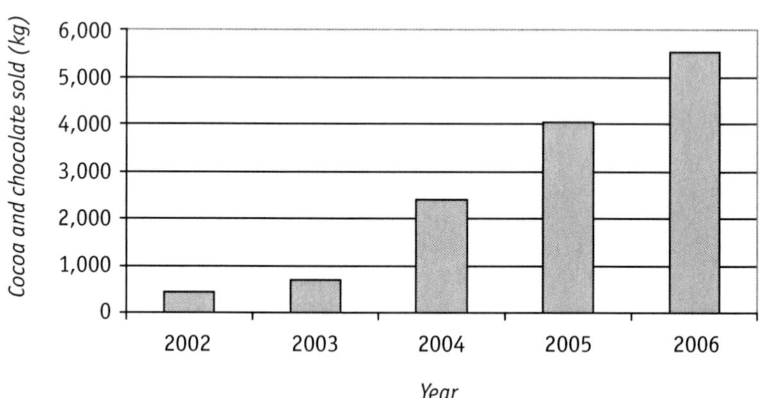

10.1.3.1.6 Rice

The reasons for the limited success (Fig. 10.6) of Fairtrade-labelled rice in Norway might be the limited distribution and high price. It could also be the packaging; the rice is sold in small amounts and in a box with an outdated design rather than in a bag. It might

13 www.co-operative.co.uk, accessed 13 July 2009.

FIGURE 10.6 Sales of Fairtrade-labelled rice in Norway

Source: Max Havelaar, Norway, Market Survey, 2006

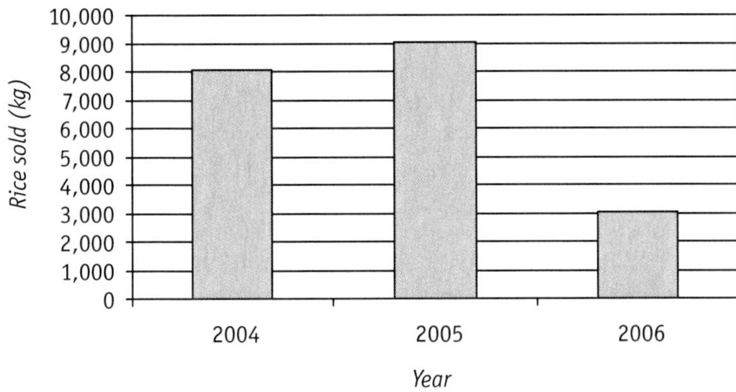

also be that the rice is of the wrong type of quality and taste. FMHN has therefore taken the original type of rice out of its product range and will be launching a new type.

10.1.3.1.7 Roses

Fairtrade-labelled roses were launched in 2005 and have been a great success. In little more than a year it has become the third biggest fair-trade product measured in retail value. This is probably because of good distribution; they are sold in all the shops of the biggest florist chain in Norway (Mester Grønn) and in some supermarkets (ICA). The price difference compared with conventional traded roses is not large. The retailer has also invested in marketing by running advertising campaigns in magazines and newspapers. In 2007 one out of three roses sold by Mester Grønn had a Fairtrade label. This example show that with good distribution, a small price premium and some promotion from the license holder fair-trade products have good market potential in Norway.

10.1.3.2 Distribution of Fairtrade-labelled products

In 1997, the same year Max Havelaar was launched in Norway, a survey showed that 64% of consumers in Norway said they had a positive attitude towards buying fair-trade products.[14] With this in mind, sales of Fairtrade-labelled goods have been lower in Norway than expected (Terragni and Kjærnes 2005). One explanation might be that the range and availability of fair-trade goods have been very limited (Terragni *et al.* 2006).

The structure of the Norwegian food market is characterised by concentration of big supermarket chains. Four main actors cover 99% of the Norwegian food market. Two chains with small product lines that compete together primarily on price have a market

14 Survey by Norsk Gallup for Alternativ Handel. See www.eftafairtrade.org, accessed 13 July 2009.

share of over 50%:[15] 'the Norwegian food market is highly centralised, concentrated and protected, leading to a limitation of the range of products available in the market, and consequently of the consumers' opportunity to make choices' (Terragni and Kjærnes 2005: 14).

Market actors, on the one hand, refer to the lack of interest among Norwegian consumers as the main cause of the scarcity of Fairtrade-labelled products in the market (Terragni and Kjærnes 2005). Terragni and Kjærnes, on the other hand, point to the consumers' lack of access to fair-trade products as a main cause for low sales figures in Norway. Through interviews with market actors, Terragni and Kjærnes (2005) found that the common view is that the Norwegian market is too small for such niche products to be profitable. A main task for FMHN is to promote fair trade to consumers. However, it is problematic to raise demand and sales volumes if consumers cannot find the products on the shelves of their local supermarkets. Therefore distribution and attractive retail space are key success factors for fair-trade products to go mainstream.

Unlike the situation in other countries there has been little competition among Norwegian supermarket chains to be positioned as the best in fair trade. In other European countries The Cooperative has been active in promoting fair trade. In Norway The Cooperative has its own fair-trade coffee called Café Futuro, which is sold as a private brand in all Cooperative outlets.[16] The best distribution of Fairtrade-labelled products has been for coffee, sugar, orange juice and bananas. For FMHN bigger licence holders could be a good strategy; the big actors in the food market have better distribution systems and easier access to more attractive retail space.

The supermarket chains Meny, Centra and Ultra have the best selection of Fairtrade-labelled products today. Norgesgruppen has recently launched more fair-trade products in their supermarket chains and made an announcement saying their goal is to sell Fairtrade-labelled products worth over NKr100 million by the year 2010. Norgesgruppen had a 59% growth in sales of Fairtrade-labelled goods from the first quarter of 2006 to the first quarter of 2007. The strongest growth was in fruit and vegetable products.[17]

10.1.3.3 Marketing and communication

FMHN has moved from network marketing by **ambassadors** (volunteers recruited mostly from the founding and supporting NGOs), focusing on front-runner groups, to more mass communication in an attempt to reach more consumers in the big market segments.

With a marketing and communications budget of just over NKr1.2 million in 2005 FMHN does a great deal of informational and educational work (publishing information, organising events and placing stands in shops). FMHN have also made extensive use of the internet to spread news and information about fair trade in general and about FMHN and Norway in particular. They run an informative internet site, distribute newsletters and run a special internet service for the fair-trade ambassadors.

15 Dagligvarefasiten 2005 (www.dagligvarehandelen.com, accessed 13 July 2009).

16 www.coop.no, accessed 13 July 2009.

17 See www.maxhavelaar.no/Internett/Nyheter, accessed 13 July 2009.

10.1.3.4 Consumer knowledge of the Fairtrade label

Although there has been a steady growth in the number of available Fairtrade-labelled goods, the label is still not very well known. FMHN's own market research shows that 40% of Norwegian consumers knew the label in 2006. This is a notable increase since 2004, when surveys showed a 30% knowledge of the label. Unaided knowledge or recognition, however, was only 13% in 2006, but this is still a doubling from the results in 2004 (6%). According to FMHN's own market research, unaided knowledge of the label was 3% in 2002, 6% in 2004 and 13% in 2006 (FMHN 2006).

In 2002 FLO launched a new international fair-trade certification mark to improve the visibility of the logo in supermarkets. A common logo also simplifies procedures for importers and traders.[18] In Norway the new logo was gradually introduced and the use of the old logo showing an elephant was phased out during 2003.[19] Use of a common mark opens up greater possibilities for cross-border trade and for introducing new fair-trade products to Norway. The retailer Smart-Club is currently importing Fairtrade-labelled goods licensed in other countries. This is facilitated by the cooperation of the national labelling organisations.

In 2006 Max Havelaar changed its name to Fairtrade Max Havelaar Norge. The English name *Fairtrade* is the international term for 'Fair Trade Labelled' and is intuitively easier to understand than *Max Havelaar*, including for Norwegians. In 2006 a market survey for FMHN showed that 61% of those who recognise the term *Fairtrade* associated it with 'better trading conditions for developing countries' (FMHN 2006).

10.1.3.5 Future challenges

The future challenge for FMHN is to reach the *mainstream* market. To improve distribution FMHN needs to maintain the goodwill and close working relationships with retailers and supermarket chains. This has to be carefully balanced with the need to promote fair-trade products to consumers as a more ethical choice. The slogan accompanying the FMHN label mark is 'creator of good stories'. Compared with many other brands Fairtrade offers a good story. Given that the quality is at least average or better, this is a unique selling point. FMHN needs to mobilise a *pull* in the market from motivated consumers feeling empowered and good about the story behind the products. It also has to ensure that price differentiation is not too high compared with conventional goods and that most of the extra costs paid by consumers go back to the producers. This points to the importance of effective certification programmes as well as the transparency of the supply chain.

18 www.fairtrade.net, accessed 13 July 2009.

19 www.maxhavelaar.no/Internett/For_presse/Pressemeldinger/2003, accessed 13 July 2009.

10.2 Results

In 2005 the combined sales turnover of fair-trade products was more than €1billion, and growth rates have been between 20% and 30% since the beginning of the 21st century (Krier 2006). The total value of products sold under fair-trade labels in 14 European countries is about €597 million (Krier 2006). The estimated retail value for fair-trade products, and the annual percentage change, for the years 2001 to 2006 (in Norwegian kroner [NKr]) are shown in Table 10.2 (see also Fig. 10.7).

TABLE 10.2 Sales of Fairtrade-labelled goods in Norway, 2001–2006

Source: Max Havelaar, Norway, Market Survey, 2006

Year	Value (NKr)	Change (%)
2001	13,355,114	–
2002	23,168,822	73
2003	33,000,000	42
2004	38,000,000	15
2005	54,518,147	43
2006	70,886,547	30

FIGURE 10.7 Estimated retail value of Fairtrade-labelled products sold in Norway

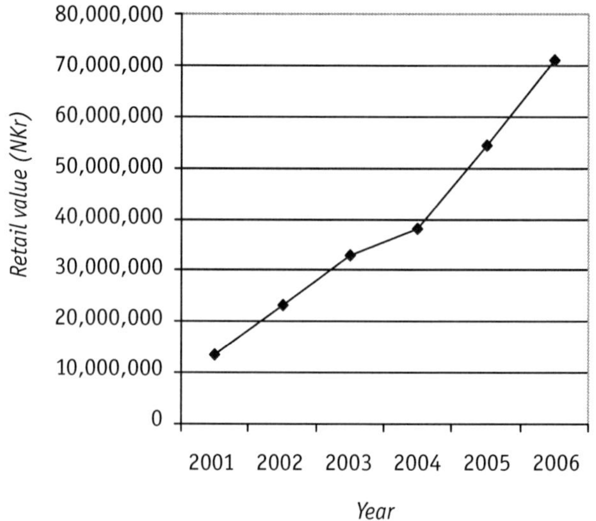

To sum up one could say that the Norwegian market for Fairtrade-labelled goods is limited and relatively immature compared with other countries in Europe. The distribution of Fairtrade-labelled products is limited, and Norwegian consumers have relatively little knowledge of fair trade. However, the Norwegian market for Fairtrade-labelled products is growing.

10.2.1 Change in social impact

It is estimated that in 2006 five million people in the world benefited from fair trade. FLO worked with 569 producer organisations, representing over a million farmers and workers. Approximately 10% of the retail value of Fairtrade-labelled goods sold in Norway is paid extra to the producers in the South.

10.2.2 Some critical points

It can be argued that fair trade in some cases can be seen as a form of cultural imperialism, where values based in Western cultures are imposed on local cultures and values (Micheletti *et al.* 2004). Labels based on social standards can also have a protectionist effect, hindering export from developing countries.

Fair trade can be criticised for shifting focus away from politicians' responsibility to solve social problems. It represents a form of privatisation of social problems that actually need to be solved at a higher political level. Another concern raised against fair trade is that it represents a form of commoditisation of empathy and caring. The producers are made dependent on consumers' concern for decent pay and working conditions.

10.2.3 Learning experiences

To gain success for a consumer labelling mark it is important to have a recognisable and self-explanatory name and logo. It is important to communicate the *good stories* behind fair trade, and positive media coverage can boost sales significantly.

To create better access for consumers to Fairtrade-labelled goods it is of great importance to work with distribution channels to mainstream supermarkets. To appeal to a bigger segment of consumers the price of a Fairtrade-labelled product should not be more than 15% higher than conventional brands in the same category. The quality standards of fair-trade goods must be good, at the same level or better than conventional products. Only a very small segment of consumers will buy a product just because it is fair trade.

10.3 Potential for diffusion and scaling up

Fair trade can still grow significantly but it is unrealistic to think that all trade could be done within the fair-trade system. Therefore the greatest impact of fair trade might emerge from the fact that it generates a focus on unfair trade. It creates pressure from different stakeholders on corporations to act more responsibly and create more 'local content': 'local content refers to the contribution to the host country and the local community by businesses who work in developing countries with regard to employment, contribution of knowledge and technology to local communities and businesses' (Ministry of Foreign Affairs 2006).

One of the by-products of fair trade is that more attention is directed to CSR. CSR can be implemented by introducing social accountability, measuring and reporting compliance with social standards. Many companies have already established ethical standards and **codes of conduct** for themselves and their suppliers. DNV has created a system standard to secure ethical supply chains, **Social Accountability 8000** (SA8000).[20] A new ISO standard, **Social Responsibility 26000** (ISO 26000), is at draft stage and is to be formally launched in 2010.[21] A central question in relation to these self-imposed standards is: how can consumers trust that these standards are being followed? NGOs such as the Clean Clothes Campaign[22] demand that this should be controlled by *independent* auditing rather than by the companies themselves. In the case of Fairtrade-labelled private brands one could argue that the FLO has in a way taken over some of this auditing function. FLO has the task of auditing producers, making sure everything is in line with the standard, and it develop standards for new products categories.

10.4 Overall conclusions

The imbalance in trading relationships between poor producers and multinational corporate buyers moved NGOs to start up alternative means of trade. Fair trade aims at empowering producers. Behind the slogan 'trade not aid' lies the idea that people can help themselves if they can get a fair chance. Consumers can therefore choose to pay a price premium to ensure that producers receive a liveable income. The retailers are the gatekeepers, ensuring Fairtrade-labelled products reach consumers. So far the range of fair-trade goods available to Norwegian consumers has been narrow and the distribution poor. Consumers' knowledge of the Fairtrade label has been relatively poor in Norway but it is rising steadily. The quality of fair-trade products must not be lower than average for the category, and the price difference compared with conventional goods should not be more than 15%. Consumers must be able to trust the labelling organisation and feel secure that the price premium will result in better payment to the producers. To motivate consumers and make them feel empowered communication should focus on the good stories created by fair trade.

20 See www.dnv.no/sertifisering/bedriftenssamfunnsansvar/bedriftenssamfunnsansvar/sa8000.asp, accessed 13 July 2009.

21 See www.etiskhandel.no/noop/page.php or www.indianet.nl/pdf/briefing_iso_sr.pdf, accessed 13 July 2009.

22 See www.cleanclothes.org, accessed 13 July 2009.

Fairtrade's potential success in Norway depends on whether or not it is seen by consumers as being meaningful. Traditionally, in Norway the consumer's role has not been particularly politicised. However, it looks as if this could slowly be changing. The impact of fair trade on market structure is still very small, but new products are being launched and sales are rising. Probably the greatest impact from fair trade will come from its influence on conventional trade to become more humane and responsible.

References

Aftenposten (200) 'Vi bryr oss: litt' ('We Care: A Little'), *Aftenposten*, 2 June 2006.

Berg, L., and L. Terragni (2006) 'Handlekurven som politisk arena?' ('The Shopping Basket as Political Arena') (Hurtig statistikk fra SIFO-survey 2006).

FMHN (Fairtrade Max Havelaar Norway) (2006) Market Survey Conducted in 2006 by Research International, Commissioned by Fairtrade Max Havelaar Norge.

FTAO (Fair Trade Advocacy Office) (2005) 'Business Unusual: Success and Challenges of Fair Trade' (Brussels: FTAO).

Krier, J.-M. (2006) '*Fair Trade in Europe 2005: Facts and Figures on Fair Trade in 25 European Countries.* (Fairtrade Labelling Organisations International [FLO], International Fair Trade Association [IFAT], Network of European Worldshops [NEWS!] and European Fair Trade Association [EFTA]).

Micheletti, M. (2003) *Political Virtue and Shopping: Individuals, Consumerism and Collective Action* (New York: Palgrave Macmillan).

Micheletti, M., A. Føllesdal and D. Stolle (eds.) (2004) *Politics, Produkts and Markets: Exploring Political Consumerism Past and Present* (New Brunswick, NJ; London: Transaction Publishers).

Ministry of Foreign Affairs (2006) 'Utenriksdepartementets rapport om norske bedrifter i utlandet og CRM' ('Report on Norwegian Businesses Abroad'); www.dep.no/ud, accessed 13 July 2009.

Nordic Council of Ministers (2001) 'Forbrugernes fornemmelse for etikk: En kortlægning af etikk-initiativener i de nordiske lande og en analyse avf perspektiverne for at styrke et etisk forbrug i de nordiske lande' ('Consumers' Sense of Ethics'), *TemaNord* 2001:583 (København: Nordisk Ministerråd).

Terragni, L., and U. Kjærnes (2005) 'Ethical Consumption in Norway: Why is it so Low? Political Consumerism: Its Motivations, Power, and Conditions in the Nordic Countries and Elsewhere', *TemaNord* 2005:517 (København: Nordisk Ministerråd).

——, E. Jacobsen, G. Vittersø and H. Torjusen (2006) 'Etisk- politisk forbruk. En oversikt' ('Ethical and Political Consumption: An Overview') (SIFO Prosjektnotat nr.1, 2006; Oslo: National Institute for Consumer Research [SIFO]).

Sætre, S. (2004) *Den lille stygge sjokoladeboka* (*The Little Ugly Chocolate Book*) (Oslo: Spartacus Forlag).

Sørensen, M.P. (2004) *Den politiske forbruger* (*The Political Consumer*) (Copenhagen: Hans Reitzlers Forlag).

Statsbudsjettet (2007) 'Etisk Forbruk' ('Ethical Consumption'); www.statsbudsjettet.dep.no/Statsbudsjettet-2007/Statsbudsjettet-fra-A-til-A/Etisk-forbruk, accessed 13 July 2009.

11

Verified sustainable agriculture: a practical experience and a significant contribution to sustainable consumption and production

Chris Wille and Joke Aerts
Rainforest Alliance

Bernward Geier
Colabora

The Rainforest Alliance is an international non-governmental organisation (NGO) with a mission to protect biodiversity and sustainable livelihoods. This is done by transforming land-use practices, business practices and consumer behaviour. The Rainforest Alliance works with people who use their land to manage a forest, grow agricultural crops or run a hotel or tour in biodiverse areas. Farmers, companies, cooperatives and landowners who participate in the programme meet comprehensive, rigorous social, economic and environmental standards.

This chapter will discuss how the Rainforest Alliance is directly supporting sustainable consumption and production (SCP) by encouraging transformative approaches to implementing best management practices in farms, forests and tourism businesses and encouraging consumers to reward these efforts by looking for the Rainforest Alliance seal of approval on products.

Over the past four centuries half of the world's forests have been cleared. In response to this significant loss, the Rainforest Alliance pioneered forestry certification in 1989 with the launch of SmartWood, the first global sustainable forestry certification programme. To encourage market-driven, environmentally and socially sound management of forests, tree farms and forest resources, SmartWood awards the Forest Stewardship Council (FSC) seal of approval to operations that follow strict standards for sustainability. The FSC is the global standard setter for responsible forestry.

Although local and indigenous people directly depend on forest resources for their livelihoods, they may lack the technical skills and tools to use them judiciously. Over the past 20 years, governments everywhere have been turning over forested areas to local communities, a trend that is expected to continue. To help local communities develop sustainable practices and earn a living from the wood and non-timber forest resources they harvest from their lands—such as palm fronds, nuts, rattan, latex and resin—the Rainforest Alliance ensures that community and indigenous forestry operations have equal access to the Alliance's certified services and markets.

With 800 million people travelling every year, tourism is not only big business but a growing source of revenue for people living in areas that are particularly rich in plant and animal species. Although tourism can produce serious negative impacts on local people and the environment, it also has the potential to provide incentives for conservation. The Rainforest Alliance is working with other organisations and experts worldwide to develop best management practices for sustainable tourism to help tourism businesses and consumers effectively contribute to biodiversity conservation and social welfare.

The Rainforest Alliance is uniting stakeholders worldwide to share information on how the tourism industry is promoting sustainable resource management. As secretariat of the Sustainable Tourism Certification Network of the Americas, the Rainforest Alliance is helping tourism certification programmes, tour operators, government agencies, civic organisations and travellers throughout Latin America and the Caribbean to develop common standards for sustainable tourism. The Alliance is also leading efforts in preparing for the launch of the Sustainable Tourism Stewardship Council, a proposed global accreditation body for sustainable tourism certification programmes to increase worldwide recognition and provide credibility for these eco-labels.

In the Sustainable Agriculture programme, the Rainforest Alliance provides farmers with incentives to meet its standards and encourages companies and consumers to support those farms making improvements toward sustainability. Consumers can look for the Alliance's green frog trust mark and be assured that the products come from a sustainably managed source.

The Rainforest Alliance together with the Sustainable Agriculture Network (SAN) set up a transparent and rigorous standard and certification programme that helps farmers 'get on the sustainability track'. The programme covers the main tropical farm commodities, including coffee, bananas and other fruits, cocoa, flowers and tea. The SAN is now working on adapting the standards for crops used as bio-fuels: namely, sugarcane and oil palm.

11.1 Case description

11.1.1 Agriculture and the link to biodiversity and sustainable livelihoods

Sustainable agriculture is at the centre of the Rainforest Alliance's efforts to conserve ecosystems by creating and/or maintaining healthy soils, rivers and wildlife and by promoting dignified living conditions for farm workers and neighbouring rural communities.

The unbridled and unsustainable growth of agriculture in recent years has encouraged rampant deforestation and careless use of agrochemicals. Today agriculture is the number one cause of ecosystem destruction and species loss worldwide. Agriculture occupies approximately a third of the Earth's total landscape and uses more land and fresh water than any other human activity in the world. Agriculture's impact goes well beyond the environment. As a major employer, agriculture affects social, economic and cultural conditions as well. Because of agriculture's widespread influence, the Rainforest Alliance works directly with those that use the land, primarily farmers, to promote economically viable, environmentally sound and socially equitable agriculture. Sustainable farms have minimal environmental footprints, are good neighbours to human and wild communities and are integral to regional conservation initiatives. Beneficiaries of sustainable agriculture include present and future generations of farmers, farm workers, consumers and wildlife.

The Rainforest Alliance certification system aims to protect biodiversity 'hot spots' or areas that boast an astonishing array of species. The *cerrado* in Brazil is one such area and, as only 1.5% is protected, its survival depends on private conservation efforts such as those made by Daterra, a Rainforest Alliance certified coffee farm. Daterra covers over 14,800 acres of this area, nearly 9,900 acres of which are natural areas that are preserved and protected.

11.1.2 The Sustainable Agriculture Network

The Rainforest Alliance does not work alone. In fact, it is the international secretariat of the SAN, a coalition of leading conservation groups based in Latin America, and looking to expand to Africa and Asia. The collective vision of the SAN is that the concept of sustainability recognises that the well-being of societies and ecosystems is intertwined and dependent on development that is environmentally sound, socially equitable and economically viable.

The SAN includes environmental groups in Belize, Brazil, Colombia, Costa Rica, Ecuador, El Salvador, Guatemala, Honduras and Mexico, with input from many associated academic, agricultural and social responsibility groups around the world.

The conservation and rural development groups that manage the certification programme and are part of the SAN understand local culture, politics, language and ecology and are trained in auditing procedures according to internationally recognised guidelines.

11.1.3 The Sustainable Agriculture Network standard

11.1.3.1 Standard setting

The activities of the SAN in terms of standard development and reviewing comply with the International Social and Environmental Accreditation and Labelling (ISEAL) Alliance 'Code of Good Practice for Setting Social and Environmental Standards' (ISEAL Alliance 2004).

ISEAL members, the Rainforest Alliance included, are collaborating in order to gain international recognition and legitimacy for their programmes, to improve the quality and professionalism of their respective organisations and to defend the common interests of international accreditation organisations while demonstrating the openness and transparency of operations that ISEAL members believe is fundamental to their integrity. The standards, certification programmes and accreditation systems being developed by ISEAL members are all global in nature and reflect a worldwide concern for the social and environmental issues being promoted. This collaboration represents a significant movement to promote the interests of workers, communities and the environment in world trade.

'The Code of Good Practice' is an international reference for setting credible voluntary social and environmental standards. It is referenced by a range of governmental and intergovernmental guidelines as *the* measure of credibility for voluntary social and environmental standards.

This compliance means that the SAN works to create standards that are developed in a transparent, participatory and multi-stakeholder process. Because of the extensive public consultations involved, the process ensures that the standard contains relevant and high-level performance criteria that create genuine social and environmental change. The process includes ensuring that there is a demonstrable need for the standard and includes measures that ensures that even the most marginalised stakeholders have a say in the standard's development (SPS 2007). The public consultation periods incorporate input from SAN members, labour and environmental experts, technical working groups, farmers and environmental and social groups.

11.1.3.2 The standard

The objective of the standard is to provide a measure of each farm's social and environmental performance and best management practices. Compliance is evaluated by audits that measure the degree of the farm's conformity to environmental and social practices indicated in the standard criteria (SAN 2005; see also Table 11.1).

Some of the criteria are critical. A farm must comply completely with all critical criteria in order for the farm to be certified or maintain certification. Any farm not complying with a critical criterion will not be certified, or certification will be cancelled, even if all other certification requirements have been met.

TABLE 11.1 The ten principles of sustainable agriculture according to the Sustainable Agriculture Network

Source: SAN 2005

	Principle	Details	Critical criterion
1	Social and environmental management system	Agriculture activities should be planned, monitored and evaluated, considering economic, social and environmental aspects and demonstrate compliance with the law and the certification standards Planning and monitoring are essential to efficacious farm management, profitable production, crop quality and continual improvement	A chain-of-custody system is necessary to avoid the mixing of products from certified farms with products from non-certified farms
2	Ecosystem conservation	Farmers are to promote the conservation and recuperation of ecosystems on and near the farm	The farm must have an ecosystem conservation programme and the integrity of natural ecosystems must be protected; destruction of or alterations to the ecosystem are prohibited
3	Wildlife conservation	Concrete and constant measures are taken to protect biodiversity, especially threatened and endangered species and their habitats	It is forbidden to hunt, gather, extract or traffic wild animals
4	Water conservation	All pollution and contamination must be controlled, and waterways must be protected with vegetative barriers	The discharge of untreated waste-water into bodies of water is prohibited, and the deposition of solid substances in water channels is prohibited
5	Fair treatment and good conditions for workers	Agriculture should improve the well-being and standards of living for farmers, workers and their families	Farms must not discriminate in work and hiring policies and procedures Farms must pay the legal or regional minimum wage or higher The contracting of children under the age of 15 years is prohibited Forced labour is not permitted
6	Occupational health and safety	Working conditions must be safe, and workers must be trained and provided with the appropriate equipment to carry out their activities	The use of personal protective gear is required during the application of agrochemicals

continued opposite →

	Principle	Details	Critical criterion
7	Community relations	Farms must be 'good neighbours' to nearby communities and be positive forces for economic and social development	–
8	Integrated crop management	Farmers must employ integrated pest management techniques and strictly control the use of any agrochemicals, to protect the health and safety of workers, communities and the environment	Only permitted agrochemicals can be used on certified farms Transgenic crops (GMOs) are prohibited
9	Soil conservation	Erosion must be controlled, and soil health and fertility should be maintained and enriched where possible	New agricultural production must be located on land suitable for that use
10	Integrated waste management	Farmers must have a waste-management programme to reduce, re-use and recycle whenever possible and properly manage all wastes	–

11.1.3.3 Scope and use of the standard

The scope of the standard covers the management of farms of all sizes and includes aspects relating to agricultural, social, legal, labour and environmental issues, in addition to sections on community relations and occupational health and safety. A farm's compliance with the standard is evaluated by observation of agricultural and labour practices and existing infrastructure plus interviews with farm workers and the management or administration team.

One of the objectives of the standard is to make certification more accessible to small farmers. During certification audits, SAN auditors concentrate on physical evidence regarding improvements and best practice 'in the field' so that documentation requirements are reduced. The results of an audit, however, may indicate the need for documentation of procedures, policies and programmes in order to guide and support the implementation of best management practices.

The SAN also has a standard for the certification of groups of organised farmers. In group certification, the administrator may suggest a document management system—policies, procedures, records and other general information—for the group members. In this way, the small producers that are members of the certified group can concentrate on agricultural practices, guided by the group administrator, as well as by the guidelines and documentation provided by the group administrator.

11.1.3.4 Certification

Under the auspices of the SAN, the Rainforest Alliance and partner organisations work with farmers to elevate their operations to the standards required to protect wildlife, wild lands, worker's rights and local communities.

The SAN awards the Rainforest Alliance Certified eco-label to farms, not to companies or products. Farmers may apply for certification for all land in production, and companies may request that all of their source farms be certified. In addition, companies may register with the Rainforest Alliance in order to begin purchasing and selling products as certified. Certification is wholly voluntary; the process begins with an application by the farmer.

Farms in the programme are fully inspected at least once a year by auditors trained by the Rainforest Alliance, and SAN groups regularly make unannounced visits. Farms must demonstrate continued progress in order to maintain their certified status. This continuous improvement is required because the Rainforest Alliance firmly believes that sustainability is never a goal achieved but a goal to constantly work towards and improve.

If it is deemed that a farm does not comply with one of the criteria, the audit team analyses this non-compliance to see whether it is a result of an isolated incident or whether it is because of a lack of a programme, policy, procedure or other element relating to social or environmental management. If the non-compliance is systemic and not an isolated incident, the auditors will conduct a more complete review of available physical evidence, supported by interviews with workers and administrators.

If the required practices of the standard are not implemented, or if elements needed to implement required practices are missing from the social and environmental management system, the audit team will apply a sanction, a non-conformity. The type of non-conformity and corrective action to be carried out by the farm depends on whether the non-conformity is an isolated incident or because of a problem in the farm's management system. In the latter case, the assigned non-conformity will focus on the need to better define and perhaps better document the policies, procedures and programmes needed to ensure conformity.

The audit team scores farm performance according to all of the criteria. In order to obtain and maintain certification, the farms must comply with at least 50% of the criteria for each principle, and with 80% of all criteria. The scoring system guides and encourages the farmers to make continual improvements with regard to all principles and criteria and allows them to compare their performance with that of neighbouring farmers and with farmers in other regions (SAN 2005).

11.1.3.5 Auditors

The credibility and relevance of the Rainforest Alliance relies not only on the standard but also on the high quality of the auditors it uses to conduct the annual farm audits. Rainforest Alliance auditors come from a wide range of technical backgrounds. They are locally based biologists, agronomists, foresters, sociologists and economists with a sustainability mission. They have all gone through rigorous training to understand and audit the SAN sustainable agriculture standard.

11.1.4 Producer partners

Rainforest Alliance certification is a voluntary, non-governmental process that is open to every kind of farm. Producers in the programme can range from large plantations, through mid-sized family-owned estates, to individual smallholder farmers who choose to organise themselves into a cooperative, around a processing mill or in a group with an exporter. Farms of all sizes all over Latin America and increasingly in Africa and Asia are starting to turn towards Rainforest Alliance certification as a driver for change on their farms and in their communities.

In Nicaragua's mountainous region, through sustainable farming practices, Finca La Bastilla perfectly complements and borders a 6,000 ha protected forested area called Datanlí Diablo reserve. Under past owners, high production pushed the soil to its limits, but La Bastilla's current owner took charge in 2004 and raised the farm's practices to Rainforest Alliance's standards for sustainability. La Bastilla is 350 ha, with a third of that area under conservation. It employs 60 people full time, but swells to a workforce of 700 during the coffee harvest. As of 2008, farm workers, their children and residents of neighbouring villages can earn up to a 10th grade education at the farm's own school. Some 250 students are currently attending, up from only 20 in 2005. The farm's ecological mill recycles water, using much less than conventional coffee mills. New workers' houses have plumbing and community recreation areas. A full-time nurse is on site, and a doctor visits twice a week to treat workers free of charge. And, ultimately, good practices lead to high quality: La Bastilla's coffee ranks very high in international cupping events (Rainforest Alliance 2007a).

Today, Colombia is one of the world's leading producers of high-quality coffee. Thanks to a collaborative effort between the Rainforest Alliance, the Colombian Coffee Federation and the Rainforest Alliance's Bogotá-based SAN partner, Fundación Natura, Grupo Kachalú became Colombia's first producer group to achieve certification. Today 110 small farms are associate members. All of them have been certified by the Rainforest Alliance for meeting a rigorous host of standards that protect the environment and the rights and welfare of workers and their families.

The Kachalú farms surround a forest preserve that is managed by Fundación Natura and is a last redoubt of an endangered species of Andean oak. Eager to be good neighbours to the park, the farmers entered their lands into the Rainforest Alliance certification programme. The farms themselves provide habitat for a rich diversity of birds and wildlife. Studies conducted by the National Coffee Research Centre in 2004 found 96 bird species in the Kachalú farm region, a population that the centre attributes to the wide variety of native trees that shade the coffee, including ingas, eritrynas, cedars, ficus, albizzias and tubebuias, among others (Rainforest Alliance 2007b).

The Ciudad Barrios Cooperative, in the Cacahuitique mountains of eastern El Salvador was founded in the late 1970s and boasts more than 1,000 participating members. The cooperative produces approximately 8.5 million pounds of coffee annually. Until 2003, all the coffee was sold as low-priced, generic coffee. The cooperative gained attention when it was recognised for its environmentally friendly practices and good workplace and social standards by being awarded Rainforest Alliance certification. Through sales of certified coffee, primarily to Kraft Foods, the cooperative has earned an average of 15 cents per pound above market price for coffee. The cooperative has used its profits to invest in its community. It has built and inaugurated a school for 125 students, most of whom are children of cooperative members and employees. They are also building a recreation centre that will be open to all Ciudad Barrios residents. Since the coop recognises that small coffee farmers are vulnerable to coffee price fluctuations, it is also investing in demonstration plots of papaya, citrus and vegetables to encourage agricultural diversification. It is considering investing in a health clinic and an eco-tourism venture. The goal is to create an environment where farmers will no longer be dependent solely on coffee for income. By reaching out to more profitable markets and using profits to reinvest in the community, members of the Ciudad Barrios Coffee Cooperative are setting an example for themselves, their children and coffee growers throughout the country (USAID 2008).

11.1.5 Business partners

Almost without exception, industries are now facing an unavoidable new array of environmentally driven issues. This new 'green wave' presents an unprecedented challenge to 'business as usual'. Actors other than the businesses themselves are now playing a prominent environmental role on the business stage. NGOs, customers and employees increasingly ask pointed questions and call for action on a spectrum of issues (Etsy and Winston 2006). Businesses involved in all parts of the agricultural supply chain, from south to north, have joined the Rainforest Alliance in order to transform their buying practices to reward more sustainable farming.

Kraft Foods, an international food and beverage company and one of the top four buyers of coffee in the world, committed to buying five million pounds of coffee from certified farms in 2003, an amount that was unprecedented at the time in terms of purchasing ethical coffee. The following year, it doubled its purchasing and has been increasing the amount since according to market demand, which has been growing quite rapidly. Kraft also partnered with the Rainforest Alliance to train and provide assistance for coffee farmers in an effort to help them reach certification standards. Kraft is blending Rainforest Alliance certified coffee into several of its popular coffee brands in the USA and Western Europe, including Yuban, Carte Noire, Jacques Vabre, Maxwell House, Jacobs, Kenco and Gevalia.

Chiquita, the international marketer, producer and distributor of fresh fruit was the first company to join forces with the Rainforest Alliance Sustainable Agriculture Programme, agreeing in 1990 to pilot the then newly developed sustainable agriculture standards on two of their banana farms in Central America (Taylor and Scharlin 2004). The certification programme grew and in 2000 Chiquita had successfully achieved certification for all the company-owned farms and promptly set about encouraging independent suppliers to pursue certification as well, by leading by example and paying a price premium per box of certified bananas. Today over 90% of its independent suppliers have a Rainforest Alliance certificate as well. In 2005 Chiquita started communicating and marketing its Rainforest Alliance certification by putting the frog trust mark—the Rainforest Alliance Certified seal of approval—on its bananas in nine countries in Europe.

Unilever, the Netherlands-based international manufacturer of leading brands in food, home care and personal care, made a significant commitment to Rainforest Alliance certification in May 2007. Although the Rainforest Alliance was only just starting to work in tea at the time, Unilever announced that it would be committing to purchase all its tea from sustainable, ethical sources. Lipton, the world's best-selling tea brand, and PG Tips, the UK's No. 1 tea, will be the first brands to contain certified tea. The company announced that it aimed to have all Lipton Yellow Label and PG Tips teabags sold in Western Europe certified by 2010 and all Lipton teabags sold globally by 2015 (Unilever 2007).

However, the Rainforest Alliance does not work only with large international companies. Licensed business partners range from small single-brand family-owned coffee roasters to boutique fruit juice companies and highly specialised chocolate enterprises.

11.1.6 Public–private partnerships

A successful model for funding work in the field has been leveraging the power of public–private partnerships (PPPs). These partnerships, with government funding and corporate sponsorship working together to assist in implementing and spreading Rainforest Alliance certification among farmers, have been key in tying together work 'in the field' and more sustainable marketplaces in buying countries in Europe, the USA, Japan and Australia.

The PPP movement is growing rapidly. There have been successes and learning experiences. With the decline in nation-state regulatory power, increasing demands for business to be more accountable amid almost routine corporate scandals, and the growing reach and power of civil society groups, stakeholders and stockholders there is little doubt that partnerships will continue to shine a light forward in the quest for sustainability (Wille 2004).

11.1.6.1 Biodiversity in the coffee conservation project

Because of the global economic importance of coffee and its potential as a conservation tool, the Global Environmental Facility, through the United Nations Development Programme, has provided the Rainforest Alliance with a seven-year grant to develop sustainable production in several countries in Central America and give them market access. A goal of the programme is to certify 10% of the world's coffee supply.

The Global Environment Facility funds go towards increasing both production and demand for coffee certified by the Rainforest Alliance. To increase demand, the Rainforest Alliance is undertaking a concerted marketing effort in collaboration with coffee companies all along the supply chain, encouraging them to buy certified coffee and helping them to promote the product by using the Rainforest Alliance certified seal.

On the ground, the Rainforest Alliance is providing Latin American farmers with the information and tools necessary to improve their management practices and make them more responsible with regard to the environment, workers and communities.

11.1.6.2 The Amazon Basin Conservation Initiative

To support the expansion of sustainable business in the Amazon, the US Agency for International Development (USAID) has funded the Amazon Basin Conservation Initiative (ABCI), a five-year, US$65 million programme involving more than 30 different organisations. The ABCI is one of the Amazon's first region-wide, multiple-country conservation initiatives. The Rainforest Alliance's role in this project is to expand the adoption of certification as a conservation tool, working in collaboration with partner groups Fundación Natura in Colombia and Conservación y Desarrollo in Ecuador.

11.1.6.3 Cocoa work in Côte d'Ivoire

In 2006 the Rainforest Alliance launched a programme in Côte d'Ivoire, the world's largest producer of cocoa, to meet the SAN standard for responsible cocoa farming. In order to accomplish this, the Rainforest Alliance is teaming up with USAID, Kraft Foods, the German Agency for Technical Cooperation (GTZ [Gesellschaft für Technische Zusammenarbeit]) and a cocoa trader in Côte d'Ivoire called the Armajaro Group. The

FIGURE 11.1 Participation in the Rainforest Alliance certification programme as a percentage of total crop produced (by area), for each crop in the areas certified

Rainforest Alliance 2008

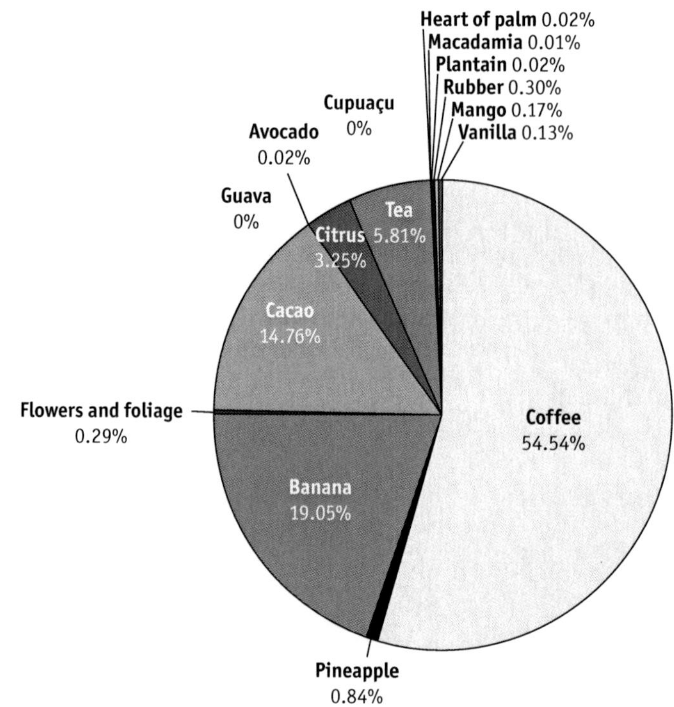

FIGURE 11.2 Total production area certified and the total area of farmland when forest refuges, conservation lands, protected zones, areas in infrastructure and other areas are included (entire bar indicates total area; black shading indicates production area)

Soucre: Rainforest Alliance 2008

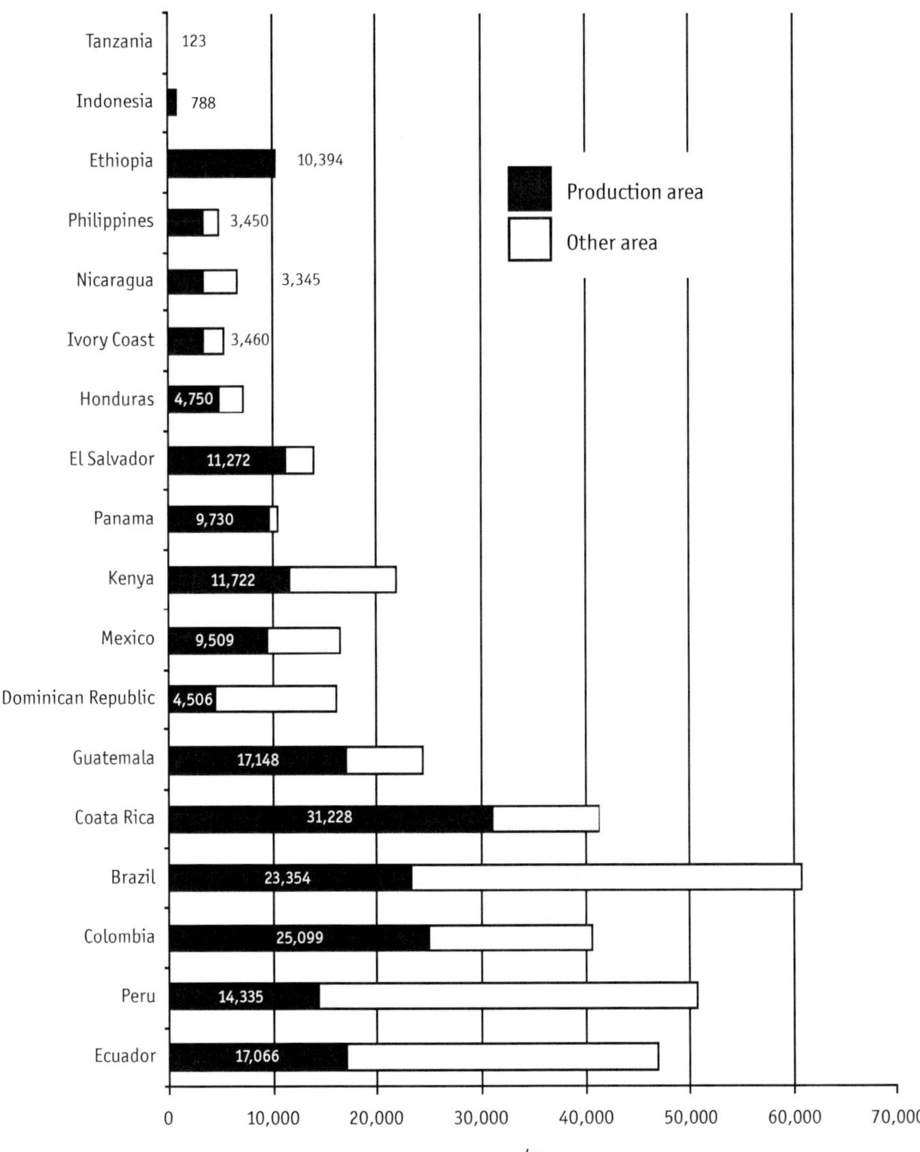

FIGURE 11.3 Total area certified, in hectares: total production area and other areas (such as conservation areas and area occupied by infrastructure) for each crop, as of 9 January 2008

Rainforest Alliance 2008

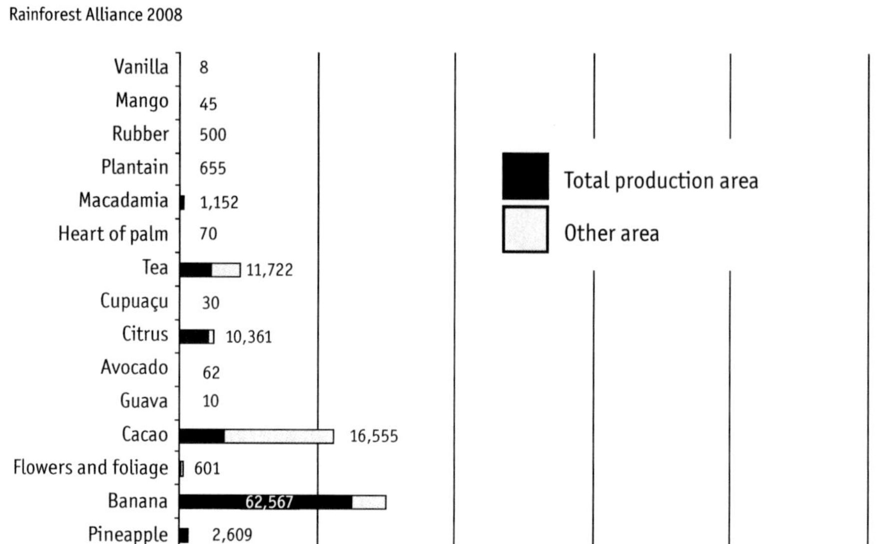

partners in the project coalition are investing nearly US$2 million to benefit the cocoa farmers, guiding them toward self-sufficiency and sustainability.

The Rainforest Alliance and its partners will provide guidance and technical assistance to about 4,000 farmers in six cooperatives in Côte d'Ivoire over three years to encourage sustainable cocoa production. Côte d'Ivoire produces about 40% of the world's cocoa, more than any other country. The country is facing the aftermath of a civil war, with land disputes ongoing and foreign mediators overseeing a peace plan. Cocoa farmers face a host of additional constraints, including poor soils, ageing cocoa trees, inefficient management practices, low productivity and lack of marketing information. The project aims to show cocoa farmers that sustainable practices can improve their productivity and their livelihoods while protecting the environment.

11.1.7 The scope of the Rainforest Alliance's work

As of 1 January 2008, some 23,921 farms, large and small, have been certified to the sustainable agriculture standard, representing 354,790 ha in 18 countries. Roughly half of that area is conservation area and the other half productive area producing crops such as coffee, bananas, cocoa, tea, citrus fruits, pineapples, avocados, hearts of palm, mangoes, flowers and foliage, guavas, macadamia nuts, rubber, cupuaçu, plantains and vanilla. In all, 1.5 million people benefit directly from the programme.

Sales of Rainforest Alliance certified coffee, bananas and chocolate exceeded US$1 billion in 2006, and the growth is projected to continue (Rainforest Alliance 2007c). The volume of sales of coffee from Rainforest Alliance certified farms has almost doubled each year since 2003, from 7 million pounds to 54.7 million pounds in 2006; sales in 2007 are projected to exceed 91 million pounds.

11.2 Benefits on the ground and for the people

11.2.1 Economic benefits

Certification is an incentive to make social and environmental improvements, but implementation of the programme also helps farms to reduce their production costs, improve the quality of the crop and increase productivity on the farm. The process of certification leads to more efficient farm management as farmers organise, plan, schedule improvements, implement better practices, identify problems and monitor progress. For large farms that have large workforces, they gain an advantage by giving their workers a wholesome, safe and sanitary place to work, reducing worker complaints and increasing worker efficiency.

Farms that meet the certification standards earn the right to sell their produce with the frog trust mark, which is a competitive advantage in the global marketplace. The seal of approval also gives them more leverage at time of sale, with premium prices and access to credit.

The objective of the Rainforest Alliance is that farmers get a sustainable price, one that covers the cost of efficient production and the social and environmental investments required by the programme, including paying proper wages to workers. The farmer should earn enough to make long-term investments and have some savings.

Higher prices are important, and most farmers in the Rainforest Alliance programme are getting significantly higher prices for their goods. But farmgate prices are not a panacea. Farmers can earn high prices but still have failing businesses. Successful farmers learn to control costs, increase production, improve quality, build their own competence in trading, build workforce and community cohesion and pride, manage their precious natural resources and protect the environment.

In the holistic approach of certification, farmers learn to navigate the global marketplace; the programme can help farmers who need it, but it is not a 'handout'. Certification and better farming practices give farmers more control over their futures. It empowers them to be better business people and to dream of a sustainable successful future. Many of the 1,500 cocoa farming families in the programme in Ecuador have doubled their income through a combination of increased production and higher prices.

Farmers in the programme in Colombia, with a total of 90,000 bags of coffee from certified farms, are selling their crops at premium prices. In addition, they have more to sell, as farmers are producing as much as 20% more coffee per hectare while controlling costs. Overall, the programme has meant US$1.6 million more income for Colombian coffee farmers.[1]

Over 10 years, with all its plantations involved in the programme, Chiquita has seen farm productivity rise by 27% and costs decline by 12%, including losses incurred by natural disasters, such as Hurricane Mitch in 1998.

11.2.2 Social benefits

Implemented standards verified by third-party certification leads to improved conditions, fair treatment and assured wages for workers. Workers, both seasonal and full time, are fully informed of their rights, with transparent and written hiring contracts and no discriminatory policies. Certified farms hire locally, pay fair wages and ensure safe working conditions and access to clean drinking water and proper sanitary facilities. In all, more than 200,000 farm families are enjoying the benefits of the programme.

Certified farms benefit the whole community, as farms join in the conservation and management of shared resources such as streams, watersheds and woodlots. Well-managed farms have a reduced environmental footprint and are good neighbours to communities and wild lands. Forested farms often serve as a resource for farm workers and neighbours, providing firewood, fruits, fibres, medicinal plants and so on. They serve as educational resources, as farmers and farm workers work on training and environmental education, and children go to school.

11.2.3 Environmental benefits

Certified farmers plant trees along roads, around housing and in areas not suitable for crops. To control erosion and to limit the need for agrochemicals, they also plant buffer zones of native vegetation along rivers and springs and allow ground cover to grow. And they manage all pollutants, from tractor fuel to coffee-mill waste-water.

Certified farmers also work on sources of contamination to fight against water pollution. They control pesticides and fertilisers, sediment, waste-waters, rubbish, fuels and so on to keep their watersheds clean. As well as protecting the water, they also use less of it, as water conservation measures are applied in washing and packing stations, housing areas and irrigation.

They experience less soil erosion, because farms implement soil conservation practices such as planting on contours and maintaining ground cover.

They reduce threats to the environment and human health as the most dangerous pesticides are prohibited, all agrochemical use is strictly regulated and farmers must demonstrate continual reductions in agrochemical use with the goal of eliminating them completely.

1 This data was gathered by the Colombia Coffee Federation as part of its routine farm surveys and confirmed by Rainforest Alliance President Tensie Whelan during a fact-finding mission where she interviewed dozens of farmers. For more information about the science and survey programmes of the Federation, see www.federaciondecafeteros.org.

More wildlife habitat is protected as deforestation is stopped, the banks of rivers are protected with buffer zones, critical ecosystems such as wetlands are protected and forest patches on farms are preserved.

The certified farm produces less waste, as farm by-products such as banana stems, coffee pulp, orange peels and un-marketable foliage is composted and returned to the fields as natural fertiliser. Other waste, such as plastics, glass and metals are recycled where possible.

The environmental benefits lead full circle to economic benefits as well. Research on Costa Rican coffee farms has shown that shade often creates more favourable microclimatic conditions for coffee cultivation by reducing coffee heat stress and enhancing coffee growth and productivity with a proper shade level in the range of 20–40% (Vaast *et al.* 2005) exactly within the range required by the SAN standard.

Ultimately, certification leads to more collaboration between farmers and conservationists. Parks alone cannot save the world's biodiversity. Concerned citizens have to ensure that wild flora and fauna find refuge outside of protected areas. Because farmers control the fate of so much land and so many critical habitats, their ideas and willing participation are essential to any local or regional conservation strategy.

11.3 Overall conclusions and outlook

In conclusion, the Rainforest Alliance programme proves that certification and sustainability can become mainstream without compromise, addressing predominantly mainstream markets yet affecting real change 'on the ground', evidenced by the participation of several leading international companies dealing in some of the world's essential food commodities and by the positive impact on farmers in Latin American and Africa.

The work of the Rainforest Alliance could be scaled up significantly as its experiences are transferable to new commodities and countries. Market demand is growing into the area of various tropical fruits, vegetables and biofuels, and, in response, the SAN is expanding its work to other continents, commodities and products. Drawing from 15 years of experience, SAN recommended best management practices are possible and applicable in new farms, given the development of appropriate local indicators.

The Rainforest Alliance believes that mission-driven, third-party, independent, voluntary, non-governmental certification is a concrete measuring tool that benefits farmers, companies and consumers. Other labelling schemes and fellow ISEAL Alliance members along with the Rainforest Alliance—such as organic labelling by the International Federation of Organic Agriculture Movements (IFOAM) and fair-trade labelling by Fairtrade Labelling Organisations International (FLO)—are complementary rather than competing approaches. Recognising that standard-setting organisations that comply with ISEAL guidelines can all have different approaches, merits and rights to existence is the work spirit in which organisations hope to be of the greatest benefit to farmers. Despite impressive growth, the majority of farmers globally have not yet had access to the benefits of working towards a certification scheme: ISEAL members have much work left to do. The circle of sustainability is a circle we must expand, not compete within.

Governments increasingly are recognising the positive link certification makes between sustainable consumption and production. In order to enable credible programmes to scale up their work, recognition in public procurement policies will greatly enable members of the public to increase their understanding and appreciation of the work and its impact. These purchasing policies should recognise that there is more than one path to sustainability, and look to ISEAL as a guide in terms of credibility for setting standards for purchasing guidelines.

The Rainforest Alliance encourages more companies to join this global effort of helping farmers to pursue certification as a concrete measuring tool towards a more sustainable supply chain. The Rainforest Alliance needs the increased support of such companies, as well as funding organisations willing to create and enter into public–private partnerships. The market demand from consumers, governments and stakeholders is growing faster than the farmers and even companies can keep up with. SAN partners need support to scale up capacity building and technical assistance training so that producers can get the assistance they need to convert to best management practices.

With the concrete support of consumers, companies and governments, the Rainforest Alliance and its partners can continue amplifying their work in making the link between sustainable production and consumption.

References

Etsy, D., and A. Winston (2006) *Green to Gold: How Smart Companies Use Environmental Strategy to Innovate, Create Value and Build Competitive Advantage* (New Haven, CT: Yale University Press).

ISEAL Alliance (2004) 'ISEAL Code of Good Practice for Setting Social and Environmental Standards', Version 3, January 2004, www.isealalliance.org, accessed 13 July 2009.

Rainforest Alliance (2007a) 'Profiles in Sustainable Agriculture: Nicaragua, Finca La Bastilla'; www.rainforest-alliance.org/profiles.cfm?id=agriculture#coffee, accessed 13 July 2009.

—— (2007b) 'Profiles in Sustainable Agriculture: Kachalú Coffee Farmers Conserve the Forest and Wildlife on Colombia's Highlands'; www.rainforest-alliance.org/profiles/documents/KachaluProfilewithhorizon.pdf , accessed 13 July 2009.

—— (2007c) 'Press Release: Rainforest Ecosystems and Sustainable Livelihoods Secured as Sales of Rainforest Alliance Certified Products reach $1 Billion Annually'; www.rainforest-alliance.org/news.cfm?id=uk_sustainability, accessed 13 July 2009.

—— (2008) www.rainforest-alliance.org; accessed 13 July 2009.

SAN (Sustainable Agriculture Network) (2005) 'Sustainable Agriculture Standard with Indicators', November 2005; www.rainforest-alliance.org, accessed 13 July 2009.

SPS (Standards and Policy Secretariat) (2007) 'Standards and Policy Development Handbook', Sustainable Agriculture Network, www.rainforest-alliance.org, accessed 13 July 2009.

Unilever (2007) 'Press Release: Unilever Commits to Sourcing all its Tea from Sustainable Ethical Sources'; www.unilever.com/ourcompany/newsandmedia/pressreleases/2007/sustainable-tea-sourcing.asp, accessed 13 July 2009.

USAID (2008) 'Success Story: Cooperative Invests Coffee Revenue back into Community: Better Coffee Yields Social Gains'; www.usaid.gov/stories/elsalvador/ss_elsalvador_coffee.html, accessed 13 July 2009.

Taylor, J., and P. Scharlin (2004) *Smart Alliance: How a Global Corporation and Environmental Activists Transformed a Tarnished Brand* (New Haven, CT: Yale University Press).

Vaast, P., R. Van Kanten, P. Siles, B. Dzib, N. Franck, J.M. Harmand and M. Genard (2005) 'Shade: A Key Factor for Coffee Sustainability and Quality', *Journal of Science of Food and Agriculture*: 887-96.

Wille, C. (2004) 'Certification: A Catalyst for Partnerships', *Human Ecology Review* 11.3.

12
Life events as turning points for sustainable nutrition

Martina Schäfer and Adina Herde
Technische Universität Berlin, Germany

Cordula Kropp
University of Applied Sciences, Germany

Worldwide, food is the most essential product for daily consumption. The production, processing and consumption of food, however, leads to much negative impact on the natural environment: for example, through the use of fertilisers and pesticides or the generation of packaging and waste (EEA 2005). Jongen and Meerdink (1998) estimate that close to half of human's impact on the environment is directly or indirectly related to food production and consumption. In Germany, the share of the food chain in energy and material consumption is about 20%, and it is responsible for 16% of greenhouse-gas emissions (Wiegmann *et al.* 2005). Yet, in addition to these ecological effects, food and its quality obviously have great relevance for human health.

A transformation to sustainable food consumption, therefore, is essential for sustainable development. Currently, no scientific consensus on the definition of 'sustainable nutrition' exists. However, in Germany, many authors refer to the definition of von Koerber *et al.* (2004), which stresses the following aspects as being most important for sustainable nutrition:

- Enjoyable and easily digestible foods
- Preferably plant-based food
- Preferably minimally processed foods

- Organically produced foods
- Regional and seasonal products
- Products with environmentally sound packaging
- Fair-trade products

Particularly in light of these criteria, present food consumption patterns cannot be called 'sustainable', as they endanger not only the carrying capacity of the Earth but human health as well. Therefore, more future-oriented alternatives have to be found and put into practice. In addition to producers, processors and traders, individual households have many opportunities to contribute towards sustainable nutrition through their consumption patterns (Spangenberg and Lorek 2001).

To get closer towards the goal of achieving sustainable development, it is necessary for the majority of people in industrial countries to change their food consumption habits. Nevertheless, it has to be taken into consideration that nutrition is a strongly ritualised and repetitive behaviour. Most decision-making in this domain follows routinised low-involvement patterns of activity that can be characterised as highly habitual behaviour with little cognitive or reflexive control (Kroeber-Riel and Weinberg 2003). These routines help in dealing with daily demands (Dahlstrand and Biel 1997; Ilmonen 2001) and create stability and security in a rapidly changing world.

Life events bring with them opportunities to change unfavourable food consumption habits, because often in these situations the need to change routines arises anyway. Among other such important life events that may prompt changes towards sustainable nutrition are the onset of a serious disease, the birth of a child, retirement or even a food or environmental scandal.

Up to now, there has not been much research undertaken in this field. Only a few types of life events have been analysed with any degree of thoroughness. In this chapter, the results of two research projects in this field will be presented. The first project, 'sustainable nutrition in the transition to parenthood', was carried out in Berlin, Germany (Herde 2007; Herde and Schäfer 2006). The second analysis, focused on 'nutritional careers', was carried out as part of a three-year project, looking at 'The Turn-around in German Agrarian Policy: New Forms of Food Consumption?', conducted in the cities of Leipzig and Munich, Germany (Brunner *et al.* 2006). Both research projects had the goal of seeking out empirical evidence for the thesis that life events bring about opportunities for changes towards sustainable food consumption habits.

The research in Berlin concentrated on the effects of the birth of a child and looked for changes concerning ecological, social, economic and health-related aspects of buying, consuming and disposing of food. The second project dealt with four different life events—the birth of a child, the onset or discovery of serious disease, retirement and food scandals. The focus of the analysis was whether these events prompted a turn towards increased consumption of organically produced foods.

12.1 Case description

12.1.1 Context

12.1.1.1 Nutritional trends

The production and consumption of food is influenced by various global trends. Conventional agriculture worldwide is still becoming more intensified, characterised by greater use of synthetic fertilisers, pesticides and technical devices and increasing average farm size. Intense forms of agricultural production have a negative impact on the biodiversity of plants and animals, on soil fertility and on the quality of groundwater as well as surface water (e.g. rivers and lakes; Jungbluth and Emmenegger 2002).

Also, globalisation is an ongoing trend, with major effects concerning food production and trade. With the consolidation of the industry through mergers and takeovers, multinational food manufacturing companies have an increasing influence, both upstream on agriculture and downstream on retailers and consumers. The globalisation of the food-processing industry is resulting in a homogenisation of the range of food products available throughout the world, especially for affluent consumers. Global branding, marketing and advertising contribute to the development of a common language or value system linked to consumption. An obvious dimension of the globalisation of food supply chains, moreover, is the expansion of the food trade and the growing need for the transport of agricultural goods (Michaelis and Lorek 2004). Because of this trend, regionality and seasonality of food are losing importance (Marshall 2001). There is a tendency toward the consumption of highly processed foods (fast and convenience foods, etc.) and use of more appliances in the kitchen, accompanied by decreasing knowledge about nutrition and food (Davies 2001; Swoboda and Morschett 2001).

More than 200 million adults in the European Union are overweight or even obese (CEC 2005), for reasons such as lack of exercise and the consumption of more food energy than is physiologically required (Michaelis and Lorek 2004). Half of women and two-thirds of men in Germany suffer from being overweight (Laberenz *et al.* 2006). Obesity in German children is the most frequently occurring disease related to nutrition (Kroke *et al.* 2004). Some common consequences of being overweight include high blood pressure, coronary heart diseases, certain types of cancer, diabetes mellitus type II, strokes, alcoholism, tooth decay and osteoporosis (Dibsdall *et al.* 2003). However, a segment of consumers are becoming more aware of health-related issues and are increasingly concerned about the nutritional content and functional value of their food (Michaelis and Lorek 2004).

12.1.1.2 Political measures

For decades, nutritional politics was of much less importance than agricultural politics. Therefore, the economic interests of the agricultural sector were for a long time more dominant than consumers' interests (Meier-Ploeger 2005). A fundamental 'reorientation and change' in German agricultural policy was proclaimed in reaction to the BSE crisis in 2001. As a result of these changes, consumer protection was supposed to become more relevant. One of the most important goals of the 'new agricultural politics' was to increase the percentage of organically cultivated area to 20% of all farmland by the year 2010. The state also aimed at adopting a new approach to agricultural politics: from the 'iron triangle' (agricultural politics, agricultural interest groups and agricultural administration) to the 'magic hexagon', comprising consumers, farmers, the fodder industry, the food industry, retailers and the state. This change of political orientation in the years 2001 to 2005 demonstrated some degree of success concerning sustainable food production and consumption. Since 2004, the demand for organic products in Germany has been growing to the extent that partial supply deficits can be observed (Rippin *et al.* 2006).

With the change of government in 2005, a programmatic reorientation in agricultural politics took place, now aiming at competitiveness on the global market. There is no longer a preference for organic agriculture, and genetic engineering, once a subject of controversy, is now seen as a future-oriented technology. The impacts of this recent political reorientation on transformation towards sustainable nutrition cannot yet be foreseen.

12.1.1.3 Challenges for nutritional communication

Communication regarding the changes in food consumption that are necessary to contribute to sustainable development faces many challenges, especially as there is no consensus concerning what the basic elements of sustainable nutrition should include. One problem with communication in this field is that it usually tries to address everyone and does not focus on selective target groups. Mass communication has not had the expected results yet because often it does not link to everyday contexts, nor does it differentiate between different lifestyles or individual situations (Lehmann and Sabo 2003). In contrast to mass communication, personal consulting allows individual problems and questions to be addressed as well as enabling the discovery of individual solutions that go beyond general advice. The classical nutritional counsellor or consultant performs an important service; however, the majority of the population cannot be reached through such means, and such services often have to be paid for out of the individual's pocket. Moreover, consulting services tend to overestimate a particular model of a rationalised lifestyle, typical only for well-educated middle-class people. Although 85% of those surveyed by Lappalainen *et al.* (1998) trust the competence of nutritional consultants, such services are generally used only in cases when severe health problems owing to unbalanced nutritional habits already exist (Keane 1997). Another problem is that nutritional consultants are not usually able to observe the food consumption habits of individuals over a long period of time (Jahnen 1998).

Another reason for the failure of many nutritional awareness campaigns is the concentration on a purely cognitive transfer of nutritional knowledge that is based on results of natural science research on the energy and nutrient content of foods. The compatibility of the given advice with the daily life requirements and culturally influenced eating styles of most people are often neglected (Rehaag 2005; Rehaag and Waskow 2006; Wilhelm and Kropp 2007).

12.1.2 Characteristics of the sample

For the project concerning sustainable nutrition in transition to parenthood, a quantitative and a qualitative analysis was carried out with first-time parents who were either still expecting their child or whose child was not older than three years. Answers to 286 questionnaires were analysed through quantitative methods, such as comparisons of means (paired-samples *t*-tests, etc.). The sample contained more women, persons with higher educational levels and higher household incomes when compared with the average population in Berlin. In addition, the number of immigrants is much lower in the sample than in the population as a whole. Consequently, because of the limited representativeness of the sample, the survey has an explorative character and should not be seen as representative data.

The qualitative survey consisted of five focus groups with 17 participants in total. Again, women and persons with higher educational levels were overrepresented, and immigrants were underrepresented. The qualitative data was analysed through the cut-and-paste technique (Lamnek 2005).

For the project concerning 'nutritional careers', the following research instruments were used: (a) a representative telephone survey with 500 consumers, (b) qualitative in-depth interviews with 140 consumers and (c) workshops with consumers, nutritional consultants and farmers. The quantitative sample included persons over the age of 16 years living in Munich and Leipzig. A comparison with the average composition of the population in these cities showed that people with a higher level of education were overrepresented. In Leipzig, the percentage of unemployed was higher than in the general population. The participants for the qualitative survey were selected according to the model of theoretical sampling; women, consumers of organic food and parents of little children were overrepresented. Participants for the focus groups were selected on scientific criteria: the main goal was to achieve a good mixture of different lifestyles, even within the group of environmentally aware persons.

Both surveys are based on self-reported data, which can be less objective than observations or, for example, detailed study of daily purchasing and eating habits.

12.2 Results

12.2.1 Main results

12.2.1.1 Parenthood

The results of Herde (2007) suggest that the transition to parenthood is a sensitive biographical period. Several statements within the focus groups made it clear that parents tend to become very interested in information about nutrition during this phase. However, some of them also feel stressed by the large amount of information available and would welcome the possibility of receiving personalised advice.

The quantitative survey shows that the parents also become more interested in 'sustainable food': they purchase a higher percentage of organic, seasonal, regional and fresh food and care more about buying products that are not genetically modified (Fig. 12.1).

FIGURE 12.1 Consumption of 'sustainable food': statistical analysis of replies to the question 'How often do you pay attention to the following food characteristics: fresh, regional, seasonal, organic and genetically unmodified?'

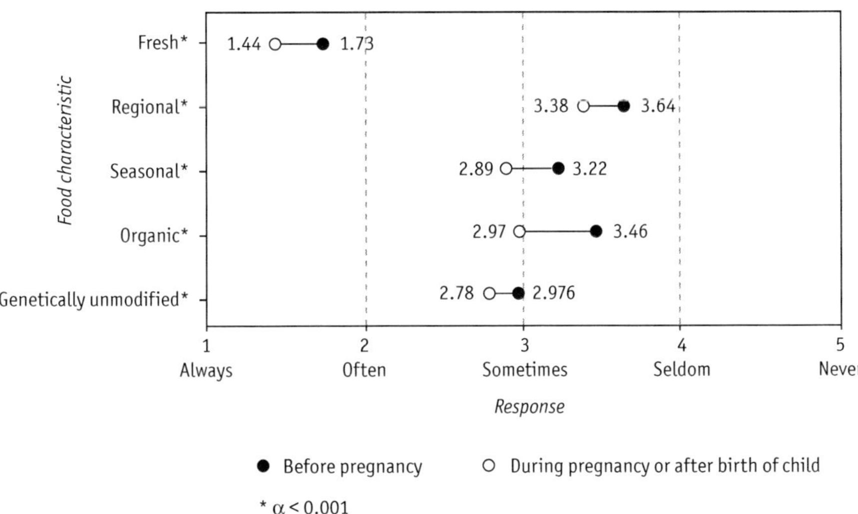

Connected to the higher interest in these types of products is the fact that organic food stores, health-food stores and food cooperatives were visited with a higher frequency than previously. Nevertheless, both normal and discount supermarkets continue to be the places where most food is bought (Fig. 12.2).

The composition of the diet tends to improve considerably during this process: not only from the health perspective but also from an ecological point of view. Unprocessed foods such as fresh fruit and vegetables, wholegrain products and plain yoghurt are consumed more, whereas the consumption of less healthy and highly processed products such as soft drinks, alcohol, sweets and fatty snacks decreases (Table 12.1).

Socio-demographic characteristics have an important influence on the impact of life events. In the project concerning sustainable nutrition in the transition to parenthood, the factors analysed were education, income and neighbourhood where participants live. The results suggest the following impacts.

- **Education.** Parents with higher levels of education buy organic, regional and seasonal food at a higher frequency than parents with lower educational levels. There are also differences regarding food composition. Parents with lower educational levels consume more products that have negative environmental

FIGURE 12.2 Shopping facilities reported to be visited, in response to the question 'Where do you buy your food?'

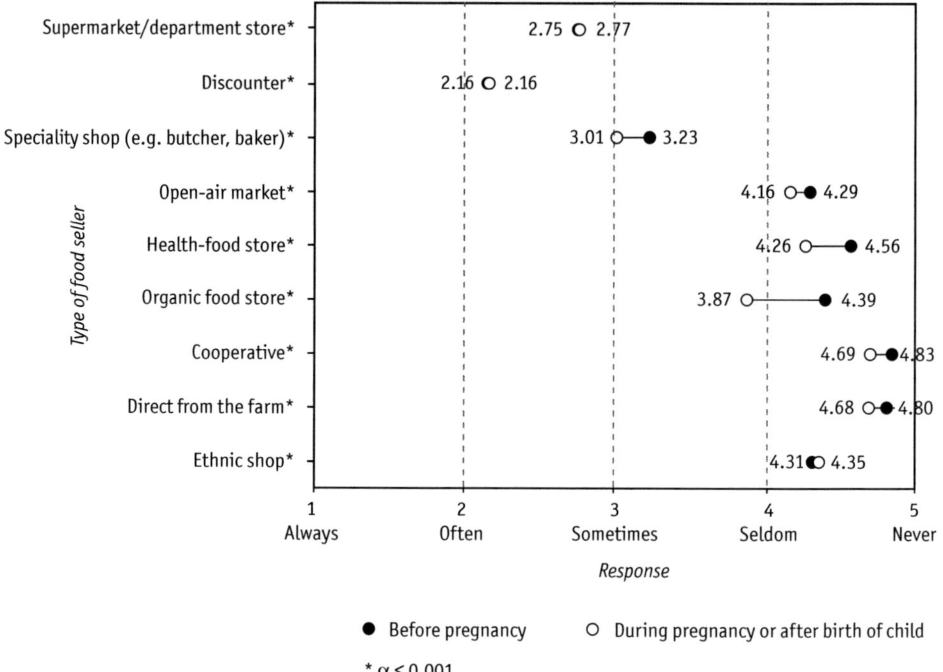

TABLE 12.1 Consumption of products: answers in response to the question 'How often do you eat or drink . . . ?'

	Mean[a]	
	Before pregnancy	During pregnancy or after birth of child
How often do you eat or drink . . . ?		
Fresh fruits**	2.41	1.86
Fresh vegetables**	2.88	2.48
Wholegrain products**	2.69	2.37
Plain yoghurt**	5.00	4.61
Lemonade**	4.86	5.24
Alcohol**	4.99	5.70
Sweets, fatty snacks*	3.80	3.97

[a] Scale from 1 to 7 (1 = more than once a day, 2 = [almost] daily, 3 = more than once a week, 4 = about once a week, 5 = two to three times a month, 6 = once a month or less, 7 = never)

* $\alpha < 0.01$ ** $\alpha < 0.001$

impacts (meat, sausages, conserved vegetables) and eat less minimally processed foods (fresh milk, cheese, plain yoghurt, water) (see also Guenther *et al.* 2005; Hart *et al.* 2006; Torjusen *et al.* 2004). They also exhibit a significantly higher price awareness regarding food items

- **Income.** In contrast to education, income does not have any impact on the preference for more sustainable products or a healthy and 'sustainable' composition of food (see also Cicia *et al.* 2002; Lockie *et al.* 2004). Differences in the readiness to pay higher prices already existed before pregnancy and remained stable during the transition to parenthood
- **Neighbourhood.** The results also show that the neighbourhood in which participants live has significant impacts on food purchasing habits (see also Block *et al.* 2004; Jetter and Cassady 2006; Weiß 2006). If organic food stores, health-food stores or open-air markets are available in the area, these facilities are used with a higher frequency than by those living in other districts without a similar supply of such shopping outlets. Concomitant with use of these facilities, the consumption of organic, seasonal and genetically unmodified products increases. However, there is no significant impact on food composition

There are few changes regarding the economic aspects of sustainable nutrition. With the birth of a child, the parents show slightly higher awareness concerning fair prices for 'third-world products' and farmers in general. Still, almost half of the parents from the sample exhibit a low level of awareness about this issue (Table 12.2).

The project concerning nutrition careers also found a higher level of awareness about and interest in nutrition accompanying the transition to parenthood. In many cases, however, the nutritional well-being of the child is of great importance whereas the parents' food consumption habits are neglected. The characteristics that are most impor-

TABLE 12.2 Willingness to pay for food

	Mean[a]	
	Before pregnancy	During pregnancy or after birth of child
I am very concerned about keeping my food bill low*	2.89	3.07
I cannot afford to pay more for higher-quality products**	3.59	3.65
It is very important to me that the farmers receive a fair price for their products**	3.52	3.40
When I buy products from developing countries, it is important to me that they have a fair-trade background**	3.74	3.63

a Scale: 1 = very applicable, to 6 = not applicable.

* $\alpha < 0.01$ ** $\alpha < 0.001$

tant to parents regarding child nutrition are fresh products that contain many nutrients and no toxins as well as recognising the preferences of the child. Brunner *et al.* (2006) identify three possible types in the period of transition to parenthood:

- **'Everything natural'.** Parenthood is seen as the main task for this type. Nutrition is a major factor. There is a general orientation towards 'natural' products and preparing them at home. However, this orientation is not automatically linked to a higher demand for organic products.
- **'I have to be completely sure!'** For this type, the responsibility regarding food safety for the child and the parent are central. There is a high degree of sensitivity regarding nutritional risks. Organic products are of special interest; however, higher prices and reduced availability are often barriers to organic consumption. Regional products are sometimes bought instead of organic products. This type dominates in households with one child or with one parent.
- **'Overstressed'.** This type is convinced that the quality of food is guaranteed by existing laws and state controls. Limited time and financial resources are the reason for preferring cheap and convenience products. There are possibilities for a change to organic convenience products if the difference in price is not too wide and if such products are easily available.

The results of both surveys support the conclusion that the greater feeling of responsibility regarding nutrition that is induced by parenthood decreases again when the children get older. Furthermore, Brunner *et al.* (2006) propose that the birth of a child is not actually the decisive factor for a turn towards sustainable eating habits; rather, the preference for or attraction to organic foods already exists beforehand, in many cases since childhood (see also Grauvogl 2005). Similar results from the project concerning sustainable nutrition in the transition to parenthood suggest that the life event 'birth of a child' is influenced by other life events (such as relocation or change of financial situation) that have additional impacts on food consumption habits.

12.2.1.2 Disease

Another life event that has the potential to initiate reflections on eating habits are the detection or onset of severe diseases and health problems, such as neurodermatitis, heart attacks, strokes, cancer, arteriosclerosis and diabetes (Brunner *et al.* 2006; see also Sehrer 2004). Brunner *et al.* (2006) differentiate between two categories of diet changes that are the result of disease:

- **'Turning to organic'.** This type dominates in the group of persons that suffer from diseases that are caused by environmental pollution (such as neurodermatitis). Because of the higher awareness of environmental problems, the preference for natural food, free of artificial ingredients and toxins, increases. This preference also carries the potential for increased consumption of organic food. However, having a disease is not a sufficient motivation to make a change to organic products. Additional important factors for this type are that they have experienced living in the countryside and with traditional agriculture. The 'turning to organic' type in this survey was dominated by women living in Munich (western Germany)

- **'Turning to a diet'.** The persons of this type often have suffered a heart attack; they are overweight or suffer from damage to bodily organs. Since explicit diets (for example, eating less fat and meat) to cure these diseases already exist, persons of this type do not suffer as much from doubts or uncertainty regarding food quality as the other type. In most cases, only absolutely necessary adaptations are made. There is no turn to organic products. The persons of this type were mostly men living in Leipzig (eastern Germany)

12.2.1.3 Retirement

Retirement is often seen as bringing about a far-reaching change in daily life routines (Brunner *et al.* 2006). Retirement is often accompanied by a new awareness of time, a higher need for social contact and by a higher degree of reflection, not only on the individual's development but also regarding societal changes. The results suggest that the newly gained time resources are not invested in the preparation of complicated meals. To the contrary, many retired people were found to have a high preference for convenience, resulting in a higher consumption of convenience products that can be divided into small portions. In the search for social contact, more time is spent, however, in purchasing food. The preference for small and customer-friendly stores and purchasing facilities has the consequence that retired persons buy more frequently in specialised stores, such as butchers and bakers and at open-air markets. Regional products are preferred, not for environmental motives but rather to secure jobs in the region and because a certain product quality is more easily guaranteed. Brunner *et al.* (2006) differentiate between two types of persons who change their eating habits as a result of retirement:

- **'Fit in old age'.** Persons of this type change their eating habits because they want to preserve or even improve the functionality of body and mind as they grow older. This motivation leads to an intensive reflection on eating habits and nutritional risks. At first, there is a change to low-calorie and health-promoting products. Later, their interest in natural or organic products may also rise
- **'Less is more'.** This type mainly wants to reduce the time and money spent on cooking and eating. There is a higher preference for spending time gardening or watching television. A turn to health-promoting food without toxins can also be observed. Innovations in food technology such as functional foods are not seen as being necessary

12.2.1.4 Food scandals

The intensive cultivation of land, mass production and intensive animal husbandry, as well as high degrees of food processing, regularly lead to food and environmental scandals, such as BSE and foot-and-mouth disease. Scandals in food production and processing are an important factor creating consumer uncertainty and suspicion regarding quality controls. The consequences of a strong sense of suspicion can be a change in food consumption habits, but less suspicious consumers may become only a little more sensitive about these issues (Brunner *et al.* 2006; see also Brunner 2006).

The consumption of organic food can be seen as a strategy to ensure security. Besides organic production, the following attributes help to confront uncertainty: minimal processing, freshness, regionality, purchasing in specialised stores and the consumption of quality products. Food scandals often result in only short-term changes to consumption habits and usually motivate people to avoid only those products affected by the scandal. Brunner *et al.* (2006) conclude from these results that scandals do not offer a sufficient basis for a long-term transformation towards organic products if they are not supported by additional motives, such as health concerns, responsibility for children and so on.

At this point, the results of two other surveys that dealt with the relation between food consumption and environmental or food scandals need to be introduced. Rohwer (1988) analysed changes in food consumption after the reactor accident in Chernobyl, which led to radioactive contamination of vegetables and milk. Two-thirds of a total of 111 persons who were questioned in the Hamburg area stated that they had changed their eating habits because of this event. Most people reduced the consumption of leafy vegetables, milk, meat and eggs, while they increased the consumption of frozen and conserved foods. These changes were motivated by fear and uncertainty about food-contamination risks. In an online survey with 767 participants, Schüz *et al.* (2005) looked for gender-specific differences in food consumption during the BSE and foot-and-mouth disease crises. The analysis showed that women reduced their consumption of meat much more than did men, at least for a short period of time.

12.2.2 Changes in sustainability performance

In the project concerning nutritional careers, the analysis of changes in food consumption habits concentrated mainly on environmental aspects. Meanwhile, the project concerning sustainable nutrition in the transition to parenthood tried to consider the three dimensions of sustainable food consumption: ecological, social and economic.

12.2.2.1 Parenthood

The quantitative data for the project concerning sustainable nutrition in the transition to parenthood showed that parenthood has the potential to bring about environmental improvement, especially because of a higher consumption of products that have a lower environmental impact. In this context, the increase in consumption of organic, regional and seasonal food is especially relevant (see Table 12.3).

Remarkable changes also occurred concerning the consumption of minimally processed foods. With parenthood the percentage of people that consumed fresh fruit and

Table 12.3 Consumption of 'sustainable food': percentage of people who, when questioned, stated they often or always consume 'sustainable products'

	Percentage		
	Before pregnancy	During pregnancy or after birth of child	Change
Organic products	22.4	42.0	19.6
Seasonal products	33.1	45.4	12.3
Fresh products	83.4	95.4	12.0
Regional products	20.0	28.6	8.6
GMO-free products	44.1	50.5	6.4

GMO = genetically modified organism

vegetables, wholegrain products and fresh milk at least once a day increased significantly (Table 12.4; see also Olson 2005; Verbeke and De Bourdeaudhuij 2007). However, the consumption of meat—another product that is of high relevance to sustainability—remained stable. Changes concerning the consumption of meat during this life period were not expected, because pregnant and breast-feeding women are often given advice to eat sufficient amounts of meat to avoid iron deficiencies.

The sociodemographic factors (education, income and the neighbourhood where participants live) appear to have no significant impact on how great the changes due to parenthood are. Only with regard to education is there a noticeable trend: the differences between persons with lower and higher educational levels appear to grow with the tran-

Table 12.4 Consumption of minimally processed products: percentage persons who, when questioned, stated that they consume the given products at least once a day

	Percentage		
	Before pregnancy	During pregnancy or after birth of child	Change
Fresh fruit	58.8	81.3	22.5
Fresh vegetables	40.6	56.7	16.1
Wholegrain products	49.1	61.6	12.5
Fresh milk	24.0	34.0	10.0

sition to parenthood. Parents with lower educational levels exhibit smaller changes in eating behaviour as a result of parenthood than do parents with higher educational levels.

The change to a slightly lower awareness of price can be viewed as an economic improvement. However, in this field there are only small changes: the percentage of persons who are very concerned about keeping their food budget down through the purchase of low-cost products increased from 65% to 67% with parenthood. Parents find it a little more important that farmers receive a fair price for their products (an increase of 4%, from 54% to 58%). There is also a small increase in persons who are ready to pay a higher price for fair-trade products (an increase of 3%, from 47% to 50%). At the same time, the percentage of people who state that they cannot afford high-quality products decreases a little (a decrease of 3%, from 49% to 46%). These changes can have a positive impact on the readiness to buy organic food.

12.2.2.2 Diseases, retirement and food scandals

As there is no quantitative data for life events concerning onset or diagnosis of diseases, retirement or food scandals, no conclusions can be drawn concerning possible environmental, social and economic improvements. There is, however, a tendency for these life events to result in higher consumption of minimally processed, regional and organic food, therefore leading to possible environmental improvements.

12.2.3 Learning experiences

It is not easy to draw conclusions concerning which life event has the highest potential for changes towards sustainable food consumption. The project looking at sustainable nutrition in the transition to parenthood showed that a significant proportion of persons questioned turned towards healthier foods, which partly lessens the impact of food on the environment. The changes are a result of a feeling of responsibility for the well-being of their child. Only a few parents, however, impose the same rigid standards that they maintain for their children on their own nutritional habits.

A severe disease can result in nutritional changes if there is scientific evidence that this will improve the person's health situation. Higher consumption of organic products is only probable if the disease results from environmental pollution. The transition to retirement is accompanied by a convenience orientation concerning the preparation of food, despite greater availability of time. A change of food consumption habits at this age is rather unusual but it can be observed if there is a strong motivation to maintain health in body and mind. Food scandals have only a low potential to prompt long-lasting changes, because reactions to them generally concentrate only on avoiding the particular products concerned.

Life events stimulate change only over a certain period of time. How long this period of time will be will depend on the life event itself, the person's circumstances and the psychological dispositions of the person concerned. Supportive factors are, for example, an upbringing that sensitises the person towards being open to making such changes.

12.3 Potential for diffusion and scaling up

In the analyses described here, research was conducted on whether certain life events have the potential to create changes towards sustainable nutrition. The results provide an empirical basis for planning concrete interventions in future projects. With an orientation towards certain target groups, the probability of carrying out successful campaigns rises. Another important factor for dissemination of the results on life events is that the majority of the population pass through life events such as the birth of a child, relocation or retirement.

The Technical University of Berlin and the University of Gießen are carrying out a follow-up project from 2008 to 2010 which seeks to use life events (birth of a child, relocation) for target-group-specific interventions made in cooperation with enterprises and associations such as environmental or consumer protection groups. Relevant consumption habits that will be addressed are nutrition, mobility and energy use. The approach of this project is to encourage developing a higher awareness among the target groups regarding certain issues by supplying them with information and offering further motivating measures. One important goal of the project is to initiate a long-lasting cooperation between the enterprises and the associations concerned, which will continue with target-specific measures after the research project has ended. From a scientific point of view, it will be especially interesting to compare the two life events (birth of a child and relocation) with regard to their impact on the three different fields of consumption patterns that are essential for sustainability (food, mobility, housing).

Whether life events can be taken advantage of as turning points for changes towards sustainable consumption depends mainly on the engagement of enterprises that offer 'sustainable' products or services. They have more financial resources to initiate campaigns aimed altering consumption habits. These campaigns allow the enterprises to address new target groups, to stabilise their position in the market and to achieve a higher profile over their competitors. Scientific support and the participation of consumer and environmental associations as well as the public sector can play an important role in assuring a certain level of quality in these campaigns.

12.4 Conclusions

The research projects described in this chapter started off with the assumption that life events bring about possibilities for alteration of food consumption habits over a certain period of time. The results show that there is indeed much potential here that could be taken advantage of by, for example, targeted marketing or information campaigns. However, since a change of habits depends a great deal on additional factors (such as the availability of products and services, sociodemographic factors, etc.) and motives (such as socialisation and upbringing) the potential of life events should not be overestimated.

References

Block, J.P., R.A. Scribner and K.B. DeSalvo (2004) 'Fast Food, Race/Ethnicity and Income: A Geographic Analysis', *American Journal of Preventive Medicine* 27.3: 211-17.

Brunner, K.-M. (2006) 'Risiko Lebensmittel? Lebensmittelskandale und andere Verunsicherungsfaktoren als Motiv für Ernährungsumstellungen in Richtung Bio-Konsum' ('Food as a Risk? Food Scandals and Other Factors of Uncertainty as a Motive for Nutritional Changes towards Organic Food Consumption') (DP-15, BMBF research project 'The Turn-around in German Agrarian Policy: New Forms of Food Consumption?'; www.konsumwende.de/downloads_fr.htm, accessed 13 July 2009).

——, C. Kropp and W. Sehrer (2006) 'Wege zu nachhaltigen Ernährungsmustern: Zur Bedeutung von biographischen Umbruchsituationen und Lebensmittelskandalen für den Bio-Konsum' ('Towards Sustainable Diets: Relevance of Biographical Situations of Change and Food Scandals for the Consumption of Organic Food'), in K.-W. Brand (ed.), *Von der Agrarwende zur onsumwende?* (Volume 1; Munich: oekom): 145-96.

CEC (Commission of the European Communities) (2005). 'Kampf gegen Fettsucht in Europa' ('Battle against Obesity in Europe'), *Ernährung im Fokus* 5.5: 145-46.

Cicia, G., T. Del Giudice and R. Scarpa (2002) 'Consumers' Perception of Quality in Organic Food: A Random Utility Model under Preference Heterogeneity and Choice Correlation from Rank-Orderings', *British Food Journal* 104: 200-13.

Dahlstrand, U., and A. Biel (1997) 'Pro-environmental Habits: Propensity Levels in Behavioural Change', *Journal of Applied Social Psychology* 27.7: 588-601.

Davies, S. (2001) 'Food Choice in Europe: The Consumer Perspective', in L.J. Frewer, E. Risvik and H. Schifferstein (eds.), *Food, People and Society: A European Perspective of Consumers' Food Choices* (Berlin: Springer): 365-80.

Dibsdall, L.A., N. Lambert, R.F. Bobbin and L.J. Frewer (2003) 'Low-Income Consumers' Attitudes and Behaviour towards Access, Availability and Motivation to Eat Fruit and Vegetables', *Public Health Nutrition* 6.2: 159-68.

EEA (European Environment Agency) (2005) 'Household Consumption and the Environment' (Report 11/2005; Copenhagen: EEA).

Grauvogl, M. (2005) 'Von der Agrarwende zur Konsumwende? Eine empirische Fallstudie am Beispiel der Zielgruppe "junge Mütter" ' (unpublished Master's Thesis; Munich: Scientific Centre Weihenstephan of Nutrition, Land Use and Environment, Technical University Munich).

Guenther, P.M., H.H. Jensen, S.P. Batres-Marquez and C.-F. Chen (2005) 'Sociodemographic, Knowledge and Attitudinal Factors Related to Meat Consumption in the United States', *Journal of the American Dietetic Association* 105.8: 1,266-74.

Hart, A., L. Tinker, D.J. Bowen, G. Longton and S.A.A. Beresford (2006) 'Correlates of Fat Intake Behaviours in Participants in the Eating for a Healthy Life Study', *Journal of the American Dietetic Association* 106.10: 1,605-13.

Herde, A. (2007) *Nachhaltige Ernährung im Übergang zur Elternschaft* (*Sustainable Nutrition and Parenthood*) (Berlin: Mensch & Buch).

—— and M. Schäfer (2006) 'Nachhaltige Ernährung und Elternschaft' ('Sustainable Nutrition in the Transition to Parenthood'), *Ernährung im Fokus* 6.4: 98-104.

Ilmonen, K. (2001) 'Sociology, Consumption and Routine', in J. Gronow and A. Warde (eds.), *Ordinary Consumption* (London: Routledge): 9-23.

Jahnen, A. (1998) 'Wechselbeziehungen ernährungsrelevanter gesellschaftlicher, gesundheitlicher und präventiver Faktoren als Bestimmungsmerkmale der Ernährungsberatung' ('Interrelations between Nutritional, Societal, Health and Preventative Factors as Determinants of Nutritional Consultancy'), in A. Jahnen, *Ernährungsberatung zwischen Gesundheit und Gesellschaft* (Frankfurt am Main: VAS): 50-166.

Jetter, K.M., and D.L. Cassady (2006) 'The Availability and Cost of Healthier Food Alternatives', *American Journal of Preventive Medicine* 30.1: 38-44.

Jongen, W., and G. Meerdink (1998) *Food Product Innovation: How to Link Sustainability and the Market* (Wageningen, Germany: Wageningen Agricultural University).

Jungbluth, N., and M.F. Emmenegger (2002) 'Ökologische Folgen des Ernährungsverhaltens: Das Beispiel Schweiz' 'Ecological Consequences of Nutritional Behaviour: The Case of Switzerland'), *Ernährung im Fokus* 2.10: 255-58.

Keane, A. (1997) 'Too Hard to Swallow? The Palatability of Healthy Eating Advice', in P. Caplan (ed.), *Food, Health and Identity* (London: Routledge): 172-92.

Kroeber-Riel, W., and P. Weinberg (2003) *Konsumentenverhalten* (*Consumption Behaviour*) (Munich: Vahlen).

Kroke, A., F. Manz, M. Kersting, T. Remer, W. Sichert-Hellert, U. Alexy and M.J. Lentze (2004) 'The DONALD Study: History, Current Status and Future Perspectives', *European Journal of Nutrition* 43.1: 45-54.

Laberenz, H., T.H. Witkowski and A. Borchert (2006) 'Lebensmittelmarketing aus Verbrauchersicht: Ein deutsch-amerikanisches Meinungsbild' ('Food Marketing from a Consumers' Perspective: A German–American Picture'), *Ernährung im Fokus* 6.12: 342-47.

Lamnek, S. (2005) *Gruppendiskussion: Theorie und Praxis* (*Focus Groups: Theory and Practice*) (Weinheim: Beltz).

Lappalainen, R., J. Kearney and M. Gibney (1998) 'A Pan EU Survey of Consumer Attitudes to Food, Nutrition and Health: An Overview', *Food Quality and Preference* 9.6: 467-78.

Lehmann, M., and P. Sabo (2003) 'Multiplikatoren' ('Disseminators'), in Bundeszentrale für gesundheitliche Aufklärung (ed.), *Leitbegriffe der Gesundheitsförderung: Glossar zu Konzepten, Strategien und Methoden in der Gesundheitsförderung* (Schwabenheim an der Selz: Fachverlag Peter Sabo): 154-56.

Lockie, S., K. Lyons, G. Lawrence and J. Grice (2004) 'Choosing Organics: A Path Analysis of Factors Underlying the Selection of Organic Food among Australian Consumers', *Appetite* 43.2: 135-46.

Marshall, D. (2001) 'Food Availability and the European Consumer', in L.J. Frewer, E. Risvik and H. Schifferstein (eds.), *Food, People and Society: A European Perspective of Consumers' Food Choices* (Berlin: Springer): 317-38.

Meier-Ploeger, A. (2005) 'Grundsatzpapier Ernährungspolitik: des Wissenschaftlichen Beirats "Verbraucher- und Ernährungspolitik" beim BMVEL'; www.uni-kassel.de/fb11cms/nue/img/publication/Grundsatzpapier.pdf, accessed 13 July 2009.

Michaelis, L., and S. Lorek (2004) 'Consumption and the Environment in Europe: Trends and Futures' (Copenhagen: Danish Environmental Protection Agency).

Olson, C.M. (2005) 'Tracking of Food Choices across the Transition to Motherhood', *Journal of Nutrition Education and Behaviour* 37.3: 129-36.

Rehaag, R. (2005) 'Ernährungskommunikation: Verständigung über eine zukunftsfähige Gestaltung unseres Ernährungsalltags' ('Nutritional Communication: Sustainable Design of Everyday Nutrition'), *Ökologisches Wirtschaften* 1: 15-16.

—— and F. Waskow (2006) 'Rahmenbedingungen von Ernährungskommunikation' ('Political Framework for Nutritional Communication'), in E. Barlösius and R. Rehaag (eds.), *Anforderungen an eine öffentliche Ernährungskommunikation* (Berlin: WZB): 21-37.

Rippin, M., A. Kasbohm, H. Engelhardt, D. Schaack and U. Hamm (2006) *Ökomarkt Jahrbuch 2006: Verkaufspreise im ökologischen Landbau* (*Yearbook of the Organic Market 2006: Sales Prices in Organic Agriculture*) (Bonn: ZMP).

Rohwer, D. (1988) 'Ernährungsverhalten nach Tschernobyl: Eine Primäranalyse' ('Nutritional Behaviour after Chernobyl'), in M.-B. Piorkowsky and D. Rohwer (eds.), *Umweltverhalten und Ernährungsverhalten* (Hamburg: Behr): 89-150.

Schüz, B., F.F. Sniehotta, U. Scholz and N. Mallach (2005) 'Gender Differences in Preventive Nutrition: An Exploratory Study Addressing Meat Consumption after Livestock Epidemics', *Irish Journal of Psychology* 26.3-4: 101-13.

Sehrer, W. (2004) 'Krankheit als Chance für nachhaltige Ernährungsumstellungen' ('Illness as a Opportunity for Nutritional Changes') (DP-5, BMBF research project 'The Turn-around in German Agrarian Policy: New Forms of Food Consumption?'; www.konsumwende.de/downloads_fr.htm, accessed 13 July 2009).

Spangenberg, J.H., and S. Lorek (2001) 'Sozio-ökonomische Aspekte nachhaltigkeitsorientierten Konsumwandels', *Aus Politik und Zeitgeschichte B* 24: 23-29.

Swoboda, B., and D. Morschett (2001) 'Convenience-Oriented Shopping: A Model from the Perspective of Consumer Research', in L.J. Frewer, E. Risvik and H. Schifferstein (eds.), *Food, People and Society: A European Perspective of Consumers' Food Choices* (Berlin: Springer): 177-96.

Torjusen, H., L. Sangstad, K. O'Doherty Jensen and U. Kjærnes (2004) *European Consumers' Conceptions of Organic Food: A Review of Available Research* (Oslo: National Institute for Consumer Research; www.organichaccp.org/haccp_rapport.pdf, accessed 13 July 2009).

Verbeke, W., and I. De Bourdeaudhuij (2007) 'Dietary Behaviour of Pregnant versus Non-pregnant Women', *Appetite* 48.1: 78-86.

Von Koerber, K., T. Männle and C. Leitzmann (2004) *Vollwert-Ernährung: Konzeption einer zeitgemäßen und nachhaltigen Ernährung (Nutrition with Wholefood: Conception of a Contemporary and Sustainable Nutrition)* (Stuttgart: Karl F. Haug Verlag).

Weiß, J. (2006) 'Umweltverhalten beim Lebensmitteleinkauf: Eine Untersuchung des Einkaufsverhaltens und der Angebotsstrukturen in sechs Berliner Wohngebieten' ('Environmental Considerations in Buying Food: An Analysis of Buying Patterns and Infrastructure in Six Residential Areas of Berlin') (Doctoral Thesis; Berlin: Mathematisch-Naturwissenschaftlichen Fakultät II, Humboldt-Universität zu Berlin; edoc.hu-berlin.de/dissertationen/weiss-julika-2006-06-06/PDF/weiss.pdf, accessed 4 August 2009).

Wiegmann, K., U. Eberle, U.R. Fritsche and K. Hünecke (2005) 'Umweltauswirkungen von Ernährung: Stoffstromanalysen und Szenarien' ('Environmental Impacts of Nutrition: Analysis of Material Flows and Scenarios') (DP-7, BMBF project 'Ernährungswende', Öko-Institut, Darmstadt and Hamburg; www.ernaehrungswende.de, accessed 13 July 2009).

Wilhelm, R., and C. Kropp (2007) 'Verbraucher zwischen Informationsflut und Informationsbedürfnis: integrierte Ernährungskommunikation als Lösungskonzept' ('Consumers between Information Flood and Informational Needs: Integrative Nutritional Communication as a Role Model'), in B. Nölting and M. Schäfer (eds.), *Vom Acker auf den Teller: Impulse der Agrar- und Ernährungsforschung für eine nachhaltige Entwicklung* (Munich: oekom): 92-104.

Part III
Conclusions

13

Sustainable consumption and production (SCP) of food: overall conclusions on SCP in the food and agriculture domain

Ursula Tischner
econcept, Agency for Sustainable Design, Germany

Eivind Stø
National Institute for Consumer Research (SIFO), Norway

Arnold Tukker
TNO Built Environment and Geosciences, The Netherlands;
Norwegian University of Science and Technology (NTNU), Department of Product Design

In this concluding chapter we discuss learning and conclusions from the previous chapters and case studies. First, we briefly summarise the structure of the agriculture and food production–consumption system and important factors influencing it (Section 13.1). Second, we discuss the potential for change in the system, including learning gained from the different case studies related to change, and introduce a vision for SCP in the domain (Section 13.2). The results of the 'food and agriculture' domain are then compared with the findings from the other two domains discussed in the SCORE! project: 'mobility' and 'housing and energy' (Section 13.3). Fourth, we summarise activities that different actor groups can contribute over the short, medium and long term and thus a kind of roadmap for change towards more SCP in the domain (Section 13.4) leading to final conclusions (Section 13.5).

13.1 Systemic description of the domain

> By 2050 we will need food for a world population that is wealthier and several billion larger. We will need to do this at the same time as adapting to a warming and less predictable climate. And, in addition, we will need to cut the greenhouse gas emissions associated with food production.
>
> Hilary Benn (2008)[1]

The above quote from Hilary Benn, Environment Secretary of the United Kingdom, launching the UK Food Strategy, summarises the landscape—that is, the meta-factors and sustainability issue—in which food production and consumption in Europe takes place and discusses the regime—that is, the food production and consumption system or value chain, with its actors and interlinked practices.

13.1.1 The 'landscape': relevant meta-factors and sustainability issues in the agri-food sector

In Chapter 2 we discussed extensively the 'landscape' in which the production and consumption of food takes place. The most important and influential developments in this framework for the agricultural and food system are described in Box 13.1.

Box 13.1 Important developments in the agricultural and food system

- The world market demand for food is increasing, because of the growing world population;[2] at the same time, because of global warming, water scarcity and soil degradation, more stress is being put on the agricultural production system
- The global tendency to consume more meat and dairy products, the need to produce more renewable energy and the increasing demand for space to live as a result of growing populations is creating more competition for land use
- As a consequence of these tendencies prices for food products are increasing; a 'world food price crisis' has occurred, associated with severe social and health problems
- Hunger and obesity are at the same time problematic threats to human health and survival in developing and industrialised countries; the demand for action has been recognised and accepted by national and international organisations

continued over →

1 Source: PoultrySite 2008.

2 Experts estimate that in 2050 there will be around 9 billion people on the planet; after that, population growth will slow down in emerging nations too (especially in India and Asia) because birth rates tend to decrease with increasing wealth.

- The current conventional agricultural production system is unsustainable because of chemical use, soil erosion and degradation; in addition, especially livestock, has a huge global warming potential
- Agriculture and food production and international trade is covered by World Trade Organisation (WTO) agreements and thus the globalisation of food value chains has become in reality associated with problems of control and need for transportation
- The search for greater efficiency in production, and the fight for market shares in international and national competition has also led to more industrialisation and concentration in the food value chains, leading not simply to higher outputs but also to increasing economic pressure on farmers and a loss of transparency in food value chains
- Demographic changes such as ageing populations, decreasing household sizes and more individualised lifestyles, as well as migration, are influencing consumer demands for food and their consumption habits
- At the same time, European consumers are losing trust in the food production system as a result of food scares and are become increasingly aware of global warming and other environmental and social issues, leading to greater demand for environmentally and socially beneficial foodstuffs
- In this political climate European politicians are reaching agreement on reforms to the Common Agricultural Policy (CAP), becoming more supportive of more sustainable agricultural production methods and initiatives to 'green' mainstream agricultural systems in Europe. National governments are putting into practice different approaches to influence national agricultural production to encourage it to become more sustainable
- ICT, GPS, tracking and tracing technologies have found a place in the food production–consumption system and are enabling precision farming and better control methods. At the same time ICT is giving consumers new opportunities such as online shopping, information exchanges and joint activities through online communities, or increasing the possibilities of local food network organisations

---

Most of the trends and developments described in Box 13.1 have an impact on the sustainability of the agriculture and food domain in Europe. As described in Chapter 2, no common definition or internationally accepted criteria system for sustainability in the food sector exists, but there are many different definitions and principles. The most relevant issues for the three dimensions of sustainability—environmental, social and economic—are summarised in Boxes 13.2–13.4. The most important **environmental** aspects related to agriculture and food are listed in Box 13.2; those relating to **social aspects** are listed in Box 13.3; and those relating to **economic aspects** are listed in Box 13.4.

Box 13.2 Important environmental aspects relating to agriculture and food

- Soil degradation and erosion results from agricultural activities
- Pollution (especially of water and soil) results from intensive agriculture, especially through use of artificial fertilisers, insecticides, herbicides, pesticides, manure, etc.
- Life-cycle impacts of food products are most important at the production stage, as a result of energy consumption in cooling processes and in transportation; also of great importance are food losses and waste—around one-third of food bought for home consumption is wasted
- The greatest contributions to global warming from food products are attributable to cheese, meat and eggs, the use of fertilisers, the use of transport, cooling processes, high processing intensity, high packaging intensity, animal-based products (particularly cattle and pigs), as a result of methane emissions and nitrous oxide from manure
- Organic products have between 6% and 15% less global warming potential than conventional products
- Animal welfare and biodiversity are especially threatened by intensive mainstream agriculture and genetic modification of crops
- Exploitation of natural resources takes place up to the point where species become extinct, e.g. the diminishing fish population resulting from over-fishing
- Organic agriculture is often used as a synonym for 'sustainable agriculture', with benefits mentioned such as:
  - Reduced environmental impact from agriculture (for example, reduced risk from pesticide residues and nitrate in groundwater)
  - The production of healthier food because of the lower content of pesticide residues and nitrates and the higher content of secondary metabolites
  - Improved biodiversity in agricultural fields
  - Food without genetically modified ingredients
  - Better-tasting products
  - More ethical animal husbandry
  - Strengthened regional development, with regional interaction along the product chain from field to table

Box 13.3 Important social aspects relating to agriculture and food

- Food security for a growing world population is threatened by unsustainable diets and agricultural production methods, by climate change and lack of water as well as by competition for land use such as from food and bio-fuel production, etc.
- Food safety needs to be guaranteed in increasingly complex production chains of food and to re-establish consumer's trust in food production and control systems
- Fair-trade issues and corporate social responsibility in globalised food value chains need to be addressed
- Cultural diversity need to be kept and supported despite the spread of unified lifestyle ideals offered mainly by multinational food producers and retailers from industrialised countries to the rest of the world
- Information and involvement, transparency and trust are needed to re-establish the connection and influence between consumers and food producers and farmers

Box 13.4 Important economic aspects relating to agriculture and food

- There is unfair distribution of income in the food chain, resulting in a North–South divide, including within Europe. Some farmers are not paid enough to survive in the long run
- Value for money, efficiency and quality are of importance. Food is relatively cheap in Europe as a result of increasing efficiency, industrialisation, rationalisation and subsidisation, but there is increasing discussion about market prices being too low to deliver high-quality, tasty, healthy and safe food products
- Long-term business, agricultural production capacity and economic stability in the system are under threat

Although there are conflicting issues in the sustainability aspects listed in Boxes 13.2–13.4, such as local and seasonal versus globalised food (the food miles discussion), organic versus fair-trade food, and sustainable food niches versus efficiency in the mainstream agricultural system, one clear conclusion from these sustainability issues can be drawn: to reach more sustainable consumption and production of food we need to reduce our consumption of meat (and dairy products), as the physicist Albert Einstein predicted, 'It is my view that a vegetarian manner of living by its purely physical effect on the human temperament would most beneficially influence the lot of mankind.'[3]

3 Translation of a letter by Einstein to Hermann Huth, 27 December 1930 (Einstein Archive 46-756).

Meat production is *the* most important contributing factor to the unsustainability of food in Western countries. We have known for a long time that in terms of using natural resources, a lot is lost by feeding animals with food that we could actually eat directly ourselves. Here are some figures (Pimentel 1997; Pasricha 2008; USDA 2004):

- Meat and dairy production account for 13.5% of total greenhouse gas emissions in the EU-25
- Animal protein production requires more than eight times as much fossil-fuel energy than production of plant protein yet yields animal protein that is only 1.4 times more nutritious for humans than the comparable amount of plant protein
- Example energy input to protein output ratios are as follows: beef, 54:1; lamb, 50:1; eggs, 26:1; pork, 17:1; milk protein, 14:1; turkey meat, 13:1; chicken meat, 4:1
- Agriculture accounts for 80% of all water used in the USA, and livestock production alone consumes 50% of all water used in the USA (this includes watering crops that are subsequently fed to livestock, watering livestock themselves, washing down stalls, etc.)

In addition, the high consumption of meat and other animal products in Western countries is taking up a large proportion of the world's vegetable food resources. Furthermore, the polluting effects of intensive farming are substantial.

The main global trend is a relatively strong increase in meat production and consumption and an increased world trade in meat. This is the case not only in Europe but also in large economies such as India and China. Their food culture has traditionally been based on rice and vegetables, and the increase in consumption of meat and dairy products represents a large challenge to sustainability.

Despite these facts, reduction of meat production and consumption is rarely on the political agenda, at the national, European or international level. There have been discussions about the differences between red and white meat, but few voices have argued for a substantial reduction in meat consumption. However, we have witnessed two relevant bottom-up 'movements' to reduce meat consumption. The first is the general vegetarian (and vegan) movement; the other is the temporary reaction to food crises in Europe, but they have not been strong enough to change the overall trend.

We are not aware of any broad public campaign arguing for consumers to reduce their consumption of meat or of any attempts to counteract forces that promote meat consumption. Even during the meat crises over the past decade the main emphasis of discussions was on safety problems, animal welfare and issues of trustworthiness rather than on addressing the impacts of meat eating on environmental and social sustainability.

It is easy to understand the main drivers for these changes. For producers there is significant added value in moving from vegetable to meat production. Historically, for European consumers one of the first signs of economic wealth was increasing consumption of meat, though there are differences between countries. Meat and other food items of animal origin generally have a high status in many of food cultures, as reflected in the central position that meat has in the structure of meals, in more ritualistic meals and celebrations and in the conceptualisation of what good and proper food is.

It is therefore difficult to imagine that the majority of Europeans will find giving up meat consumption an easy change. Yet changing the relative proportions of meat, vegetables and staples is not a big deal in terms of practical effort. The 'grammar' of a meal (for example, meat; potatoes, rice, noodles or pasta; and vegetables) does not need to be changed. Most consumers would probably also agree that having a few meat-free days per week would not hurt, representing a more healthy diet anyway.

However, there is no reason to believe that a substantial reduction of meat consumption will result from decreased consumer demand alone. Consumers will not be the main drivers in this necessary change. Actions of other actors in the triangle of change, policy–business–consumer, are needed too.

13.1.2 The agri-food production–consumption chain and interlinked practices: the 'regime'

The agricultural and food production–consumption chain or value chain, with its main actors and activities, was described in Chapter 2, leading to the definition of a triangle of relations and influences between the three main groups of actors in the 'regime':

- Actors in the food provisioning system
- Households and consumers
- The state and civil society, including policy and legislative or regulatory bodies, intermediary organisations, non-governmental organisations (NGOs), media, researchers and so on

General descriptions of the primary actor groups in the agro-food system are provided in the following, with a focus on the questions: what change can they influence, what change are they unable to influence significantly and where might they have room to act?

13.1.2.1 A general dilemma: the need for clear definitions and guidelines

One general dilemma in the agriculture and food system is that it is still unclear what sustainable agriculture and food is and where different regions, countries and continents should set priorities for the development of such a system. Even the definitions of 'organic food' in different countries vary:

- European definition: 'organic farming favours renewable resources and recycling, returning to the soil the nutrients found in waste products. Where livestock is concerned, meat and poultry production is regulated with particular concern for animal welfare and by using natural foodstuffs. Organic farming respects the environment's own systems for controlling pests and disease in raising crops and livestock and *avoids the use of synthetic pesticides, herbicides, chemical fertilisers, growth hormones, antibiotics or gene manipulation*. Instead, organic farmers use a range of techniques that help sustain ecosystems and reduce pollution' (CEC 2007)

- US definition: in the US definition, chemicals are allowed in 'organic' farming, and, 'in addition to these practices and materials, a producer may apply a crop nutrient or soil amendment included on the National List of synthetic substances allowed in crop production'[4]

Therefore, for politicians, company managers and consumers it is very difficult to decide whether they should promote organic food or fair-trade food, local food or only vegetarian food, or if the 'greening' of mainstream agriculture with no-till and low-till, less chemical use and more efficient production techniques should be the goal—or all of this. In this book, Dewick, Foster and Webster and Cooper for the UK (Chapters 3 and 4) as well as Schäfer, Herde and Kropp for Germany (Chapter 12) show how agriculture and food policy has been changing over recent decades to promote different priorities of farmers, sometimes being contradictory from one government to the next.

There is strong evidence that producing and consuming less meat and dairy products is an important strategy to move the agri-food system towards greater sustainability, but how can we convince consumers to eat less meat when meat consumption emerged in European history as a synonym for wealth and well-being and at the same time huge economies such as China and India are continuously increasing their demand for meat and dairy products? However, if for health and sustainability reasons Europe moves in the direction of a diet low in meat and dairy products it might also be able to influence consumption patterns in other nations.

13.1.2.2 Policy

Most politicians behave strategically so that they have a chance of being re-elected. In most of their decisions they reflect on who is in favour of and who is against specific decisions and measures.

Some important influences on agriculture and food policy at the national and European level can be distinguished:

- Strong agricultural lobbies: these are fighting to keep subsidies for agriculture as this policy is best for larger conventional farmers and processors, more dominant in lobby organisations

- Strong industry lobbies for use and promotion of genetically modified organisms (GMOs) and advanced biotechnology: these strategies are promoted by very large multinational companies

- Significant concerns among consumers about food safety and health issues and more and more mistrust in the regulatory and control systems for food in Europe: there is a strong consumer movement against GMOs, chemical use in

4 This quote is taken from the national organic standards rule published in the Federal Register on 21 December 2000. The law was activated 21 April 2001; see the National Organic Program's website, www.ams.usda.gov/nop, accessed 13 July 2009.

agriculture and industrialised livestock production, which have been cited as factors in food scares over recent years

- Growing lobbies for organic, fair-trade, regionalised and other alternative agricultural and food production systems: these are gaining prominence in the media and the public imagination and are aiming to gain more influence in Brussels

Thus politicians at the national and European level, manoeuvring between these different lobbies, suffer from a lack of certainty about the 'right' direction for sustainable agriculture and food and, in addition, have to balance national and European strategies with international issues such as globalised (WTO) trade issues and international initiatives to generate food and water security for all.

13.1.2.3 Farmers

We can distinguish two types of farmers: those who run larger conventional industrialised farms and those who run (usually) smaller organic and other alternative farms. Both are embedded in their own systems of production that are relatively slow to change (for example, conversion from a conventional to a certified organic farm takes around three years; changing the type of crops takes at least one season), and in the mainly conventional production and processing systems, such as production cooperatives and large food processors.

As power in the production system is concentrated in a few hands (large processors and large retailers) farmers have two main options:

- Becoming more efficient by increasing yield and farm size to increase output and thus income despite low prices
- Shifting to high-quality and specialised products (such as organic vegetables, fruit or milk) to achieve higher prices per output and thus survive; for this, networking is often necessary to create new organic farming cooperatives or to join existing cooperatives in order to increase the efficiency and power of the alternative farmers in the whole production–consumption system

Furthermore the European CAP influences the prices of agricultural products in Europe; thus farmers look closely at Brussels and try to influence developments in the CAP not only through their lobbies but also in self-organised groups of farmers. In a European survey 64% of European farmers said the CAP does not favour them but tends instead to favour the food processing industry and consumers (a few mentioned also the environment as benefiting from the CAP; EOS Gallop Europe 2000). Most of the CAP payments go to the biggest farmers: it has been calculated that 80% of the funds go to just 20% of EU farmers, while at the other end of the scale 40% of farmers share just 8% of the funds (BBC News 2008).

As the CAP is being reformulated to give more freedom to farmers to decide what they grow and to stimulate more sustainable agriculture in Europe, there is a good chance that farmers slowly but surely will find it more beneficial and are motivated to produce more sustainably.

13.1.2.4 Consumers

Consumers feel trapped in a system they cannot change: they feel unable to influence how agriculture takes place, how food products are processed or how policy is made. They only have the choice to vote every *X* years for a political party, to buy a product or not, to decide where they buy it and how they process and prepare it themselves—some may even grow their own food.

If consumers are very concerned, very active and find a supportive network around them they can set up activities or experiments such as community gardens, slow food, fair-trade initiatives or cooperatives to demand and purchase organic food, organise organic catering in their school or company canteens and so on. By using the internet and other media they can speak out about their concerns and demands and can organise protest groups. These very active consumers might constitute 5–10% of the population.[5] Other less-active citizens, but nevertheless very interested in the issue, join the first-movers and strengthen the movement. Thus experiments and niches are created. However, the majority of consumers are bound in their daily routines and it needs a good reason and a starting point to change their behaviour. Such reasons can be life event, as Schäfer, Herde and Kropp discuss in Chapter 12. A 'life event' may be an illness, pregnancy, retirement or individual concerns and motivation to protect nature and animals and fight global warming. They may also be external 'events' such as food scares, rising prices for food products and so on.

If there are no such internal or external drivers then education or the peer group moving in a specific direction might motivate an individual person or family to follow suit (for instance, children coming home from school and discussing sustainable food issues with their parents may bring about change); otherwise, it is difficult to change highly routinised behaviour such as food shopping and food preparation.

13.1.2.5 Food processors and retailers

These companies have to 'play the game' of the market, fight for market shares and beat their competitors. All too often in the case of food this is done by price policy alone (that is, by lowering prices); for some agricultural products, this might mean prices are below the levels that are feasible for farmers, resulting in disastrous consequences for the agricultural production capacity in Europe. However, more recently, large supermarket chains have started fighting for market shares by using a quality strategy, listing more quality and sustainable food products, more seasonal and regional food, starting cooperatives with local farmers and working with the supply chain to reduce packaging. Even discounters are introducing more and more organic vegetables and fruit, organic dairy products and bread and so on. One reason for this is that organic food products are one of the few products with amazingly high growth rates compared with conventional food produce, a low market share and with large potential as a result of increased consumer interest in sustainability and the health aspects of food. Another reason might be that it is becoming obvious that continually lowering prices is not a long-term business strat-

5 For example, in the German ecobiente project, 5–10% of the German population were identified as front-runners in sustainable consumption choices (ecobiente 2007).

egy and—in a situation where products are the same in every supermarket—product differentiation, innovation and creation of a positive brand identity for the supermarket, which leads to a close and positive customer relationship, are very important for market success.[6]

A system-inherent problem for processors is the difficulty of dealing with agricultural products of different quality in the same production facilities; for example, from the organic point of view, conventional and organic ingredients should not be mixed, as organic products are not to be 'contaminated' by conventional ingredients to guarantee the quality that is necessary to display the European Bio-label. Thus production lines and supply chains have to be separated or be extremely well controlled. This is even more so for genetically modified (GM) food ingredients. The same is true in the supply chain to farms, such as in the fodder industry or producers of other agricultural inputs.

13.1.2.6 Preliminary conclusions: there is momentum for change

It seems there is increasing momentum for change in the otherwise relatively inert agriculture and food system. On the one hand, the change is driven by the dissatisfaction of European farmers with the prices they are able to achieve for their products, especially when recognising global market developments in food consumption and prices. On the other hand, change is driven by consumers and their growing distrust in the food and agricultural system—the image of the 'good farmer' and food industry has been damaged in recent years by food scares and scandals.

In addition, discussion of health-related issues and food (such as fast food, obesity and diabetes) and discussions on global warming regarding food-related aspects (such as meat consumption) are bringing a climate of change and are providing a fruitful situation for emerging concepts. Consequently, an influential driver for change is consumer interest and thus the economic success of alternative organic and fair-trade products in mainstream food markets.

13.2 Potential for change in the agri-food domain

In this section we summarise the findings of the case studies in this book and other sustainable initiatives and cases regarding SCP models in the food domain. We then discuss factors that support and hinder change and how they lead to stability and windows of opportunity for change in the system. Finally, a vision for a more sustainable consumption and production system in the food domain is presented.

6 An example of the negative effect of competing on price alone is that of Wal-Mart's failure in Germany because the US supermarket chain did not really understand and adapt to the German market, its consumers or their values. In 2006 Wal-Mart sold all its German markets to Metro AG (see e.g. Handelsblatt 2006: www.handelsblatt.com/unternehmen/handel-dienstleister/wal-mart-zieht-sich-aus-deutschland-zurueck;1113795, accessed 26 May 2010).

13.2.1 Sustainable consumption and production models in the agriculture and food domain

As we can learn from the case studies in this book and other developments in the agriculture and food domain there are already numerous activities in the domain that are worth being called 'more sustainable' than the conventional mainstream production and consumption system. Some of them are focused more on production and some more on consumption; some deal more with environmental issues whereas others deal more with socioeconomic issues. The main approaches are summarised below.

13.2.1.1 'Greening' mainstream agricultural and food production

As the predominant agricultural system in Europe is industrialised intensive agriculture, with very negative environmental impacts, measures to increase the sustainability of the current mainstream production and consumption system of food are already under development (for example, see the publications listed by the German Cleaner Production initiative on agriculture: Cleaner Production Germany undated; see DG Environment 2008).

Promising concepts to tackle the negative environmental impacts of agriculture in the primary production stage include: better resource management, including water management; cleaner production measures (reduction of waste and emissions and minimisation of chemicals used; for example, see JRC 2003, 2006); precision farming; no-till and low-till agriculture; crop rotation; biological pest management; and biotechnology as historically used in agriculture.[7] Regarding genetic modification of crops to become more resistant to pests and more suitable for extreme weather conditions, as expected under global warming, expert opinions are still contradictory regarding the potential benefits against the high risks and uncertainty of free field experiments and their application more widely.

With some of the (new) methods such as no-till and low-till farming additional positive aspects can be generated such as the potential for carbon sequestration in no-till crop fields. By reducing tillage, leaving crop residues to decompose where they lie, and growing winter cover crops such as grains or alfalfa, a farmer can slow carbon loss from a field while contributing to carbon transfer from the atmosphere to the soil. This might be a carbon sequestration method also of interest to energy companies.

Some techniques applied in conventional agriculture, such as crop rotation and biological pest control, are already being applied in organic farming; some of them, such as no-till agriculture and precision farming methods, could be of interesting to organic agriculture too if they were developed so that they are consistence with organic farming principles (for example, pattern-recognition technology has potential for use in the development of automated weed management in organic agriculture; see MSTI 2003)

To improve a farm's economic sustainability strategies such as vertical integration between the stages of production and processing, further product differentiation at the processing stage and more local production and consumption are often suggested. How-

7 Biotechnologies such as hybridisation and artificial selection have been applied in agriculture historically and do not change the DNA of the crops.

ever, as Dewick, Foster and Webster show in Chapter 3 for the example of milk production in the UK, there can be considerable trade-offs between economic and environmental sustainability resulting from all three strategies. For the case of milk production in the UK they conclude that (Section 3.1.1):

> in the absence of radical new technologies for efficient small-scale, differentiated processing and new energy and water efficiencies or new transport or fuel technologies to reduce the impact of more localised food distribution to individual consumers (likely, given current retail structures in the UK, to involve more car and small van transport *vis-à-vis* large lorries), a shift away from the current regime may lead to greater environmental impacts, whatever its economic and social consequences.

They furthermore identify technological change and behavioural change in industry as the most important factors to improve overall sustainability in the sector.

To reach sustainability, the British government has introduced a 'Strategy for Sustainable Farming and Food' (SSFF) (Defra, 2002) with the aim of increasing the economic sustainability of farmers but not at the expense of environmental sustainability. A combined approach of legislative and subsidy reform was introduced centrally but with close liaison between government, farming and environmental bodies at a *regional* level to gain acceptance by industry. Evaluation of one of the regional initiatives created under this strategy showed positive results in long-term behavioural changes within government: The regional initiative was able to support networking of different stakeholders and to help government bodies to target their support for sustainable behaviour in farming and food and to pilot initiatives involving significant change within industry.[8] The regional initiative was not so successful in changing the attitudes and behaviour of farmers. Reasons for this included the overwhelming commercial pressures and financial constraints on farmers and the deeply entrenched attitudes and behaviour of many farmers. Overall, the regional initiative was successful in identifying and shaping targeted interventions (for example, projects and campaigns); it enabled arguments about sustainability to be brought into mainstream thinking at the relevant farming and food industry organisations. Moreover, through the development and exploitation of better government–industry relationships it was able directly to change public-sector food procurement (Chapter 3).

13.2.1.2 Food self-sufficiency of nations and local production and consumption of food

The lack of a harmonised international or even European understanding of what sustainable consumption and production in the agriculture and food domain means is nicely documented by Cooper's chapter in this book about Britain's self-sufficiency in

8 For example, F4C was able to help bring together public-sector procurement managers and wholesale meat suppliers. By liaising between them and providing funds for a small pilot project, changes were made to the specifications given to the wholesalers, resulting in an increased use of local (in contrast to imported, often from outside the UK) foods and a higher specification of nutritional quality of the food provided to schoolchildren.

food production (Chapter 4). While self-sufficiency has long been associated with sustainability, and many experts see decentralised power and smaller-scale, localised production as intrinsic elements in a more sustainable future, the environmental facts show a different situation for food. Arguments for reducing food miles rightly refer to the environmental impacts related to the transportation of food, especially of products that need cooling and air transportation. However, in life-cycle assessments for many food products the transportation, including packaging, accounts for only around 8% of the overall environmental impacts of the food produce, and very significant here are the last transport distances by truck in the country where the food is finally sold (Defra, in the UK, states that most environmental damage is done by internal traffic rather than imports; Defra 2006). That means local production makes sense environmentally only if the domestic production, transportation and distribution can be as eco-efficient or even better than production in foreign countries. This might not be the case, for example, for greenhouse products in countries with colder climates compared with free field products in countries with warmer climates.

Thus the food-mile discussion is rather a political and sociocultural one and based more on social and economic aspects than on environmental facts. But even the argument that a country's self-sufficiency would create independence from global factors such as climate change, peak oil prices and international terrorism is opposed by the fact that a nation relying on own food production would be more vulnerable to internal threats, such as a harvest failures, natural disasters or diseases.

In addition, from a socioeconomic perspective it also has to be taken into account that food production in developing or emerging countries might bring more sustainability gains as a result of improvements to the socioeconomic basis in regions where poverty is an issue and can go hand in hand with protection of natural areas such as rainforests (see also the fair-trade cases in this book: Chapters 9–11).

Thus from a sustainability perspective it is especially interesting to increase local production and consumption of those foodstuffs that can be produced *efficiently* and *sustainably* in the respective region. If this is the goal the following factors are important (see Chapter 4):

- **Policy.** Restriction of imports and subsidisation of local food production is no longer a viable strategy to support regional food production because of WTO regulations and the unification of the single European market.[9] Instead, more market-based strategies for sustainable farming and food, focusing on better food quality and sustainability and cooperation and consultation of farmers on a regional level, should be developed (see also Chapter 3)

- **Labelling.** One pre-condition and one of the most important measures for improving transparency in food chains and potentially increasing demand for local products is better and more transparent labelling of the 'region of origin' and 'place of production' of food

9 Although there are some voices suggesting we should exclude food products from the WTO regulations because of their importance for the survival of the population.

- **Consumer choices.** As Cooper (Chapter 4) argues, for British consumers choosing locally produced food is not a very high priority.[10] This might be different in other European countries, such as in Italy, the birthplace of the Slow Food movement, where more consumers are in favour of regional and local food and distinctiveness in cultural identities (see Chapter 8). Generally, consumers got used to freedom from regional and seasonal constraints and enjoy exotic products. Furthermore, in a multicultural world, consumers rooted in foreign cultures might have an understandable desire to purchase food items that cannot be grown in the nation in which they currently live. Thus consumers still have to be convinced of the benefits of 'buying local'
- **Diets.** Here a change in diet to eat less meat and dairy products and more products in season would be in favour of greater self-sufficiency of a nation because of reduced land use by agricultural production
- **Retail.** Hand in hand with more locally generated food products and food choices go local retail structures such as farmers' markets (see Chapter 6) or supermarkets campaigning for local foodstuffs
- **Networking.** Local networks promoting greater involvement of consumers in local food production such as community-supported agriculture (CSA; Chapter 7) support a region's self-sufficiency regarding food

13.2.1.3 Organising the growth of organic food products for niche markets and substitution of unsustainable products at a feasible pace

Even though organic farming and food or fair-trade food still has relatively small market shares their growth rates and projected market potential are high (see Chapter 2). Examples such as the German product Bionade® show how successful an organic food product can become, with growth rates at around 300% per year. In many European countries organic farming and food are already on their way out of niche markets, as the successful case of Denmark shows.[11] However, in Denmark the growth in sales of organic products has recently slowed down and some products such as meat have not had the same success as that of the organic milk. Jørgensen (Chapter 5) sees the reasons for this in the following barriers:

- There is too much of a 'contradiction' between the practices of highly specialised conventional farms and those of organic farming
- There has not been a substantial change in the Danish diet

10 In a survey 89% of respondents wanted British food to remain widely available, only 18% said that they always tried to buy British food even if it is more expensive, 29% preferred to buy British food but only if at the same price and quality as other food, 51% did not mind where food comes from and the remaining 2% tried to avoid buying British food (IGD 2005).

11 The development in Denmark towards around 6% of agricultural land in use for organic farming and a market share for organic products of around 25%, especially within the dairy sector and the vegetable sector, can be seen as a success story (Chapter 5).

- Organic food production has not been broadly accepted as the way forward for sustainable food production in Denmark
- The present market-based policy strategy may have reached its maximum potential for encouraging voluntary conversion from conventional to organic farms

The following measures can help to overcome these types of barriers and increase the organic market share above the current threshold (see also Chapter 5):

- **Policy.** This can act on three levels, the first in combination with the second two:
  - Restrictions on the conventional production paradigm, such as further restriction of use of chemicals in agriculture, to be combined with the following two measures
  - Support for the development of more sustainable agricultural practices, such as a national support scheme for organic farmers
  - Market-based approaches, such as higher taxes on pesticides, activities to generate consumer confidence in organic food and public purchasing programmes in order to develop a bigger market for organic food
- **Institutional change.** This is achieved through a combination of the creation of new and the re-use and reshaping of existing institutions and structures, such as processing plants, the role of the supermarkets and agricultural advice and research within organic food
- **Interaction of actors.** The different actors in the food production–consumption system are shaping each other. For example, organic food offers on the market are shaping the strategies of the retail sector, and vice versa. Organic agriculture has in some cases inspired conventional agriculture, such as the adoption of strategies to use healthier types of fodder, etc. Organic agriculture and food can be developed as a multi-stakeholder strategy where all stakeholders in favour of and benefiting from organic food production work together, from regional politicians, to cooks, to health professionals and so on
- **Changing diets.** A strategy based on high-quality products—for example, inspired by the principles of the Slow Food movement and the so-called 'new Nordic food' concept, with a focus on high-quality products with a good taste—could be a way to change food consumption towards less food but of a higher quality

Important success factors for marketing organic food products derived from cases of successful organic food products can be summarised as follows:

- The products meet more than just the environmental expectations of consumers and are also healthy and support well-being, are products that fit into people's lifestyles and sell for a price that is acceptable and easy to understand for 'normal' consumers

- There is a nice and credible story behind the product that creates public sympathy for it and generates free advertising, because media and peer groups start talking about it
- There is good expertise in branding and design during product development
- The advertising strategy fits the product and target group; for example, with Bionade® it was more word of mouth at the beginning and, later on, 'viral marketing' (that is, a marketing strategy that uses existing social networks to attract attention to brands and products, by spreading information like a virus within these network) that advertised the product
- Attraction of the so-called 'cultural creatives' or LOHAS (lifestyles of health and sustainability) as first-mover consumer groups can help
- The growth can be organised and supported by a regional network of suppliers that are willing and able to grow with the product's market success
- The product can be more than 'hype' or a short-term trend and survive over the long run or have follow-ups with product innovation or diversification

13.2.1.4 SCP bottom-up initiatives: from niches to mainstream

Several different bottom-up strategies for the production and consumption of food in Europe have emerged: that is, activities started by consumers at the local level. Most of these focus on re-establishing a closer connection between farmers and consumers.

The concept of community-supported agriculture (CSA) as described by Vadovics and Hayes (Chapter 7) includes several approaches such as 'share farms' (where consumers buy 'shares', paying for part of the harvest in advance), food subscription and delivery schemes or farmers' markets (see Chapter 6). Close to the philosophy of CSA is the Slow Food movement, which started in Italy but spread throughout the world and from Slow Food to Slow Cities and general 'Slow Ways of Living' (see Chapter 8). Slow food should be good, clean and fair food—that is, the food people eat should taste good, be produced in a clean way that does not harm the environment, animal welfare or human health, and food producers should receive fair compensation for their work.

All of these strategies have in common the aim of reducing negative social and environmental issues related to industrial agriculture and food production and distribution by bringing together producers and consumers in local networks and strengthening the local economy and local culture.[12] Thus producers and consumers get to know each other (again), farmers gain additional local direct retail channels, consumers gain more influence on and knowledge about the production of their food (for example, recovering recipes based on regional and seasonal fruits and vegetables) and may even become co-producers. Often, organic farming and food is supported, and food miles are reduced.

12 Examples of negative effects and issues are increasing dependence on energy and fertiliser inputs as well as on transportation, fewer jobs in the countryside, increasing levels of environmental pollution, growing consumer distrust in food and economic insecurity for farmers.

This is especially environmentally beneficial if sustainable transportation is used for goods and people (for example, the use of bicycles).

Normally, these activities are started by a group of like-minded people, often with academic backgrounds, detecting a problem and trying to find alternatives to the mainstream markets. NGOs or other non-profit organisations can help to set up local activities. At a later stage of development the CSA organisations are likely to change from a not so professional non-profit structure to more market-oriented professional organisations. This often goes together with a shift from an ideologically very strict concept to a more pragmatic and customer-oriented solution, (for example, in farmers' markets organic and non-organic products might be sold together). Essential issues for the success of these kinds of concepts are:

- **Consumer choice.** Even though the systems are based on local and seasonal products, the diversity should not be too restrictive in terms of what is available via the system— product diversity should to be created at a local level or perhaps organic food from other regions may be imported
- **Organisational adaptability.** A combination of non-profit (to attract funding and gain more credibility) and for-profit parts of the organisation (to become more professional and efficient) has proven successful
- **Strategic pricing.** It is necessary to set prices right—not too high for consumer acceptance and not too low, to allow the system to survive
- **Partnerships and networking, creating trust.** Interaction between producers, consumers and the organisers of the system (through events, visits, newsletters, etc.) are used to create trust between actors and to attract new members; the involvement of further supportive local organisations such as schools or other public bodies can be very beneficial
- **Local context and participation.** Regional and local strategies must always be adapted to the specific regional context and circumstances; a participatory approach is essential
- **Additional services.** The Slow Food organisations also campaign to protect traditional foods, organise food tastings and seminars, encourage chefs to source locally, nominate producers to participate in international events, work to bring taste education into schools, set up their own universities and so on

13.2.1.4.1 Are these concepts sustainable?

While the socioeconomic benefits of these local food production–consumption networks are obvious, studies and data for an environmental evaluation are still missing. If local systems are based on organic production and processing then the environmental benefits of organic agriculture can be achieved (see Chapters 2 and 5).

However, Dewick, Foster and Webster (Chapter 3) point out that for the UK dairy industry the smaller scale and more localised production and retail strategies (based on vertical integration by farmers producing more differentiated products with local processing and distribution) very likely lack some of the efficiencies of the large-scale industrialised production and retail structures and thus might become less eco-efficient. Only if this alternative strategy is accompanied by the development, adoption and diffusion of new technologies such as efficient small-scale processing and storage (refrigeration) technologies, and by the emergence of local retailing and logistics models that are not reliant on shopping by cars, might they become positive in terms of environmental aspects.

13.2.1.5 Trade Not Aid: fair-trade concepts

In contrast to local food production–consumption networks another especially socio-economically beneficial strategy of food production and retailing is represented by fair-trade initiatives, such as the international Fairtrade organisation and label, as discussed by Osmundsvåg (Chapter 10), the 'verified sustainable agriculture' concept of the Rainforest Alliance, presented by Wille, Aerts and Geier (Chapter 11) and an individual case of marketing the açaí berry, produced in a sustainable way in the Brazilian Amazon, mainly to the USA, as discussed by Tunçer and Schroeder (Chapter 9). Also, these initiatives are generally driven by NGOs (such as Christian organisations) or small groups of like-minded people—in the açaí berry case it was three young Californians who 'discovered' the berry on a holiday in Brazil.

The central aim of fair-trade strategies is to support producers, particularly those in less-developed and poorer regions of the world, in their socioeconomic development by establishing cooperatives between (non-profit) importers and retailers in industrialised countries and small-scale producers in developing countries and by increasing the efficiency and sustainability of their agricultural and food production (also, trade in non-food products occurs but is not discussed in this book). The slogan 'Trade Not Aid', expresses the aim that trade will help poor countries develop better and faster than if they were to receive only development aid. In the case of the Rainforest Alliance the trade activities were a response to massive clear-cutting of half of the world's forests over the past four centuries and the starting point of the projects were people who use their land to manage a forest, grow agricultural crops or run a hotel or tourist trade in biodiverse areas. The Rainforest Alliance is active in supporting sustainable forestry, tourism and agriculture. Rainforest Alliance-certified farms hire locally, pay fair wages and ensure safe working conditions and access to clean drinking water and proper sanitary facilities; in addition, farmers and farm workers receive training and environmental education and their children go to school.

With most fair-trade activities *economic success* can be created: in 2005 the combined sales turnover of fair-trade goods was more than a billion euros, and growth rates have been between 20% and 30% since the beginning of the 21st century. The total value of products sold under fair-trade labels in 14 European countries in 2005 was about €597 million (Krier 2006).

Sales of Rainforest Alliance-certified coffee, bananas and chocolate exceeded US$1 billion in 2006, and further growth is expected. The sales of coffee from Rainforest Alliance-certified farms has almost doubled each year since 2003, from 7 million pounds (weight) to 54.7 million pounds in 2006; sales in 2007 are projected to exceed 91 million pounds.

As prices paid for fair-trade products to primary producers are higher than in conventional supply chains, where primary producers often do not earn enough to make a living, fair-trade products tend to be a bit more expensive than conventional products. This also means that market-driven fair trade depends on the will of consumers to pay more for a product. To motivate this, consumers should feel that they are making a difference. Therefore it is important to report back to the market and tell the **good stories** about the improved quality of life for real people. Compared with many other brands, fair trade offers a good story; given that the quality is average or better, this is a unique selling point. For consumers, fair-trade choices are a form of political consumerism. However, only a very small segment of consumers will buy a product *only* because it is fairly traded.

Important factors for the success of fair-trade concepts and products are as follows:

- **Distribution in mainstream retail stores and cooperation with mainstream companies.** Kraft Foods, one of the top four buyers of coffee in the world, was committed to buying 5 million pounds of coffee from Rainforest Alliance-certified farms in 2003 and has increased its commitment in subsequent years
- **Price.** Prices should not be more than 15% higher than those for conventional brands in the same category
- **Quality.** The quality of fair-trade goods must be good, either at the same level or higher than that of conventional products
- **Labelling.** The product should have a clear, credible and understandable certification scheme or label and a good story behind the product; for example, in 2005 Chiquita started communicating and marketing its Rainforest Alliance certification by putting the frog trust mark—the Rainforest Alliance-certified seal of approval—on its bananas in nine countries in Europe
- **Target groups.** The right target groups must be addressed, and these should be as broad as possible; for example, Sambazon, the brand behind the açaí berry products, emerged through grass-roots marketing at music and athletic events, letting word of mouth about Sambazon spread among surfers, skateboarders, world-class athletes and health-conscious consumers, making sustainability something not far off and 'crunchy'[13] but approachable, desirable and cool
- **Funding.** The Global Environmental Facility, through the United Nations Development Programme, has provided the Rainforest Alliance with a seven-year grant to develop sustainable production in several countries in Central America

13 Ideologically green. See www.slangcity.com/email_archive/4_01_2004.htm (accessed 26 May 2010).

- **Supporting measures.** In the UK fair-trade towns have been established, which have been a great success, and many see the 150 fair-trade towns as a key factor for understanding the boost to fair trade in the UK
- **Media attention.** In a popular Norwegian TV programme on consumer issues a special report on the living and working conditions on banana plantations in Latin America boosted the sales of fair-trade-labelled bananas; similarly, business for Sambazon really picked up after Dr Nicholar Perricone named açaí the 'number 1 superfood' in his 2004 book, *The Perricone Promise*
- **A non-profit and for-profit mix.** A good combination of and cooperation between non-profit and for-profit organisations is beneficial
- **Combining forces.** A joining of forces among different organisations is also beneficial; for example, fair-trade actors can join the International Social and Environmental Accreditation and Labelling (ISEAL) Alliance, which formulated the Code of Good Practice for Setting Social and Environmental Standards, with members such as the Rainforest Alliance, the International Federation of Organic Agriculture Movements (IFOAM), offering organic labelling, and the International Fairtrade Labelling Organisation (FLO), all of which are complementary rather than competing approaches

13.2.1.5.1 Are fair-trade concepts sustainable?

From an environmental point of view the positive effects of fair-trade activities are seen in the protection of important natural areas such as rainforests and in reducing the environmental impacts of the primary production of food in developing countries.

The Rainforest Alliance has a strong but pragmatic approach to sustainable farming—sustainable farms are to be certified by the Sustainable Agriculture Network (SAN), which awards the Rainforest Alliance certified eco-label to farms (not companies or products) that have minimal environmental footprints, are good neighbours to human and wild communities and are an integral part of regional conservation initiatives. The certification is based on 10 principles of sustainable agriculture but does not follow strictly organic agricultural rules: that is, specific agrochemicals are allowed. Farms must demonstrate continued progress in order to maintain their certified status.

A trade-off can occur between fair trade from developing and emerging countries to industrialised countries and transportation distances and thus the food miles involved. There is no simple solution to this: it has to be evaluated on a case-by-case basis whether production, distribution and consumption of an exotic fair-trade product is environmentally more or less beneficial than of a local product with the same nutritional (and emotional) value. Also, in fair trade, further environmental aspects in the supply chain such as energy consumption, transport, packaging and waste need to be considered.

In the end it often comes down to a (political) decision as to whether socioeconomic development in developing countries, with global fair trade as an instrument, is more important than environmental efficiency and reduction of food miles. In an ideal world all fair-trade products would be environmentally sustainable and all eco-products would be produced and traded in a fair way.

13.2.1.5.2 Problems of growth

In addition to the above uncertainties regarding environmental impacts, fair trade faces two other problems. First, with growing demand larger companies that do not follow the same sustainability principles as the fair-trade organisations might enter the market and set up unsustainable monocultural plantations. Second, a social problem can occur as a result of increasing demands for fair-trade products—the attainment of higher prices can mean that local people in the region can no longer afford to buy a product grown or manufactured in their own neighbourhood.

13.2.1.6 Changing diets and consumer behaviour

> If all the grain currently fed to livestock in the United States were consumed directly by people, the number of people who could be fed would be nearly 800 million
>
> *David Pimentel, Professor of Ecology at Cornell University's College of Agriculture and Life Sciences (Pimentel 1997)*

Different theories, methods and tools exist regarding how to convince and motivate consumers to shift to more sustainable food consumption habits. But as food consumption is a strongly ritualised, routinised and repetitive behaviour, it is not trivial to change and, despite efforts and some developments, mainstream food consumption habits are unsustainable most of the time. Some approaches to support behavioural changes towards more sustainability are discussed below.

13.2.1.6.1 Life events as windows of opportunity

Schäfer, Herde and Kropp (Chapter 12) discuss life events as opportunities to change unfavourable food consumption habits, because often in these situations the need to change routines arises anyway. A serious disease, the birth of a child, retirement or food scandals have been researched in two German projects to find out if they may prompt changes towards more sustainable food consumption. The following was found:

- **Diseases.** A severe disease can result in nutritional changes if there is scientific evidence that this will improve the health situation; increased consumption of organic products is probable only if the disease results from environmental pollution
- **Retirement.** Retirement is accompanied by a tendency to look for convenience regarding the preparation of food, despite more time generally being available to the retired; a change of food consumption habits at this age is rather unusual but can occur if there is a strong motivation to maintain functionality in body and mind
- **Food scares.** Food scandals usually motivate people to avoid only the products affected by the scandal and often result only in short-term changes to consumption habits unless they are supported by additional motives, such as health concerns, responsibility for children and so on

- **Parenthood.** Parents may become more interested in 'sustainable food', purchase a higher percentage of organic, seasonal, regional or fresh food, care more about buying products that are not genetically modified and shop more often in health-food stores and eco-supermarkets; this especially the case for parents that have a higher level of education

Although **income** is not a relevant factor in dietary changes, the availability of eco-products in nearby shopping facilities is an important factor. Also, the birth of a child is often accompanied by other life events (such as relocation or change in financial situation), and these have additional impacts on food consumption habits. Thus we can conclude that life events stimulate change only over a certain period of time. A more permanent change of habits depends much on additional factors (such as the availability of products and services, socio-demographic factors, etc.) and motives (such as socialisation and upbringing).

13.2.1.6.2 Targeted communication measures

In communication campaigns, companies or governments can take advantage of life events as turning points for change towards sustainable consumption. However, such communication must be targeted, aimed specifically at consumer groups and their everyday contexts. So far such targeted communication have not been done sufficiently well. Often, communication in the field of sustainable food is designed to address everyone. Mass communication has not had the expected results (yet) because it often does not link to people's everyday contexts, nor does it differentiate between different lifestyles and individual situations (Lehmann and Sabo 2003). If the information is aimed specifically at certain target groups, the probability of a successful campaign increases.

13.2.1.6.3 Labels as a tool for sustainable food consumption

Labels are mass-communication tools that play an important role in the business–consumer dialogue in the market of food and agricultural products in Europe. Labels are chosen both as an alternative and as a supplement to other political and economic instruments. The reason for this seems to be that it is easier to agree on information instruments than on juridical and political decisions. The political process and discussion concerning GM foodstuffs illustrates this clearly: the result has not been a moratorium but a label.

In theories about consumer behaviour there is a tension between models explaining consumer behaviour as being based on rational decisions, and models linked to ordinary and routinised consumption (Stø *et al.* 2008). However, labels in general fit in very well with both approaches. Within rational theories, labelling schemes function as a tool for consumers to make choices in the market for goods and services. Without this kind of information, consumer choices would be substantially more difficult. In many ways this argument is similar to that regarding the role of labels within theories assuming routinised everyday consumption. Modern consumers do not have the time or knowledge to make specific choices every time they do their regular shopping. If consumers trust the labels, they are able to make the necessary selection in their everyday life as part of their routine shopping behaviour, in line with their own values. Research has shown that consumers trust labels that are scientifically based and run by independent bodies or consumer or environmental NGOs (Rubik and Frankl 2005).

Labels function as symbols of rational choices and as signals of political, social or environmental consciousness. When buying fruit and vegetables with organic labels, value-for-money and concern for one's own health and the environment can be signalled at the same time.

In the European consumer market a large number of food labels exist (see Box 13.5).

Box 13.5 Overview of labels for sustainability aspects of food

- At least one organic label is available in all European countries. Such labels are in line with the Council Regulation EEC 2092/91.[14] Following the recommendation to reduce the diversity of labels generally, one interesting suggestion is to use the European Bio-label as the overarching label for organic products all over Europe, particularly because of its success with consumers. There is a danger that the entire system will collapse if the numerous organic labels are not integrated.

- In addition there is also a large number of fair-trade and other social and political labels. Many of the fair-trade labels belong to the International Fairtrade Labelling Organisation (FLO).

continued over →

14 www.organic-europe.net/europe_eu/eu-regulation-on-organic-farming.asp, accessed 13 July 2009.

- Furthermore we see an increasing number of nutrition labels within many product categories. The UK has developed a red–amber–green traffic light label to indicate the nutritional value of the product; in Scandinavia, the label is a green keyhole.

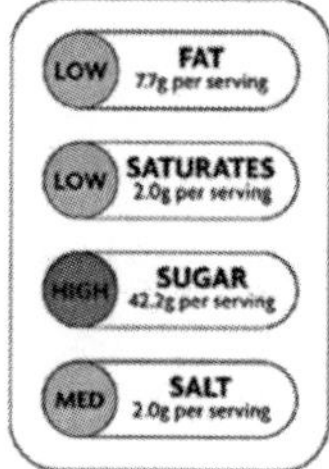

- Some countries (or regions) have strong quality labels, often linked to special local or regional products. The Slow Food labels are also, to some degree, regional labels. The EU has regulated PDO (protected designation of origin), PGI (protected geographical indication) and TSG (traditional speciality guaranteed) labelling.[15]

- The recent focus on climate change has also put food miles and carbon dioxide emissions on the labelling agenda, by developing carbon footprint labels (through, for instance, the Carbon Trust[16]) or air-freight labels (such as by Tesco and by Marks & Spencer[17]).

15 ec.europa.eu/agriculture/quality/logos/index_en.htm, accessed 13 July 2009.

16 www.carbon-label.com/business/label.htm, accessed 13 July 2009.

17 See www.igd.com/index.asp?id=1&fid=1&sid=5&tid=52&cid=340 and corporate.marksandspencer.com/howwedobusiness/our_policies/climate/commitments, accessed 18 April 2010.

There are reasons to believe that the large number of labels may on the one hand increase the possibilities of consumers to choose products that respond to their wants and values. However, on the other hand, it may be confusing and make it even more difficult to choose. There is no simple answer to this dilemma but it seems very likely that a stronger integration of environmental criteria with social criteria in the labelling schemes may be an important measure to increase the effectiveness and acceptance of the labels. European consumers should not be forced to choose between social and environmental sustainability. The EU could take steps to introduce a kind of 'meta-label' that would combine several aspects into one scheme. Generally, consumer information on food products should include at least the following dimensions:

- Information on whether it was produced organically or non-organically
- Fair trade aspects and social standards
- The origin of the product (its main parts or ingredients) and producer
- The carbon dioxide (CO_2) footprint (including transportation) or 'ecological rucksack' of the product
- Date of production
- Nutritional factors
- A full declaration of all ingredients, including whether any of them were genetically modified, and warning of allergenic ingredients

Labels can also have negative effects as they could function as barriers to trade, especially for agricultural products from developing and emerging countries. This could be even more so if environmental and social aspects are combined into one label—that is, specifying that products should not only be organic but also satisfy the fair-trade criteria, be in line with human rights and the trade union charter and avoid child labour. However, this is not an argument for not including environmental and social aspects in the same label, only that such an integration process has to be organised and planned in order to avoid trade-barrier effects.

13.2.1.7 Preliminary conclusions

In most European countries we see the co-existence of separate, parallel supply chains for niche local and regional food and for the more globalised vast majority of the food we eat. We see activities to 'green' mainstream agriculture and at the same time the growth of sustainable niche concepts. We see fast-food chains starting to use organic ingredients and local food delivery systems integrating exotic fair-trade foodstuffs in their range. We see large supermarket chains setting up cooperatives with local and organic producers and creating own organic brands for their markets. Conventional food producers are becoming very successful with new organic food products in the marketplace and consequently setting up projects to convince local farmers to convert to organic practices to secure their supply of organic raw materials in a growing market. Demand for organic and fairly traded food is increasing and sometimes the pace is too fast for the agricultural production system. It seems that the majority of conventional farmers do not (yet) trust this increasing demand and need to be convinced with long-term contracts and policy measures that this is an economically 'safe' way to go. And all these developments are happening at the same time.

13.2.2 Factors supporting and hindering change: stability and windows of opportunity in the system

In this section we try to draw more general conclusions about which factors are supporting and hindering change in the agri-food sector, whether we can extract any general patterns for change happening in the domain and where the most promising windows of opportunity are to influence the domain towards more sustainability.

13.2.2.1 Factors supporting or hindering change

There are some issues not only blocking change in the agri-food domain but also creating dissatisfaction among actors in the system with the potential to lead to such great frustration that concerted action will follow. Table 13.1 lists these factors and points out how they are blocking or supporting change towards greater sustainability in the food domain.

13.2.2.2 How and why does change happen?

As we can see from most of the case studies, very often change is driven by motivated individuals. They may be consumers starting CSA projects (as in the Open Garden project in Hungary), individuals (as in the Sambazon case) or people in NGOs initiating fair-trade activities (which is how the fair-trade movement started). Also, in companies often individuals with a specific motivation to change their or their company's situation are the drivers for change or are change agents (as in the case of Bionade®). But where does this motivation to drive change come from? Often, it is a mix of being aware of a societal problem, having a personal motivation (such as being concerned with one's own health or self-esteem) combined with a more altruistic or humanistic mission to be good to nature, planet and fellow citizens. These kinds of mind-sets are more often present with people with a medium to higher level of education, as shown by many studies (e.g. Deloitte and SGMA 2009; UNESCO 2000; Bentley 2001). The motivation for change becomes extremely strong when pressures such as a food-related disease, food scares or economic problems for the company such as decreasing sales figures or a shrinking market share are also present. However, if there is too much pressure this can also block changes because actors often are too scared or have no scope and financial means to act. Table 13.2 gives an overview of actors, typical SCP driving measures and motivations.

Table 13.1 Important issues in the agri-food domain and how they support or hinder change towards more sustainability

Issue	Supporting change	Hindering change
The European CAP Rural development policy National agricultural legislation	A reformulated CAP and rural development policy could support sustainable farming UK Strategy for Sustainable Farming and Food, with regional bodies to support sustainable agriculture	Unsustainable production quotas and subsidies favour mainly large companies In some countries subsidies for organic agriculture are being reduced
A lack of a clear and harmonised international or European definition of sustainable agriculture and food	There is room for openness to discussion and development of new approaches	Absence of clarity leads to a lack of understanding so that decision-making is very difficult and conflicting issues are left unsolved
Increasingly larger consumer groups are focusing on sustainability issues	They create demand for healthy, tasty, environmentally friendly and socially fair food products, increasing market shares for more sustainable food offers	–
Increasingly, there are more consumer groups with a low income and high price sensitivity	–	This creates a demand for cheap food Retail and food-processing companies need to provide cheap food products and thus put cost pressure on the supply chain
Institutions and companies in agriculture tend to be conservative	Alternative movements have to fight against the mainstream agricultural system and cooperate to set up alternative institutions	Current actors do not want to change anything and lobby to keep the status quo
There is power concentration in the value chain	This can be supportive if the companies that are important players in the chain support more sustainable food products	This can hinder progress if the companies that are important players in the chain do not support more sustainable food products
(Conventional) smaller farmers are under economic pressure	Farmers and food processors are motivated to shift to alternative farming and food production practices such as organic production, where income often is better and more stable, funding opportunities are better and market demand increases However, initial investment is needed	These actors do not have enough economic leverage to change anything They are fighting for sheer survival and may choose to leave the business

continued over →

Issue	Supporting change	Hindering change
More sustainable niche food offers such as organic and fair-traded food and alternative retail channels such as farmers' markets are achieving increasing market success	More companies, especially conventional companies, are becoming interested in such strategies and start offering them as well More consumers get access to these kinds of offers and become interested in them	–
Globalisation of food production	This leads to the discovery of new food products, with possibly more (eco-) efficiency of production and socioeconomic benefits for developing countries There is an increasing counter-movement to support and rediscover local food products and traditions	Less transparency means it is more difficult to control, involves unsustainable transportation and is produced under possibly lower social and environmental standards
Food scares	These encourage policy-makers to respond, consumers and companies to act to avoid food scares and change the system towards more control and better quality standards of food products, increased motivation of consumers to eat less meat (but only small long-term changes)	–
Global warming, peak oil discussion and greater demand for renewable energy and so on put pressure on food production and consumption systems worldwide	This creates discussion about the right way forward, combining economic, environmental and social issues and encourages more demand and support for research and for new more sustainable food solutions	It creates fear and short-term demand for immediate and mainly efficiency-based solutions, which might not be sustainable in the medium and long term
Diet	A growing part of the population (especially the middle and upper class) are adopting more healthy and more sustainable diets	An increasing majority of consumers (especially in the lower classes) have adopted the typical 'Western' diet with too much meat, fat, dairy products and sugar, leading to obesity and health problems
Multinational companies are pushing policy, the market and consumers to accept GMOs	A growing anti-GMO movement in Europe is forcing policy to formulate stricter legislation	Political decisions in favour of GMO are being made despite consumers' refusal to accept the technology

CAP = Common Agricultural Policy

GMO = genetically modified organism

TABLE 13.2 Actor groups, typical actions driving sustainable consumption and production (SCP) and motivations

Who?	Typical measures	Level	Altruistic motivation	Personal motivation
• Government • Top-down	• Regulation • Legislation • Standardisation • Subsidies • Research programme • Sustainable procurement • Public–private partnerships	• Regional • National • European • International	• Tackling larger problems such as climate change, food security and public health	• Being re-elected • Good conscience
• Business	• Cleaner or sustainable food production • Marketing of sustainable food products • Influence on food supply chain to become more sustainable • Development of more sustainable food services • Sustainable choice editing by retailers • Education of consumers towards more sustainable food consumption	• Market-inherent • Regional • National • European • International	• To have a healthy business but be better to the planet and better than competitors • To survive in a changing environment	• Personal success • Good conscience • Media attention • Economic benefits
• Interest groups • Civil society	• Fair-trade movement • Slow Food movement • Animal welfare movement • Environmental groups	• Regional • National • European • International, normally non-profit	• Specific religions • Beliefs • Values • Shared in a group	• Fulfilment • Being useful • Position in society

continued over →

Who?	Typical measures	Level	Altruistic motivation	Personal motivation
• Consumers • Bottom-up	• Community-supported agriculture • Green purchasing groups • Bottom-up community projects	• Regional • Usually starts as non-profit	• Protection of natural and social environment	• To feel empowered • To be healthy while doing something good • Economic benefits

Furthermore we can identify three typical levels for mechanisms driving change in the agriculture and food sector:

- There are bottom-up or grass-roots initiatives that are started by individuals or groups of like-minded people and often grow and become successful because they meet the desires, needs and wants of other peer groups and are able to activate a larger group of actors to work together, being they consumers, businesses, NGOs or political actors such as municipalities. If these experiments become niches and are economically successful, often larger companies become interested in the concepts and consider offering them as well. This can take place only if the concepts are not too different from their own philosophy
- Changes are started by companies. Here especially the role of retailers as the bottleneck for the market success of a lot of food products and as editors of consumer choices must be mentioned; however, individual companies are also starting new more sustainable business ideas and launching new more sustainable products, conventional farmers are converting to organic farming and so on. Also, if the new products or business strategies are successful it is very likely that other companies will follow suit
- Change is driven by policy, because regulations, standards and labels, subsidies and support for research or market-based instruments such as green procurement are influencing market actors (mainly business and lesser consumers). Compared with changes driven by consumers and business, policy is able to put much more pressure on a much larger group of actors, as when formulating legislation to phase out specific chemical substances in food production or when subsidising only specific types of (more sustainable) agriculture in Europe

In most other cases the initiatives started by few actors or one company or company association has to happen in a supportive socioeconomic environment, with appropriate infrastructure, to become successful, continue and even grow. Often, networking and cooperation are important support functions.

Two more change mechanisms are present but there are not (yet) very many such case studies available in the food sector:

- Research and academia develop new more sustainable agricultural or food production methods and food products and set up spin-off or start-up companies and institutions to offer them on the market. One such case is the permaculture method initiated by researcher Bill Mollison in Australia[18]
- Popular persons, role models or leading thinkers in society may start a movement, foundation or similar to support change towards more sustainability. These may take the form of educational and informational campaigns (Al Gore and his movie *An Inconvenient Truth*)[19] or foundations (from social and environmental foundations from Bill and Melinda Gates[20] and Bill Clinton,[21] to Brigitte Bardot[22] for animal rights). In the agri-food domain for instance the W.K. Kellogg Foundation supports and reports about more sustainable agriculture and food initiatives and projects[23]

13.2.2.3 Preliminary conclusions

There are many opportunities for all actors in the agri-food domain to becoming more involved in driving change towards more sustainability in food production–consumption systems. However, most of the changes cannot be started by one actor alone but requires more actors and cooperation in the system to create larger-scale impacts. Only European and national policy actors (if they manage to find agreement) or big players in the market have the power to initiate change in mainstream markets; most other initiatives start in smaller niches and groups and have to be amplified to become significant for the whole sector. Nevertheless, in niche experiments and activities a greater number of radical new concepts can be developed than in mainstream markets. Thus they are very interesting as fields of experimentation and inspiration and also for consideration if upscaling or transfer, multiplication and reproduction of the niches elsewhere may be possible (strategic niche management).

In Section 13.4 we discuss more in detail activities for different actors to promote more sustainable agriculture and food production and consumption—a kind of action plan. But before that we introduce a vision of a sustainable system of consumption and production of food in Europe.

18 Bill Mollison, Australian researcher, author, scientist, teacher and naturalist developed, together with David Holmgren, 'permaculture', an integrated system of design that encompasses not only agriculture, horticulture, architecture and ecology but also economic systems, land-access strategies and legal systems for businesses and communities (for example, see Mollison and Holmgren 1978).

19 See www.climatecrisis.net, accessed 13 July 2009.

20 See www.gatesfoundation.org, accessed 20 April 2010.

21 See www.clintonfoundation.org, accessed 20 April 2010.

22 See www.fondationbrigittebardot.fr, accessed 20 April 2010.

23 See wkkf.org/default.aspx?LanguageID=0, accessed 13 July 2009.

13.2.3 A vision for sustainable consumption and production of food in Europe

What could a more sustainable system of production and consumption of food in Europe look like and what would be important elements for this? Figure 13.1 presents an overview of an improved system, the main elements and strategies. From all case-study research and expert discussions taking place in the SCORE! project the following directions for future developments of SCP of food can be extracted. We need to:

- Change our diets towards less meat and dairy products and change our food shopping and food processing behaviour
- Influence mainstream agricultural production to become more sustainable
- Amplify, multiply and upscale the more sustainable niche strategies
- Take the best of both two sides (mainstream and sustainable niches) and add other functions such as energy production into a new and radically multi-functional agricultural production system
- Reduce and deal with agricultural and food waste (and methane emissions) throughout the whole value chain in a much more intelligent and efficient way

The following scenario tries to picture a vision of a sustainable food production and consumption domain in Europe including these strategies.

FIGURE 13.1 Elements and strategies of a more sustainable agri-food system

Source: Ursula Tischner

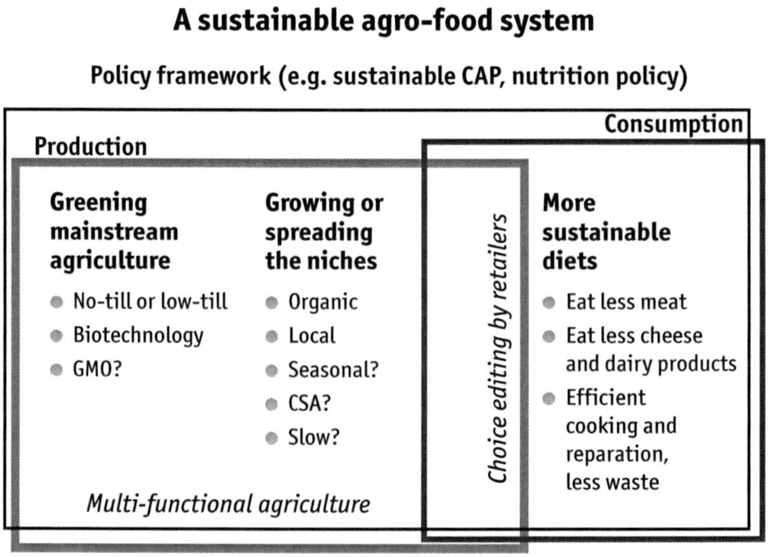

Scenario 13.1 Sustainable consumption and production of food in Europe in 2020

- In 2020, as the prices for oil and transportation have continuously increased, the local production of food has become much more popular in Europe—and more reasonable because of imported foodstuffs being more expensive
- Owing to rising world market prices for food and the increasing demands for meat and dairy products, these types of animal-based foodstuffs have become more expensive and the demand for meat and dairy products in Europe has decreased. This was supported by campaigns by the World Health Organisation (WHO), the UN Food and Agriculture Organisation (FAO) and the EU, which was titled 'Three no-meat days per week to get healthy and sleek', suggesting that consumers each week should have at least three days without meat in their meals. This was a concerted action of health experts and doctors, national governments, educational institutions, the churches, NGOs and others. Main arguments for the campaigns were the negative health effects of eating too much meat and dairy products and the cost induced by that on public health systems, climate-change issues and animal welfare. In addition, in public purchasing strategies and public canteens all over Europe (schools, hospitals, municipalities, etc.) the reduction in meat consumption was implemented. Thus the demand and production volume of livestock in Europe went down by half, making more land available for production of food that is eaten directly by consumers and for production of renewable energy
- Driven by prospects of higher incomes farmers were teaming up with local municipalities and energy producers and added renewable energy production to one of the functions of agriculture. They developed multi-functional agricultural systems where the land is used mainly to grow food products, then the agricultural and food production waste that is not needed in agriculture as fertilisers for the fields is used in bio-gas plants to create energy. In 2020 Europe's complete demand of natural gas was substituted by bio-gas[24] thus creating independence from gas deliveries from external nations. Also, around 15% of European cars are driven by bio-methane instead of mineral oil or natural gas. Bio-gas has been proven to be the much more efficient fuel compared with bio-ethanol or other bio-fuels, because farmers can produce four times more bio-gas than bio-fuel from the same amount of land.[25] The European law on renewable energy translated to national legislation made it possible for producers of climate-friendly renewable energy to be guaranteed sales for a fixed price and receive tax reductions

continued over →

24 See for example the German study that in 2020 natural gas could be completely substituted by biogas produced within Europe. The potential of bio-gas production by European agriculture is calculated as 500×10^9 m^3 annually; see www.umweltschutz-news.de/123artikel1693rss.html, accessed 13 July 2009.

25 For example, see the bio-gas activities in Sweden, described at www.svenskbiogas.se/sb/english, accessed 13 July 2009.

- At the same time by adopting low-till and no-till agricultural practices European farmers are generating more sustainable soil conditions, same or better yields and additional income from energy companies, which are using the agricultural land as a carbon sink for power generator emissions. Thus farms are becoming zero-carbon companies
- Furthermore farmers, supported by research, are adopting several new sustainable farming practices such as: permaculture techniques, to revive depleted natural areas; aquaculture techniques, especially fish production, to reduce overfishing in combination with algae production as CO_2 sinks and feed for the fish or to produce bio-fuel;[26] and sustainable greenhouse production (Nichols and Christie 2002) based on solar energy, heat pumps, re-use of water and waste and so on
- Owing to ongoing public criticism—caused by a recurrence of negative impacts on bio-systems, and media news stories about unfair pressure put on farmers in developing countries by multinational companies who sell them GMO crop systems and chemicals—major parts of the European population have been strongly opposed to genetic modification in production of foodstuff. Consequently, GMO production in Europe and the use of GM ingredients in food sold in European markets has not been successful. Through a redesigned labelling system for GM content in foodstuffs and the production chain (see below), consumers are able to identify and avoid such products in the supermarkets
- A new meta-label has been created by the European Commission in cooperation with agricultural and food producer organisations, NGOs and large European supermarket chains in a bottom-up, multi-stakeholder process. This label includes information on: nutrition; GM content in products and processes; origin of the largest components by weight; and location of production, method of production (conventional, eco-efficient conventional, semi-organic, strictly organic) and social aspects. It took some time to develop, but, because the most important stakeholders were involved, were supportive of the final version of the label, and promoted it to consumers from their various perspectives, it found broad market acceptance
- Driven by new research results and better cooperation between actors in organic and conventional agriculture, new models of eco-efficient conventional, semi-organic and strictly organic agricultural production systems are emerging. For instance no-till and precision farming practices have been added to organic farming methods, organic fodder is more widely used in conventional livestock production, and cleaner production methods are widely used in conventional farming and food processing, etc.

continued opposite →

26 For example, see www.algaelink.com or www.oilgae.com, both accessed 13 July 2009.

- A zero-waste strategy for agricultural and food products has been developed as well, based on the assumption that there should simply be no waste in agriculture and food production. All organic waste is re-used in natural cycles or to produce energy in bio-gas plants throughout Europe; furthermore, all packaging is made from bio-plastic, produced from agricultural resources, and thus is biodegradable and can be fed into the same re-use infrastructure as the food waste itself. All food packaging that is not biodegradable is made out of recyclable and recycled materials (glass and cardboard) and fed back in the recycling infrastructures. Of course, this kind of cyclic economy requires 'clean production practices', i.e. no use of any hazardous substances that can accumulate in the natural environment or human body
- In line with these developments, in the CAP reform of 2015 it was decided that subsidies are paid only to these three forms of agriculture (on a scale from 'eco-efficient conventional' to 'semi-organic' to 'strictly organic' with the last being awarded the highest subsidies); and more advanced minimum standards for livestock production and animal welfare have been formulated. In parallel, European legislation related to chemical use (REACH *et al.*) has been advanced to include restrictions on the use of biocides, fertilisers, antibiotics, hormones, etc.
- Thus the multifunctional agricultural sector has gained increasing economic importance in Europe, with more and better jobs in agriculture and a growing share of European GDP. At the same time, the sustainability of the agriculture and food sector has increased considerably and nutrition-related health problems of European citizens has been reduced (for example, obesity, diabetes, lactose intolerance and food allergies have been contained by changed diets)
- These European approaches have had an influence elsewhere: for example, the European meta-label for food became mandatory for food producers worldwide, if they wanted to sell food products in Europe—therefore many countries (and companies) chose to adopt a similar label. The trend in Europe of eating less meat and dairy products (notwithstanding strong opposition from meat and dairy producers) is also becoming evident elsewhere, especially in countries where the population has been enduring similar nutrition-based health problems. Large food processors have developed new and popular plant-based convenience products and snacks, and even multinational fast-food chains have changed their menus substituting animal-based fast-food products with plant-based ones

13.3 A brief comparison with the mobility and housing domains

The SCORE! project has also provided analyses of two other priority consumption domains: mobility, and housing and energy use. The result of this work has been summarised in two other books that have been published alongside this book in the System Innovation for Sustainability series (Geerken *et al.* 2009; Lahlou *et al.* 2010). This section gives a short comparison of the main differences and similarities across domains, using some key concepts introduced in the first SCORE! book (Tukker *et al.* 2008) and elaborated in Chapter 2 as well as the conclusions of this book. It concerns:

- Characteristics of the system: is there agreement on the sense of urgency of change, the means of realising change and the key players in the system that may drive change?
- Potential role of each actor in the 'triangle of change': what could consumers, producers and governments do?
- What factors contribute to windows of opportunity for change, and what are the main challenges?

Table 13.3 gives this comparison, summarising information from Chapter 2 and the concluding chapter of each book. We see here that in the different domains rather different strategies for change may have to be chosen, since the three fields have rather different characteristics.

When it comes down to agreement on the sense of urgency and the means to realise change, the food area appears to be a complicated one. Some would pursue biological and local food production, whereas others point at the fact that precision-farmed food produced at the place where local conditions are optimal can be sustainable given the relatively low impact of (say, ship) transport. GMOs are outright controversial: seen as a means to enhance food production sustainably by some, and seen as wholly unnatural by others. The sense of urgency in the areas of mobility (energy use, congestion) and housing (energy use) are much more widely shared; here, the only dispute is to what extent one could pursue an outright reduction of mobility or housing area needs. The housing area is also the domain where a possible direction for solving problems seems clear (energy-neutral housing), whereas the other areas still need more convergence in terms of insight into what approaches would work.

When it comes down to the potential role of actors in the triangle of change (an issue that will be elaborated in the domain central in this book in the next section), one sees important differences too. Unlike in the case of mobility and housing or energy-using products, in the food area consumers have a relatively high degree of freedom with regard to consumer choice.[27] Although they are obviously constrained by habits, routines and so on, they can easily switch food choices and even diets in radical ways. Such

27 In the area of housing and energy use consumers in most EU countries now also have a free choice of energy supplier, but the practical hurdles in this field seem still to be significant.

TABLE 13.3 **Comparison of important issues for sustainable consumption and production (SCP) in the three SCORE! domains: food, mobility, and housing and energy**

	Food		**Mobility**		**Housing and energy**	
	Importance	Comment	Importance	Comment	Importance	Comment
The system						
Agreement on goals and sense of urgency	Low	There is no full consensus what a sustainable food system looks like There is dispute about the sustainability benefit of GMOs compared with regular food, biological compared with precision-farmed food, and local grown compared with imported food	Medium	Reducing congestion, CO_2 emissions and emissions of other pollutants are generally accepted goals. The need for an absolute reduction of mobility is much less accepted	High	Zero-energy houses and low-energy products are suggested as answers. Limiting the housing area per person to reduce energy use is not, however, feasible
Certainty or agreement about means	Low	See above	Low	There are few success stories of change; the growth in mobility demand seems unstoppable	Medium	Many successful illustrative projects on zero-energy houses are available, but mainstreaming seems difficult

continued over →

	Food		Mobility		Housing and energy	
	Importance	Comment	Importance	Comment	Importance	Comment
The system *(continued)*						
Geographical characteristics	Global	Most modern food chains span the globe	Global and local	There are just a handful of car manufacturers left. However, the local situation highly determines mobility patterns and the availability of alternatives	Local	The building industry is rather locally organised and has local characteristics
Power nodes in the system	–	Retailers Major food companies	–	Car producers Infrastructure providers		Varies per country: Social housing agencies Landowners Developers Customers
The triangle of change						
Potential for consumer-driven change	Medium	Consumers make daily purchase choices and can change diets, though in practice they are locked in by routines, habits and availability of food alternatives	Low	Consumers have often limited choice with regard to their pattern and means of mobility, except for during holidays	Low	In most countries consumers have little power over how houses are built

continued opposite →

	Food		Mobility		Housing and energy	
	Importance	Comment	Importance	Comment	Importance	Comment
The triangle of change *(continued)*						
Potential for producer-driven change	Medium to high	All of the following have proven to be able to drive change: – Retailers (for instance, through choice editing) – Major food companies (for instance, via certification schemes such as the MSC) – Smaller companies (as in the case of Bionade)	Low to medium	Automotive companies at best will realise incremental change Alternative mobility providers tend to stay in niches	Medium	Many illustration projects are available, but mainstreaming seems difficult
Potential for government-driven change	Medium	Government can support change by public procurement and regulation, but the latter will work mainly to set minimum standards rather than driving radical change	Medium to high	Government has a major role in spatial planning, the availability of public transport and encouraging a more sustainable modal split	Medium	Via energy labelling and minimum standards governments can ensure a basic level of energy performance, but radical change is possible only if zero-energy housing is mainstreamed

continued over →

	Food		Mobility		Housing and energy	
	Importance	Comment	Importance	Comment	Importance	Comment
Factors providing windows of opportunity						
	–	Consumer expectations about health, safety and social issues Reformulated CAP	–	High energy prices Congestion	–	High energy prices
Main challenges						
	–	Negotiating a view on sustainable food and implementing this Various views and related production structures may co-exist	–	Overcoming various lock-ins (regional planning, physical infrastructure, the importance of the car industry to many industrialised countries)	–	Mainstreaming of proven practices with regard to low-energy houses Dealing with the existing housing stock

choices can be made on an almost daily basis, whereas when it comes down to mobility and housing the physical lock-ins greatly reduce consumer choice. Compared with the building sector, it seems that the power at the production side in the domains of mobility (car and infrastructure providers) and food (retailers, major food producers) is probably much more concentrated. Interestingly, in the food area one sees that such mainstream players seem more active in pursuing sustainability goals and related management systems than in the car industry, most probably because of a greater flexibility and implicit consumer expectations with regard to health, environment and social responsibility.

The driving forces, main challenges and, with these, the role of government seem rather different in each domain too:

- In the **building area** the main challenge seems to be to mainstream solutions focused on realising zero-energy housing. For new houses, this implies mainstreaming proven solutions, whereas for the existing housing stock the question of how to tackle the problem is still unsolved. A smart innovation and diffusion policy seems the main challenge here
- In the **food area**, one sees an implicit societal negotiation going on about what implies 'sustainable food'. There seems to be, however, a relatively high degree of flexibility in the food system and a supportive role in the background from government may be sufficient
- In the **mobility area**, the problems and lock-ins seem most problematic. Given the important role of infrastructure, spatial planning, the need to provide alternative transport modalities and the fact that the few success stories rely heavily on government incentives (such as the London congestion tax, discussed in the mobility book: Geerken and Borup 2009: ch. 8) a strong integral government role seems inevitable

13.4 Actions by actor group and time-horizon

We have grouped the most important actors in and around the agriculture and food domain into three main clusters (the triangle of change actors): (a) consumers and households, (b) the production and distribution system, including farmers, the food industry and retailers and (c) civil society, including policy, NGOs, the media and intermediate organisations. Another potential view of the system is that change can happen from the production side, from the market or from the consumption side. All these changes can be influenced by policy, media, research, intermediate organisations and actors in the other clusters. Box 13.6 gives an overview about the more proactive actors regarding sustainability in the agriculture and food domain.

Box 13.6 Proactive actors in the domain agriculture and food

Business

- Conventional companies in the food industry and retail that include sustainability in their unique selling proposition (USP)
- (Grass-roots) sustainability-oriented companies in the food industry and retailing that are becoming more professional in their operations
- Online retailers and brokers of sustainable food
- Companies integrating consumers more actively into innovation, design and communication[28]

Consumers and citizens

- Cultural creatives and LOHAS, celebrities and role models, TV cooks and doctors promoting sustainable cooking and eating
- People experiencing life-changing events, and people with special food needs (such as those with allergies as well as vegetarians and vegans)
- Active consumer groups taking action and starting movements towards more sustainable food

Policy

- Progressive politicians who like to profile on these issues across party political interests and who speak out in public
- Governments and municipalities stimulating more sustainable food production at a local level, such as by providing start-up subsidies that enable low-risk pilot projects to get off the ground or allow learning-by-doing and by entering into public–private partnership and green purchasing
- Governments, on an intermediate level, that strive towards goals that are difficult to realise through partnerships of (commercial) actors, such as fighting obesity by promoting healthy food and exercise
- Governments on a global level, jointly active to foster fair trade and ensure world food security

Research

- Scientists that speak out in public, such as experts from the Intergovernmental Panel on Climate Change (IPCC)

continued opposite →

28 New (sustainable) products and services must accurately respond to user needs if they are to succeed in the marketplace. Thus, integration of consumers and customers in the development of new products and services can be a promising option, especially if sustainable innovations are to increase their chances of market success.

- Research and knowledge-exchange networks, joining forces across EU, trying also to gain influence on policy (for example, SCORE!)
- Researchers developing new technologies for enabling efficient agriculture on a small scale and with diverse products, including renewable energy
- Researchers identifying more sustainable options in the agri-food domain and clarifying some of the potential areas for trade-offs

Other stakeholders

- Strong NGOs, test organisations and media that focus on sustainable food issues, inform and mobilise consumers and set up and get involved in projects
- Lobby organisations for sustainable food in Brussels
- Consumer consulting organisations that promote sustainability
- Educational organisations not only teaching the subject but also setting up cooperatives in practice, such as sustainable food in schools
- New internet communities concerned with sustainable food, such as in blogs

In the following we summarise measures that different groups of actors can and should take to move the agriculture and food system towards more sustainability. To introduce a time-horizon in this action plan short-term, medium-term and long-term measures are discerned.

Usually, short-term measures are in line with the meta-trends and factors described above: that is, the 'rules of the game' or of the landscape that cannot easily be influenced by market actors.

Medium-term measures are based more on common views (that is, a general consensus in society that food should be healthy and fair) where it is not really clear which means will reach the goal in the best way. Thus some research and experimentation might be needed.

The more long-term measures are long-term because, so far, the goals and means are somewhat controversial. Thus it might need some changes in paradigms and perceptions in society to reach common agreement on the measures needed (for example, regarding the question of the paradigm of growth or of the consumption of meat).

13.4.1 Short-term measures: means and ends are not too controversial

Short-term time-horizons in the agricultural and food sector should cover around five years because the agricultural system naturally and historically is slow to adopt and follow changes and trends in the marketplace. So what can different actors do in the short term to move the agri-food system towards more sustainability?

13.4.1.1 Conventional farmers

Conventional farmers could:

- Adopt cleaner production methods, using decreasingly fewer environmentally problematic chemicals
- Explore no-till and low-till methods, precision farming and the option to shift to organic agriculture
- Consider the multi-functions of agriculture, such as for energy production

13.4.1.2 Organic farmers

Organic farmers could:

- Consider whether growth is possible
- Explore whether vertical integration of production makes sense (integrating more production steps at the farm)
- Set up and/or use local food networks and local retail structures such as farmers' markets or sell from their own farms
- Consider the multi-functions of agriculture, such as for energy production
- Use networking to improve market power
- Become more professional and efficient

13.4.1.3 Conventional food producers

Conventional food producers could:

- Introduce clean production methods in food production
- Use organic and fairly traded ingredients as much as possible
- Use Bio-label, Fairtrade and other labels if possible and appropriate
- Bring back processing and value creation to the region or even to one producer
- Consider whether production that has been outsourced to other continents can be brought back to Europe to avoid transport and transportation costs and improve freshness

13.4.1.4 Sustainable and/or organic food producers

Sustainable and/or organic food producers could:

- Professionalise and scale up their sustainable food business
- Gain new and broader consumer groups for sustainable offers

- Tackle first-mover groups as well as groups paying special attention to sustainable food, from LOHAS to people in life-changing phases
- Improve the design, branding and communication of their sustainable offers to become more successful outside niches markets and to gain broader target groups and use credible labels for this purpose (see Chapter 10)
- Be aware that 'being sustainable' is not the only or main reason for consumers to prefer sustainable food products—there should be a bundle of expected qualities embedded in the offer (price, aesthetics, function, taste), with sustainability being an additional quality: sometimes it is the decisive factor in consumers preferring the product; often it is only a 'nice thing to have'
- Seduce rather than force customers to shift from 'conventional' to 'sustainable' products; it can be a successful strategy even to increase the bundle of qualities of sustainable products by adding more customer benefits such as savings in time and effort, enabling new experiences and so on
- The bundle of specific qualities of an offer has to be carefully designed to fit sustainability aspects, the specific product or service and company image and the specific target group(s) addressed[29]

13.4.1.5 Farmers (cooperatives), energy companies and municipal waste collectors

Farmers (cooperatives), energy companies and municipal waste collectors could:

- Explore and set up infrastructure to use food and other organic waste to create energy
- Explore the carbon sequestration possibilities of agriculture

13.4.1.6 Large (conventional) retailers

Large (conventional) retailers could:

- Practise sustainable choice editing as much as possible, suitable to their image; for example, list more organic, local, seasonal and fair-trade food products and use this as a means of quality competition instead of focusing only on price competition
- Consider introducing sustainable food services, such as delivery and food subscription schemes for organic, local and seasonal food
- Set up new partnership with local (organic) food producers to increase consumer identification with the offer and freshness of the food products

29 For example, see the ecobiente Project: www.econcept.org, accessed 13 July 2009.

13.4.1.7 Smaller local retailers

Smaller local retailers could:

- Professionalise, multiply and upscale their activities
- Improve logistics and customers' individual transportation issues to become more sustainable, such as by encouraging consumers not to use their car to get to markets
- Introduce efficient food delivery and subscription systems for local, seasonal and organic offers
- Use sustainable means of transportation
- Team up with other supportive actors in civil society, municipalities and other similar networks

13.4.1.8 Restaurants and other public food services

Restaurants and other public food services could:

- Offer sustainable, organic and vegetarian dishes and recipes in a high-quality and delicious way[30]
- Educate consumers about healthy, delicious and sustainable food and recipes

13.4.1.9 Consumers

Consumers could:

- Practice ethical and political consumerism
- Consider how to change their food shopping, preparation and eating habits to become more sustainable, especially at life events when habits are changing anyway
- Select shops with sustainable food offers
- Travel to food shops by bike or public transportation
- Consider labels when selecting products
- Adopt more sustainable diets, especially by eating less animal-based food (less meat, dairy products and eggs)
- Prefer more organic, local, seasonal and fairly traded products
- Try to reduce cooking time and prefer eco-efficient kitchen appliances

30 For example, see the case of Bux in Munich and De Kas in Amsterdam, at www.restaurantdekas.nl, accessed 13 July 2009.

- Use leftovers of meals for the next day or for other purposes
- Compost food waste
- When going out for food, select restaurants with delicious and sustainable meal offers
- Become more active and set up purchasing cooperatives, buying sustainable products collectively to achieve economies of scale
- Start or join CSA schemes, or set up a community garden, where consumers share pieces of land to produce their own food[31]
- Talk to local politicians, neighbours, friends and colleagues and discuss the issues of sustainable food
- Join internet communities on the issue
- Mobilise other people on the issue
- But, still, enjoy food!

13.4.1.10 Producers, consumers (and policy) together

Producers, consumers (and policy) together could:

- Set up local CSA schemes, bringing together producers and consumers in local food networks
- Pay attention also to the environmental efficiency of such networks

13.4.1.11 Policy-makers

Policy-makers could:

- Launch research programmes and concerted actions to get more clarity and agreement about effective strategies and measures for more sustainable consumption and production of food in Europe and globally
- Develop a sustainable agriculture and food programme at the regional, national or EU level
- Agree on European goals for sustainable agriculture and food and harmonise or standardise national and regional legislation and regulation in line with this
- Develop a stronger concept for the European CAP to support more sustainable farming and include it in the reformulation in 2015

31 See the case of Les Jardin de Ceres, in Palaiseau, a small town close to Paris (EMUDE 2005).

- Influence WTO discussions to tackle food as a special issue, which is not just about global trade but critical to the survival of human kind—issues such as food security and sustainability should receive a higher priority than trade issues
- Continue and increase and redesign funding and support schemes for more sustainable farming and food production: for example, include principles of sustainable agriculture in start-up grants for young farmers who want to take over farms or start their own businesses[32]
- Support and secure the availability of independent, credible and clear labels for the most important aspects of the sustainability of food
- Start, continue and strengthen sustainable public purchasing programmes, such as by integrating in public purchasing the preference for more sustainable and less animal-based food or by establishing long-term contracts between municipalities and local farmers
- Launch a sustainable public food programme, as in the UK, where the Department of Health is launching a new 'healthier food mark' to get the public sector providing healthier, more environmentally sustainable food; this will help consumers to access healthier choices when eating out
- Provide information that considers both the health and environmental aspects of food (see Cabinet Office 2008)
- Start regional implementation groups for sustainable agriculture and food, such as the Framework for Change (see Chapter 3), which was one of many regional implementation groups established following the publication of the UK Strategy for Sustainable Farming and Food

13.4.1.12 Policy and all other actors

Policy and all other actors could:

- Improve sustainable food education by starting and supporting educational programmes for sustainable food and agriculture, such as educational programmes for farmers about more sustainable production methods and for consumers about more sustainable diets and cooking and so on. In many studies consumers with higher education levels are reported as being more aware of food-related health and sustainability problems and of changing their behaviour as a consequence; these consumers could be motivated to move towards more sustainable food choices (see Chapter 12). Thus education generally and about the connections between personal behaviour and (food) choices, well-being, health and sustainability seems to be an important factor for change

32 The EU rural development policy for the period 2007–13 foresees that member states will offer young farmers installation aid as a single payment or interest payments up to €55,000.

- Improve sustainable food communication. It is very important to address specific target groups in such communication. A 'one message for all' approach often does not work. Preferably, the most important and trend-setting target groups should be identified and addressed. It is not sufficient only to transport rational information. Consumers need to understand the story behind the product or issue and like to be addressed in an emotional way. The message should be connected to their own values and dreams. Still, most consumers today have a rather romantic idea about agriculture, with cows, pigs and hens running free over green fields. This is harshly contradicted by the reality of industrialised agriculture when it is shown in the media. This dilemma in consumer perception has to be acknowledged and solved

13.4.1.13 Non-governmental organisations and media

Non-governmental organisations and media could:

- Inform, educate and mobilise people towards sustainable food
- Cooperate with and consult consumers as well as producers
- Lobby for sustainable food, especially in the field of sustainable food labels and sustainable diets
- Discuss the credibility of labels and support the credible and trustworthy labels
- Function as testing and watchdog organisations, such as the German *Öko-Test* magazine

13.4.2 Medium-term measures: agreement on direction of change, means uncertain

Medium-term measures are more difficult to implement because larger changes are necessary in the system and there is still uncertainty about the best measures to reach the goal of more sustainable agriculture and food. The medium-term time-horizon can be around 10 years from the time of writing.

13.4.2.1 Farmers generally

Farmers could:

- Apply multifunctional agricultural strategies, diversifying the functions of agricultural land, the products that are produced, and the way in which income is generated
- Select agricultural products in line with climatic changes with the goal to use all parts of the crops as food or raw materials, such as for bio-plastic production, building and insulation materials, cosmetics, medical substances and so on

- Use waste not needed as fertiliser for bio-gas and energy production
- Use soil and crops for CO_2 sequestration and adopt agricultural production methods suitable for this
- At the same time create enjoyable cultural landscapes and living spaces for humans, flora and fauna
- Consider introducing sustainable agri-tourism services, including educational programmes for consumers about sustainable agriculture and food
- Invest in sustainable greenhouses and other more sustainable infrastructure
- Consider setting up a zero-energy (or even plus-energy) and a zero-waste farm: for example, by installing bio-composters and/or bio-gas plants that can even treat organic waste from other sources
- Team up with municipalities and energy producers in these activities
- Help to develop and install technologies to use methane emissions in livestock production for energy generation

13.4.2.2 Large (conventional) producers

Large (conventional) producers could:

- Advance cleaner and eco-efficient production technologies and distribution
- Consider if it is possible to shift from larger to smaller, more local and more diversified businesses, as a socioeconomic sustainable strategy for one country or region; to avoid environmental inefficiencies this has to be done in an efficient way or by using, for example, renewable energy and manual labour to reduce the negative environmental impacts of 'inefficiencies'. This would also need radical new technologies for efficient small-scale farming, differentiated processing, new energy and water efficiencies and new transport and fuel technologies to reduce the impact of more local transport and so on (see also Chapter 3)
- Demand and support development of this kind of eco-efficient technology

13.4.2.3 Small (sustainable) producers

Small (sustainable) producers could:

- Scale up and multiply sustainable niche concepts but take care to keeping the sustainable qualities
- For the above, work with researchers and large producers and team up to develop efficient small-scale production, processing and logistics technologies, principles and means

- Apply no-till or low-till, precision farming and other techniques, if interesting and feasible

13.4.2.4 Policy

Policy could:

- Implement and advance a more sustainability oriented CAP and rural development and agricultural policy
- Introduce dynamic and progressive performance indicators for agriculture, asking for increasing sustainability
- Evaluate, improve, continue and strengthen all successful measures from legislation and regulation to incentives and market-based instruments at the European, national and regional level
- Support the export of more sustainable agricultural production methods, products and lifestyles
- Influence international discussions and agreements on agriculture and food to become more sustainable
- Create national as well as international task forces on the future of food and agricultural production, as these fields are essential for the survival of humans; for instance, there could be a food and agricultural task force within the UN Marrakech Process[33] and a specific chapter on food and agriculture in the European Action plan on Sustainable Consumption and Production (Makela 2008)[34]

13.4.2.5 Policy and all other actors

Policy and all other actors could:

- Create one sustainability label for food products, combining the most important aspects of sustainability in one scheme: nutritional factors, origin of the components of greatest weight and location of production, method of production (conventional, organic, etc.), content of GMOs in product and process and social aspects

33 See esa.un.org/marrakechprocess, accessed 13 July 2009.

34 See also ec.europa.eu/environment/eussd/escp_en.htm, accessed 13 July 2009.

13.4.2.6 Consumers, producers and architects/project developers

Consumers, producers and architects/project developers could:

- Re-introduce food production in cities, for example, by vertical agricultural concepts

13.4.3 Long-term measures: controversy over means and ends

The long-term measures are difficult to implement over the next ten years because they are not necessarily in line with some major trends or paradigms predominant in European society. Thus more time is needed to discuss these issues, and they may or may not be implemented over a longer time-horizon. However, some of the topics discussed below are essential for the fundamental shift of European society needed for it to become truly sustainable.

13.4.3.1 Policy and all stakeholders

- Policy and all stakeholders need to question the growth paradigm: with a shrinking population in Europe and the adoption of healthy and sustainable diets by consumers there will be less demand for food products in Europe. The opposite is the case in developing and emerging countries, where the population and demand for food products is still growing and will continue to do so. Thus the European food and agricultural industry either has to produce less food over the long run, which would give space for other 'agricultural' products such as bio-fuels and bio-gas products, or the industry has to increase the export of food products, generating higher demands for packaging, transportation and energy consumption as well as more emissions
- Policy, together with industry and other stakeholders, should develop roundtable discussions on the future of agriculture and food industry within a shrinking European population, globalisation, climatic changes, the situation in developing countries, food and water scarcity and so on. Most of the niche experiments described above that are based on regional approaches are good examples of how small businesses can survive in a region over the long run without any growth but based on a regional network of producers and consumers and on cooperation instead of competition
- There is also a need to solve conflicting issues. When is it more sustainable to aim to reduce food miles and increase local production and self-sufficiency of regions, and when is it more sustainable to support fair trade with developing countries? How is it possible to generate food security for all? Is GMO technology needed for this and, if so, how can it be controlled so that the risks are minimised and the power distribution be skewed more towards farmers and consumers rather than towards large multinational companies?

- Concepts and definition of sustainable agriculture and food need to be customised to different regions, climates and cultures. Furthermore, a concept of pricing for natural resources and natural areas needs to be developed where nature has a price and there is an external cost so that the consumption, pollution and degradation of nature can be internalised in economic activities. This may be reflected in taxation and subsidies

13.4.3.2 Consumers and producers

The concept of production and consumption needs to be reformed. The niche experiments of CSA and community gardens are pointing toward a new understanding of production and consumption. They are the translation of the old system of subsistence, where everybody would produce or hunt their own food, into a modern system of cooperation between landowners, people having the right knowledge and skills to produce food on the land, and people without land, time and skills to invest but having shares in the system. They can together organise the agricultural production and share the results and consume the products. A fair system of investment and payment has to be set up and should be organised by a central and neutral organisation.

13.5 Conclusions

At the end of this book and the European project SCORE! we would like to draw some overall conclusions regarding sustainable consumption and production in the agriculture and food domain.

First, it has to be acknowledged that the agri-food area has received a lot of attention and is getting even more attention because it is very much related to environmental, cultural and health issues in society (from obesity, through food safety, to security); it is also of great relevance to political and economic issues, such as agricultural legislation at the European level and the huge budget invested in agricultural subsidies in the EU, despite the fact that agriculture makes a relatively small contribution to GDP in most European countries.

Furthermore, we must state that there is still controversy and a lack of clarity about what the term 'sustainable food and agriculture' means. There is a lack of a common and harmonised definition beyond the diverse definitions of organic agriculture and food, or the different approaches to define sustainable agriculture—and there are many contradictory issues in the domain. This ranges from the discussion about whether local food is more sustainable than imported food, whether social issues such as fair trade are more important than ecological issues, such as eco-efficiency and energy consumption through transportation (food miles), whether it is better to 'green' mainstream agriculture, including issues of genetic modification and biotechnology, or whether we should shift as much as possible to organic agriculture, including a shift in diets towards less meat and to multi-functional agriculture and so on.

Here, we clearly need more research regarding different food products, different types of agriculture, in different geographical and climatic conditions, covering the whole life-cycles and production–consumption chains of the products and a complex set of criteria in social, environmental and economic dimensions of sustainability. And sometimes political decisions by policy actors, industry, consumers and other stakeholders might be needed when there is no clear evidence to hand.

One way forward in this open and unsure situation is a pragmatic approach: there should be a robust policy framework developed at the European level, such as a reformulated CAP and rural development policy to support more sustainable agriculture as well as restrictions on chemical use in agriculture in line with this. Then each country and region could define its own national or regional strategies for sustainable production and consumption in agriculture and food, tackling questions of self-sufficiency and environmental efficiency but keeping the global situation (globalised and fair trade as well as food security issues) in mind.

The whole process should involve the most important stakeholders from the triangle of change—policy, producers, civil society (consumers, NGOs, the media, etc.)—in a participatory way and should be transparent, well communicated and covered by the media. The special roles that each of the partners in the triangle of change can play in this shift towards a more sustainable production–consumption system in the short, medium and long term has been described in Section 13.4.

Analysing the developments in the agri-food domain in recent decades it became clear to us that when there is a kind of equilibrium between these three groups (consumers, producers, and policy and civil society) the situation in the domain is relatively stable, but when one or more of the groups or sub-groups are dissatisfied with the situation then there is momentum for change. This individual or joint dissatisfaction initiates significant change only if a critical mass is reached, such as several individuals jointly expressing their dissatisfaction or acting to overcome it. How big the critical mass is will depend on several factors, such as the size and power position of the actors. It also depends on how the public domain, such as mass media and leading persons in society (role models), take up the issue and publicly talk about it and amplify it.

Overall, it is clear that considerable inertia exists in the agriculture and food sector not only because of its institutional structure but also, perhaps, because most of the agricultural sector is dominated by males over 50 years old.[35]

A more sustainable food consumption and production system is achievable but requires a combination of technological and industrial structural change, on the one hand, and behavioural and attitudinal change on the other. Making changes to behaviour and attitude is difficult for consumers and producers alike, particularly in a highly political sector where individuals and institutions hold deeply entrenched views and where the subject is so connected with the survival of human kind. Here societal as well as regional dialogue and concerted action is needed to develop transition strategies that reflect the starting point of different actors in the production and consumption system

35 The age distribution of farmers in Europe is as follows: 52% of all EU farm holders are over 55 years old, and only 7% are under 35 years old (Eurostat 2008a). In Finland the average age of farmers is 49 years old (MLF 2006); in Austria in 2004 the average age of farmers was 48.5 years old (Larcher *et al.* 2007).

in terms of knowledge, values, motivation, infrastructure and scope of action. In addition, changes to the initial conditions (regime) to become more supportive for sustainability in the domain are required.

It is also very important to note that we have to avoid discussions about sustainable production and consumption in the agriculture and food domain (but also generally) that question whether we need to convert to organic farming and sustainable niche concepts *or* whether we need to 'green' mainstream business and consumption. It is very simple: we need to do both and we need to act fast. There is a sense of urgency emerging in almost all stakeholder groups through the far-reaching discussions about global warming, global inequity and unfair development as well as personal health issues. And, while we are working to develop an overall goal and strategy for sustainable consumption and production regionally Europe-wide and even globally, at the same time we need to start acting quickly and at several levels: in the mainstream *and* in experiments and niches, politically as well as through business initiatives and consumer activities. The short-term measures as suggested above can be implemented *right now*. While the experiments and niches can develop much more radical SCP strategies the 'greening' of mainstream systems can be more efficient in terms of scale effects.

In all discussions and strategy development related to food we need to acknowledge cultural issues. Nutrition is not just about the intake of calories and vitamins. Eating habits are deeply rooted in the culture and history of a society and influenced by family, education, peer groups, advertisements, religion and other external as well as internal aspects. In the end, how and what we eat is an expression of our culture, and thus an approach to sustainable food must leave room for a great deal of diversity: there will never be a one and only sustainable food strategy and sustainable diet, but multifaceted different models that we hope can be realised within one (political and economic) framework.

An understanding of sustainable food as suggested above adds at least one dimension to the triple-bottom-line approach. It is about:

- People
- Planet
- Profit
- Pleasure

References

BBC News (2008) 'Q&A: Common Agricultural Policy', BBC News, 20 November 2008; news.bbc.co.uk/2/hi/europe/4407792.stm#where, accessed 13 July 2009.

Bentley, M. (2001) 'Consumer Trends and Expectations: An International Survey Focusing on Environmental Impacts', in *UNEP Industry and Environment Review* 23.4 (October–December 2000).

Cabinet Office (2008) 'Food Matters: Towards a Strategy for the 21st Century' (London: Cabinet Office, Strategy Unit).

CEC (Commission of the European Communities) (2007) 'Council Regulation (EC) No 834/2007 of 28 June 2007 on Organic Production and Labelling of Organic Products and Repealing Regulation' (EEC 2092/91; 20 July 2007; Brussels: CEC).

Cleaner Production Germany (undated) 'Agriculture'; www.cleaner-production.net/index.php?id=73&L=1&tx_exozetcpgproject_projects[cat]=2417, accessed 13 July 2009.

Defra (Department for Environment, Food and Rural Affairs) (2002) 'Strategy for Sustainable Farming and Food: Facing the Future' (London: Defra).

—— (2006) 'Food Security and the UK: An Evidence and Analysis Paper' (London: Defra).

Deloitte and GMA (Grocery Manufacturers Association) (2009) 'Finding the Green in Today's Shoppers: Sustainability Trends and New Shopper Insights'; www.gmabrands.com/publications/greenshopper09.pdf, accessed 18 April 2010.

DG Environment (2008) 'LIFE on the Farm' (European LIFE Programme; Commission of the European Communities).

ecobiente (2007) 'ecobiente: Successful Design of Sustainable Products'; www.econcept.org/index.php?option=com_content&task=view&id=61&Itemid=44&lang=en, accessed 13 July 2009.

EMUDE (2005) 'Les Jardins De Ceres, Palaiseau, France', 20 June 2005; www.sustainable-everyday.net/EMUDE/?p=70 accessed 13 July 2009.

EOS Gallup Europe (2000) 'Farmers' Attitude towards the CAP' (Eurobarometer Flash Survey 86 for DG Agriculture; EOS Gallup Europe).

Eurostat/CEC (Commission of the European Communities) (2008a) *Agricultural Statistics: Main Results, 2006–2007* (Eurostat Pocket Books, 2008 edition; Luxembourg: Eurostat).

—— (2008b) *Food: From Farm to Fork; Statistics* (Eurostat Pocket Books; Luxembourg: Eurostat).

Geerken, T., and M. Borup (eds.) (2009) *System Innovation for Sustainability. 2: Case Studies in Sustainable Consumption and Production—Mobility* (Sheffield, UK: Greenleaf Publishing; www.greenleaf-publishing.com/SCP2).

IGD (Institute for Grocery Distribution) (2005) 'Connecting Consumers with Farming and Farm Produce' (Watford, UK: IGD).

JRC (Joint Research Council) (2003) 'Integrated Pollution Prevention and Control (IPPC) Reference Document on Best Available Techniques for Intensive Rearing of Poultry and Pigs'; eippcb.jrc.es/reference, accessed 13 July 2009.

—— (2006) 'Integrated Pollution Prevention and Control Reference Document on Best Available Techniques in the Food, Drink and Milk Industries'; eippcb.jrc.es/reference, accessed 13 July 2009.

Krier, J.-M. (2006) 'Fair Trade in Europe 2005: Facts and Figures on Fair Trade in 25 European Countries' (Fairtrade Labelling Organisation International [FLO], International Fair Trade Association [IFAT], Network of European Worldshops [NEWS!] and European Fair Trade Association [EFTA]).

Lahlou, S., M. Charter and T. Woolman (eds.) (2010) *Case Studies in Sustainable Consumption and Production. 4: Housing/Energy Using Products* (Sheffield, UK: Greenleaf Publishing; www.greenleaf-publishing.com/SCP4).

Larcher, M., S. Vogel and R. Weißensteiner (2007) 'Einstellung und Verhalten von Biobäuerinnen und Biobauern im Wandel der Zeit: Ergebnisse einer qualitativen Längsschnittuntersuchung', DP-28-2007, Instituts für nachhaltige Wirtschaftsentwicklung, Universität für Bodenkultur, Vienna.

Lehmann, M., and P. Sabo (2003) 'Multiplikatoren', in Bundeszentrale für gesundheitliche Aufklärung (ed.), *Leitbegriffe der Gesundheitsförderung: Glossar zu Konzepten, Strategien und Methoden in der Gesundheitsförderung* (Schwabenheim an der Selz, Germany: Fachverlag Peter Sabo): 154-56.

Makela, T. (2008)'Sustainable Production and Consumption Plan', paper presented at the *ESDN Conference*, 30 June 2008; www.sd-network.eu/pdf/doc_paris/ESDN%20Conference%202008%20-%20Plenary_MAKELA.pdf, accessed 13 July 2009

MLF (Ministerium für Landwirtschaft und Forsten) (2006) 'Studie Agrar und Lebensmittelwirtschaft in Finland' (Helsinki: MLF).

Mollison, B., and D. Holmgren (1978) *Permaculture One* (London: Transworld Publishers).

MSTI (Ministry of Science, Technology and Innovation) (2003) *Grønt teknologisk fremsyn* (*Green Technology Foresight*) (Copenhagen: MSTI).

Nichols, M., and C.B. Christie (2002) 'Towards a Sustainable "Greenhouse" Vegetable Factory', *Acta Horticulturae* 578 (www.actahort.org/books/578/578_17.htm, accessed 13 July 2009): 153-56.

Pasricha, A. (2008) 'Global Warning: Climate Change and Farm Animal Welfare', Compassion in World Farming (CIWF); www.fao.org/fileadmin/user_upload/animalwelfare/GlobalWarningExecutive Summary1.pdf, accessed 16 April 2010.

Perricone, N. (2004) *The Perricone Promise: Look Younger, Live Longer in Three Easy Steps* (New York: Warner Books).

Pimentel, D. (1997) 'Livestock Production: Energy Inputs and the Environment', reported in *Cornell University Science News*, 78 August 1997; www.news.cornell.edu/releases/aug97/livestock.hrs.html, accessed 16 April 2010

PoultrySite (2008) 'DEFRA: 21st Century Challenges for Food Set in the UK', ThePoultrySite News Desk, 8 July 2008; www.thepoultrysite.com/poultrynews/15351/defra-21supst-sup-century-challenges-for-food-set-in-the-uk, accessed 16 April 2010.

Rubik, F., and P. Frankl (eds.) (2005) *The Future of Eco-labelling: Making Environmental Product Information Systems Effective* (Sheffield, UK: Greenleaf Publishing; www.greenleaf-publishing.com/catalogue/ecolabel.htm).

Stø, E., H. Throne-Holst, P. Strandbakken and G. Vittersø (2008) 'Review: A Multi-dimensional Approach to the Study of Consumption in Modern Societies and the Potential for Radical Sustainable Changes', in A. Tukker, M. Charter, C. Vezzoli, E. Stø and M. Munch Andersen (eds.), *System Innovation for Sustainability: Perspectives on Radical Changes to Sustainable Consumption and Production* (Sheffield, UK: Greenleaf Publishing; www.greenleaf-publishing.com/SCP1): 248-68.

Tukker, A., M. Charter, C. Vezzoli, E. Stø and M. Munch Andersen (eds.) (2008) *System Innovation for Sustainability. 1: Perspectives on Radical Change to Sustainable Consumption and Production* (Sheffield, UK: Greenleaf Publishing; www.greenleaf-publishing.com/SCP1).

UNESCO (2000) 'Youth, Sustainable Consumption Patterns and Lifestyles' (Paris: UNESCO).

USDA (United States Department of Agriculture) (2001) 'Part V: Conservation and the Environment', in *Food and Agricultural Policy: Taking Stock for the New Century* (www.usda.gov/news/pubs/farmpolicy01/chapter5.pdf, accessed 13 July 2009): 72-87.

—— (2004) 'Irrigation and Water Use'; www.ers.usda.gov/Briefing/WaterUse, accessed 16 April 2010.

About the contributors

Joke Aerts coordinates tea certification and other projects for the Rainforest Alliance. She is based in London and Amsterdam; her fieldwork is mostly in India and Africa.

Tim Cooper was appointed as Professor of Sustainable Design and Consumption at Nottingham Trent University in 2010. His interest in self-sufficiency originated in the late 1970s, when he wrote an undergraduate dissertation on the topic. He graduated from the University of Bath in 1978 with an Economics degree and worked in industry for 15 years before a short spell as a researcher at the New Economics Foundation. In 1995 he was appointed Senior Lecturer in Consumer Studies at Sheffield Hallam University, where he managed the Centre for Sustainable Consumption. He was awarded a doctorate in 2001 for research into the life-span of household goods. He has served as a Specialist Advisor to the House of Commons Environment Committee and was an inaugural member of the EU research network SCORE! (Sustainable Consumption Research Exchange). He manages the Research Network on Product Life Spans and is contributing editor of *Longer Lasting Products* (Gower, 2010).

Paul Dewick is an economist at the Manchester Institute of Innovation Research, Manchester Business School, University of Manchester, UK. He researches and teaches in the area of innovation and environmental sustainability.

Chris Foster is an environment consultant and Research Fellow in the Manchester Institute of Innovation Research, Manchester Business School, University of Manchester, UK.

Bernward Geier is the founder and president of Colabora, a consulting agency based in Germany that draws on his decades of experience in the organic farming movement. Geier represents the Rainforest Alliance in Germany and neighbouring countries.

Matthew Hayes began working in practical organic agriculture in 1984, and trained in environmental biology at Wye College, University of London. He gained an MSc in sustainable agriculture and rural development from Imperial College, London. He moved to Hungary in 1995, and currently manages the organic demonstration garden of Szent István University and teaches related subjects. Further to this, he is a founding board member of Open Garden Foundation, of which he was the managing director for ten years. He is interested in supporting the development of sustainable local organic food systems in the region.

Adina Herde has undertaken research on life events and sustainable consumption in her PhD thesis 'Sustainable Nutrition in Transition to Parenthood'. She currently works as an online editor at the Hochschule für Technik und Wirtschaft (University of Applied Sciences) in Berlin.

Michael Søgaard Jørgensen has an MSc in Chemical Engineering and a PhD in Technology Assessment. He is an associate professor in user participation in technology development at DTU Management Engineering within the section on Innovation and Sustainability. He works within the research tradition of community-based research and action research and is active within the international network of science shops, Living Knowledge. His research areas are environmental management in companies and product chains, sustainable transitions, technology assessment and technology foresight, developmental work, and system innovation and governance in the food sector in relation to organic food, food at work and obesity. He has been member of a national panel on green technology foresight and member of international networks on technology foresight. He is the chairman of the Society of Green Technology within the Danish Society of Engineers and a member of the Danish Board of Technology.

Unni Kjærnes is Senior Researcher at the National Institute for Consumer Research (SIFO), Oslo, Norway. She is coordinator of the Trust in Food project. She has published widely on food consumption, consumer concerns and food policy and is co-editor of *Regulating Markets: Regulating People: On Food and Nutrition Policy*.

Ingrid Kjørstad is a sociologist and researcher at SIFO (National Institute for Consumer Research), Norway. Her research within this area has centred on consumers and animal welfare, participating in the Project Welfare Quality®: Science and Society Improving Animal Welfare in the Food Quality Chain—within the 6th Framework Programme. Kjørstad is currently Research Fellow at SIFO, working in the ICT and Digital Media research group, where her primary interest lies with ICTs and gender.

Cordula Kropp is professor at the Department of Applied Social Sciences in Munich, where she specialises in social future and innovation studies. Essentially asking how societies can prepare themselves to deal with the challenge of sustainable development, her research activities span topics including environmental discourse, science policy interfaces, risk governance and processes of open innovation.

Dr **Oksana Mont** is an Associate Professor at the International Institute of Industrial Environmental Economics at Lund University, Sweden. She has a PhD in Technology, an MSc in Environmental Management and Policy and an MSc in Biology and Chemistry. Her research interests lie in sustainable consumption and production, product-service systems, sustainable lifestyles, innovation studies and waste management. She works in close contact with policy-makers, business and consumer organisations, and teaches at undergraduate, master and PhD levels at universities and in 'educate the educators' programmes.

Helen Nilsson is an environmental researcher based in Sweden. She has an MSc in Environmental Management and Policy. Her research interests lie in sustainable production and consumption of food, especially at the local level, as well as bioenergy crops grown on agricultural land. She has been conducting research on local farmers' markets in southern Sweden, and how they can benefit the local community.

Ingri Osmundsvåg was a researcher at the National Institute for Consumer Research (SIFO), Norway, at the time of writing for this volume.

Martina Schäfer works as a professor of sustainability research at the Center for Technology and Society, Technische Universität Berlin. Her main research areas are sustainable consumption, sustainable regional development and sustainable land use.

Patrick Schroeder works as consultant for the UNEP/Wuppertal Institute Collaborating Centre on Sustainable Consumption and Production (CSCP) and the EU's SWITCH-Asia Programme, supporting projects for sustainable consumption and production in Asia, particularly China. He is also international advisor at the China Association for NGO Cooperation (CANGO) and the China Civil Climate Action Network (CCAN), a position supported by the Centre for International Migration and Development (CIM) in Germany. He holds a BA Hons in Chinese (University of Westminster) and an MA in International Relations (Victoria University of Wellington). He will complete his PhD in Environmental Studies in 2010.

Eivind Stø is the Director of Research at the National Institute for Consumer Research (SIFO), Norway. He gained a Mag.art. in political science from the University of Oslo in 1972. He was an assistant at the Norwegian Election Programme from 1972 to 1976; and from 1976 to 1998 he worked for the Norwegian Fund for Market and Distribution Research. Since 1989, he has been a researcher at SIFO, as Head of Research from 1990, and Director of Research from 1998. He has initiated, participated and coordinated several European projects. His research interests include consumer complaints, consumer policy, sustainable consumption, energy use and nanotechnology.

Professor **Ursula Tischner** studied Architecture and Industrial Design and specialised in Eco- and Sustainable Design of products and services. Following her master's graduation in Germany, she worked for over four years at the German Wuppertal Institute for Climate, Environment and Energy, where she was involved in research and developing concepts such as eco-efficiency, MIPS, Factor 4 and Factor 10, Ecodesign, Eco-Innovation, etc. In 1996 she founded econcept: Agency for Sustainable Design, in Cologne (www.econcept.org). With econcept she carries out research and consulting projects with small and large companies and other organisations on sustainable and eco-design and innovation. She has organised numerous training and educational courses and programs, such as the Sustainable Design Program at Design Academy Eindhoven, The Netherlands, Ecodesign training courses for the Electronics Industry in Korea and the Furniture Industry in Australia. In 2010 she became professor of Sustainability Design at Savannah College of Art and Design (www.scad.edu). She publishes books and organises conferences and networks around Eco- and Sustainable Design, is member of design juries and standardisation bodies such as ISO, and is an evaluator in European research programs.

Dr **Arnold Tukker** joined TNO in 1990 after some time working for the Dutch Environment Ministry. Over time, his focus shifted from life-cycle assessment, material flow analysis and risk assessment to interactive policy-making and sustainable system innovation and transition management. In 1998 he published a book on societal disputes on toxic substances, for which he was awarded a PhD from Tilburg University. He has published about 40 peer-reviewed papers, 5 books, 10 book chapters and 150 other publications, and is frequently asked as invited speaker worldwide. In his career, he has been awarded over €15 million in mainly international research grants. He currently manages the research programme on Transitions and System Innovation within TNO Built Environment and Geosciences, Business Unit Innovation and Environment. This programme was evaluated as one of TNO's top-ranking programmes during the 2006 scientific assessment exercise. Arnold is the initiator and manager of the SCORE! network. From 1 April 2010, Arnold is also part-time professor of sustainable innovation at the Industrial Ecology Program/Department of Product Design, NTNU, Trondheim, Norway.

Burcu Tunçer works as a researcher and communication expert at the SWITCH-Asia Network Facility, which is the single largest Europe-Aid programme focusing on sustainable consumption and production (SCP). She holds masters' degrees in engineering, business administration as well as environmental management and policy. She has eight years of international professional experience in SCP research, and specifically worked on capacity building and environmental assessment within global value chains. Recently, she has researched potential contribution of sustainable consumption and production practices to poverty alleviation, especially through market-based approaches within base-of-the-pyramid markets.

Edina Vadovics has a background in education, environmental management and sustainable consumption. She worked in environmental and sustainability management, and delivered training courses in the field both for companies and students in higher education. She is currently working on her PhD at Central European University, her research focusing on sustainable consumption, low-carbon communities, and new models of well-being. Ms Vadovics is also co-founder and president of GreenDependent Sustainable Solutions Association, a non-profit organisation active in research and awareness-raising related to sustainable consumption and well-being.

Steve Webster is Principal Consultant at Delta Innovation Ltd, a specialist rural and agricultural research consultancy.

Chris Wille is the chief of sustainable agriculture for the Rainforest Alliance. From his base in Costa Rica, he helped organise the Sustainable Agriculture Network, which manages the leading standard for environmentally friendly and socially responsible farming.

Index

Page numbers in italic figures refer to illustrations

SPECIFICATION WRITING

for Architects and Surveyors

Ninth Edition

Christopher J. Willis
FRICS, FCIArb
and
J. Andrew Willis
BSc, ARICS

BSP PROFESSIONAL BOOKS
OXFORD LONDON EDINBURGH
BOSTON MELBOURNE

First edition published in Great Britain by Crosby Lockwood & Son Ltd 1953
Reprinted 1954
Second edition published 1958
Third edition published 1962
Fourth edition published 1966
Fifth edition (metric) published 1968
Sixth edition published 1971
Reprinted by Crosby Lockwood Staples 1973, 1975
Seventh edition published 1979
Seventh edition (revised) published by Granada Publishing 1981
Eighth edition published 1983
Ninth edition published by BSP Professional Books 1989

British Library
Cataloguing in Publication Data
Willis, Christopher J. (Christopher James)
Specification writing: for architects and surveyors - 9th ed.
1. Building specifications. Preparation - Manuals
I. Title II. Willis, J. Andrew III. Willis, Arthur J. (Arthur James) Specification writing
692'.3
ISBN 0-632-01998-0

BSP Professional Books
A division of Blackwell Scientific Publications Ltd
Editorial Offices:
Osney Mead, Oxford OX2 0EL
(Orders: Tel. 0865 240201)
8 John Street, London WC1N 2ES
23 Ainslie Place, Edinburgh EH3 6AJ
3 Cambridge Center, Suite 208, Cambridge, MA 02142, USA
107 Barry Street, Carlton, Victoria 3053, Australia

Set by DP Photosetting, Aylesbury, Bucks
Printed in Great Britain by
Hollen Street Press Ltd, Slough

Whilst every care has been taken to ensure the accuracy of the contents of this book, neither the authors nor the publishers can accept any liability for loss occasioned by the use of the information given.

Contents

Acknowledgements

We are most grateful to Mr David Juniper DipArch, RIBA for permission to use his drawing and for his help in preparing the copy for reproduction and to Mr Tony Allott BArch, FRIBA for his assistance with the chapter on the NBS and to National Building Specification Ltd for permission to reproduce samples of the NBS text.

Specify To speak of or name something definitely or explicitly, to set down or state categorically or particularly

Shorter Oxford English Dictionary

'There's no place like home
If it's all to specification'

Comic song Arthur Askey, 1938

Preface to the ninth edition

When the eighth edition of this book was published the work of the Co-ordinating Committee for Project Information (CCPI) was just beginning and only passing reference could be made at that time to the then forthcoming Common Arrangement of Work Sections for Building Works (CAWS) and the accompanying codes of procedure for Production Drawings and Project Specification. In the five years that have elapsed the work of CCPI has been completed and now all three documents are published for all to use.

Common Arrangement has been adopted for structuring both the National Building Specification (NBS) and the National Engineering Specification (NES), as well as for the PSA General Specification and the Seventh Edition of the Standard Method of Measurement (SMM7) and in consequence used in the PSA/RICS/BEC sponsored Library of Standard Descriptions. In addition it is being considered for adoption in other areas in the industry. Following the precepts of Co-ordinated Information it is intended that it should be used in project documentation generally so that the ideal of information on the drawings equalling that on the specification and in the bills of quantities descriptions will be achieved.

In consequence the ninth, and silver jubilee, edition of this book has been restructured to reflect the impact of Common Arrangement on specification writing. It opens as before with three chapters of a general nature and a chapter on the NBS and its use. There then follow the Common Arrangement sections and the book ends with an example small works specification rewritten and updated in CAWS order.

This book should be looked upon as an adjunct to the CCPI Specification Code. It is intended to be both an introduction to the subject and a desk top reference for specification writers but like previous editions it concentrates on the simpler forms of building, although the basic principles remain the same whatever the complexity.

The distinguished architect Sir Edwin Lutyens once said that a working drawing was merely a letter to a builder telling him precisely what was required of him and not a picture to charm an idiotic client and, whilst not being over complimentary to his clients, this is a succinct and apt definition of the role of drawn information. The same sentiments can be said to apply equally to specification writing. It is in that spirit that this book has been written.

The doctrine of co-ordinated project information must in the main be accepted by architects, engineers and surveyors, in fact by all who specify, as being a wholly worthwhile objective. How quickly it will be taken up, and how soon the clients of the industry will realise the undoubted benefits to both their plans and pockets, remains to be seen. If this book helps in any way to hasten that process we shall have achieved our aim.

September 1988

C J W
J A W

Abbreviations used

BEC	Building Employers' Confederation
BS	British Standard (of the BSI)
BSI	British Standards Institution
CAWS	Common Arrangement of Work Sections for building works
CCPI	Co-ordinating Committee for Project Information
CI/SfB	Construction Indexing Samarbetskommitén for Byggnadsfrägor
CP	Code of Practice (published by BSI)
CPI	Co-ordinated Project Information
JCT	Joint Contracts Tribunal
NBS	National Building Specification
NES	National Engineering Specification
PSA	Property Services Agency
RIBA	Royal Institute of British Architects
RICS	Royal Institution of Chartered Surveyors
SMM	Standard Method of Measurement

1 Purpose and use of the specification

INTRODUCTION

A specification writer should not require model clauses to know, for instance, what size of fascia to use, how to specify kitchen cabinets or what weight or width of flashings should be used. These are matters which must be decided from a knowledge of building construction, the fundamental aspect, the aesthetic effect required and the cost. Specifications must not be drafted in a haphazard fashion just as the items are thought of, but must follow a system whereby those using a specification can visualise the whole building in advance, whether to measure its constituent parts for an estimate or find definite instructions for erecting the building.

A large proportion of the items in every specification concern matters common to all, varying only in detail to suit the ideas and designs of the individual architect: these items form the backbone of the lists given in Chapter 5 for each section. It is not possible to anticipate every item in every specification. Each building will have its own peculiarities which are fairly certain to present themselves clearly to the specification writer. If, however, a specification properly covers all the points mentioned which are relevant, it will stand high in intrinsic merit and, what is more important, practical value.

The scope of this book is limited to simple building work. Engineering works are so varied in their nature that their special requirements must be considered on their merits. One can, perhaps, speak of an average or simple building, but it is difficult to apply such terms to engineering work and in consequence only passing reference is made. Road work in this book is limited to probable requirements for site roads, courtyards etc., and drainage, excluding the special needs of sewers.

The best test for a specification is for the architect who wrote it to go on an extended holiday whilst quantities are being prepared or, if there are to be no quantities, whilst the building is erected, leaving to an assistant routine replies for the quantity surveyor or the inspection of the works. How many architects would feel confident in so doing?

A building begins as a concept in the mind of the architect. The conception must be elaborated in the drawing office, adapted to practical considerations and then conveyed to a number of people who will co-operate in the erection of the building. Until the time comes when telepathy is a certain science, intentions can only be conveyed by specific instructions. Whether the instructions are conveyed to the contractor direct or through the quantity surveyor is immaterial: unless they are specific and complete without gaps and overlapping, there will be delays and mistakes, to the detriment of the client's interests. The writer of a specification is assumed to know what is wanted. Once it is known what is wanted all that is required is for it to be expressed fully, clearly and systematically.

WHAT IS A SPECIFICATION FOR?

Unless the writer grasps the purpose of the document, he cannot prepare it sucessfully. The specification may have three purposes, in each case in conjunction with the drawings:

- To be read by the contractor's estimator as the only information he has on which to prepare a competitive tender.
- To be read by the quantity surveyor to enable him to prepare a bill of quantities as a basis for such competitive tenders.
- To be read by the clerk of works and the contractor's agent during the progress of the contract as the architect's instructions for carrying out the work.

THE SPECIFICATION AS A BASIS FOR TENDERS

In smaller contracts, usually those under about £100,000 in value, contractors prepare their tenders from drawings and specifications only. Estimators take their own measurements of the work from the drawings and build up their estimates relying on the specification for a full description of quality, materials and workmanship. Besides this, drawings and specifications, when read together, must indicate everything required to be included in the estimate. If anything is omitted, something that is required is not mentioned or shown, or very obviously necessary or implied, such work will not be part of the contract. If its carrying out is insisted upon, the contractor will be entitled to extra payment.

The writer of a specification for this purpose will, therefore, realise the importance of the work necessary. Instructions must be specific and crystal-clear and complete in detail. The specification will be one of the contract documents and is not to be scribbled hurriedly. It must have all the preciseness of an agreement (it will, in fact, be part of one) conveying exactly to the contractor what is wanted and protecting the building owner from claims for extra payment which would arise from vagueness and uncertainty.

THE SPECIFICATION FOR THE QUANTITY SURVEYOR

For contracts of a value of over about £100,000 it is usual for a bill of quantities to be supplied on behalf of the client if competitive tenders are being invited. The measuring work, which in the previous case would be done by all the tenderers, is in these circumstances done for them by the quantity surveyor who puts before them the facts, but each tenderer is left with the estimating, this being largely a matter of individual judgment.

In order that the quantity surveyor may prepare his bill, instructions must be given by the architect as complete as those required by the contractors when taking their own measurements and these instructions are conveyed in a specification. In this case the architect's specification is not usually a contract document, it can be less formal and convey the information in the form of notes. For certain standard clauses reference may be made to another contract. Whilst such a specification should be as complete as possible, omissions are not as vital as in the first case. The quantity surveyor will find the gaps, as in preparing the quantities every stage in the erection of the building has to be visualised and questions will arise whenever further information is required to complete the specification preambles which should be in CAWS order to enable easy reading with the measured items. It should be noted that where the architect's specification is not a contract document (as in the JCT 'with quantities' forms) the preambles must convey the specification information to enable the bills to 'fully describe and accurately represent the quantity and quality of the work' as required by the SMM.

THE SPECIFICATION FOR SITE AGENT AND CLERK OF WORKS

When the erection of a building starts the work will be supervised on behalf of the contractor by his site agent. On large projects a clerk of works will be employed to inspect on behalf of the employer, since constant inspection will be necessary and the architect is not expected to be continuously on the site. Both site agent and clerk of works require instructions and they take these, subject to any variations ordered by the architect, from the contract documents, i.e. drawings and specification or drawings and bill of quantities. Where quantities have been prepared, the quantity surveyor will have incorporated the specification in the descriptions or in the preambles in the bill. There is however, certain information required by the site agent and clerk of works, which will have been excluded from the bill. The location of the items, for instance, will not usually be mentioned in the bill because they do not normally affect price; however the site agent must have this information when it comes to erecting the building. There are also matters which must be conveyed as instructions to the site agent. For instance it may be required that floor joists shall not be spaced more than 300 mm apart. This the quantity surveyor will have duly noted in taking the measurements, but, having taken them accordingly, there is no reason to state the requirement in the bill. When it comes to erection, however, the site agent must have this direction which in this case would probably have been shown on the drawings.

THE CONTENTS

The specification is written to convey to the reader all the information about a proposed building which the architect cannot easily show on the drawings. He cannot, for instance, show the quality of a material to be used nor the way pipes are to be jointed together, except by adding such full written notes or large-scale details on his drawings that their main purpose is obscured. On the other hand without this supplementary information the estimator, working without quantities, cannot prepare the tender. If quantities are being supplied, the quantity surveyor is in equal difficulty in preparing them as the architect's wishes are not known.

ESSENTIALS IN WRITING

There are two essentials in writing a specification:

(a) To know what one requires
(b) To be able to express it clearly

Many specifications fail for lack of (a). Such lack of knowledge has two causes:

(1) Insufficient thought
(2) Insufficient knowledge of building construction

If the architect has not thought about what is wanted, the estimator or quantity surveyor can hardly be expected to know. It must be remembered that every step in the carrying out of the work has to be visualised in advance by the estimator (or the quantity surveyor). If some requirement of the architect is not made clear it may quite possibly be missed and consequently result in a claim by the contractor for extra payment. The estimator in competitive tendering has not the time to work out what the architect or quantity surveyor has left undone, nor can the estimate be loaded to cover the item when the competitor's estimate may not. The quantity surveyor may realise the possibility of a claim arising and measure something to cover, but, may quite likely, guess wrongly. This will involve unnecessary adjustment of variations and extra expense to the employer.

Even if full thought has been given to the job it will still be unsatisfactory if the architect is lacking in knowledge of building construction. The students can only acquire this knowledge by hard work, constant visits to buildings under construction and by keeping up to date with information on new materials. The specification writer put in the place of the reader should be able, from the information available, to direct the operatives constructing the building. There should be no occasion to stop and say 'How am I to get over this difficulty?' or 'What do I put on top of this?'

The architect must have a knowledge of materials. If some branded article, e.g. fibreboard, plaster, plastic sheets, paint etc., is to be used, the architect must be thoroughly acquainted with the maker's recommendations, because what may apply to one brand may, quite possibly, not apply to others. Whilst one must remember that the anxiety of manufacturers to sell their products may make them optimistic, unless their recommendations are followed architects may find themselves blamed when defects appear.

In short, the writing of a specification by the architect is nothing more than setting down what is required when it has been fully thought out. If it is known, say, that a timber fascia 175 mm wide would cover the ends of the timbers and be satisfactory aesthetically, that if 16 mm thick it will probably curl or split, if 40 mm thick money is being wasted, that it can be securely fixed to the particular roof construction by driving nails through it, that the nails must be driven into the holding material for a depth at least equal to the thickness of the fascia, that nails should be punched in and stopped or they will rust through the paint – then, if put in plain English, it is possible to explain precisely what is required.

Granted that the writer can specify the fascia referred to above the next important thing is completeness. Have all the varieties of fascias and all the members of the roof construction been specified? all carpentry items? and so on? Chapter 5 sets out the main sub-divisions of each work section with the essential points to be mentioned for each. The writer in following through the lists must consider which items apply to the particular case and then say what is wanted about each.

FOUNDATIONS

A possible uncertainty, which needs special consideration in specifications for tender where no quantities are supplied, is depth of foundations. The drawings and specifications read together will determine the extent of the contract work. They should either be definite, when any variation would be adjustable, or it should be stated that the whole will be measured and valued as done up to a specified level. The attention of the tenderer may be specifically drawn to the need for a schedule of rates to be prepared with a view to such measurement. The schedule will naturally follow the tenderer's method of estimating, so it is difficult to lay down particular requirements as to its form.

2 Subject-matter

MATTER FOR DRAWINGS

Some of the information required of the architect is best shown on the drawings, and being so shown, is not required in the specification. The CPI Code for Production Drawings sets out the links that can be achieved by common arrangement between specification and drawings. The architect should see that the following are all shown on the drawings:

- Ground levels of the site
- Floor levels related to ground levels
- Figured dimensions:
  - (a) walls over all on plans
  - (b) inside dimensions of rooms
  - (c) heights
  - (d) (in the case of steel-framed buildings) centre to centre of stanchions and relation of steelwork lines to wall and floor surfaces
  - (e) setting out of openings
- Position of steps in concrete foundations
- Room titles (or serial numbers)
- Opening portions of windows
- Direction of opening of doors
- Runs of service pipes
- Position of cupboards or other fittings
- Position of electric points

It will also simplify reference in the specification if windows and doors are given serial numbers in one or separate series. The numbering should start in one corner of the plan (usually top left) and continue clockwise.

It is not usually necessary to incorporate sketches in a specification, as anything that needs illustration in that way would be shown on the drawings and reference to the drawing number would be sufficient.

SUBDIVISION

The traditional subdivision in the case of a new building was into recognised 'Trade Sections'. Now with the introduction of Common Arrangement they will be structured in CAWS order. Work should be specified in its proper section, but where several sections are involved cross-references to clause numbers should be given.

In the case of alteration work, whilst Common Arrangement has sections for Demolition (C10) and Works on Site and Spot items (C20) it is sometimes more useful to subdivide the specification according to rooms or parts of the building, so that everything in connection with that room or part is together. Or a combination of the two systems may be adopted to save unnecessary repetition, namely, to specify the alteration work room by room but keep in a separate section all new work in connection classified into work sections (the description of which might otherwise have to be repeated many times). Each alteration job must be considered on its merits and that system adopted which seems to suit that particular scheme best.

WORKS OF ALTERATION

In specifying works of alteration by rooms or by parts of a building one should take each room in a sequence which can conveniently be followed walking round the building, thus putting oneself in the position of the contractor's estimator and consider what information is necessary for quoting a price. The impossible should not be attempted and if it is not reasonable for the estimator to anticipate the work required a provisional sum should be given which will in due course be adjusted against actual cost.

A point often not visualised sufficiently by specification writers is the amount of damage that will be caused by traffic, dumping of materials, vibration etc. One of the commonest arguments over alteration jobs is how much plaster should have been renewed under the contract. Inherent defects, such as a bad key, are apt to reveal themselves unexpectedly during the progress of the work and the contractor will disclaim liability for them.

PC PRICES AND PROVISIONAL SUMS

There is nearly always certain work required on a building which is outside the sphere of the normal contractor and necessitates the introduction of a specialist on the site. Such work may vary from plastering or plumbing which some contractors do undertake with their own staff, to electrical work which is only occasionally part of a contractor's business. Such work is let by the contractor as a sub-contract. In the former more ordinary trades it is usually left to the contractor to make his own arrangements, but in such matters as electrical work, central heating, lifts etc., the selection of the firm will probably be made by the architect. Under the terms of the standard forms of contract firms carrying out this latter work become 'nominated sub-contractors' ('named sub-contractors' in the case of the JCT Intermediate Form of Contract) and, though the contractor remains responsible for their work, the architect controls the appointment. In the specification the tenderers are told to provide a PC sum of £x for the work, and the sum so included in the tender will be subject to adjustment against the specialist's account.

Such nominated, or named, sub-contractors are essentially firms doing work on the site. It often happens, however, that the architect wants to select certain materials or goods which the contractor is to fix. What is wanted may be known but it may be necessary to defer decision as to the exact model or make until the contract is let. If so, again, tenderers will be given a figure to include subject to adjustment as their services are restricted to supply only: such firms are known as 'nominated suppliers'.

Sums so included in a tender subject to adjustment are termed 'provisional sums' or 'prime cost sums'. Provisional sums are lump sums given provisionally i.e. subject to adjustment, to include all profit, attendance etc., which, when the adjustment is made, will be allowed in accordance with the terms of the contract governing variations. Prime cost sums (usually abbreviated to PC sums) are sums deemed for the purpose of the tender to be the prime cost (basic cost before addition of overhead charges and profit) to the contractor: if the actual cost is more or less, adjustment will be made accordingly, both of the expenditure and of any profit, attendance on sub-contractors etc., added in the tender.

The architect is responsible for obtaining the quotations from nominated sub-contractors and suppliers.

BUILDING REGULATIONS

Responsibility that the work shown and specified complies with the regulations of controlling authorities is for the architect. Existing local bye-laws were replaced as from 1 February 1966 by Building Regulations (Building Regulations – HMSO) which were to apply to the whole of England and Wales, except the Inner London Boroughs where building was still to be controlled by the London Building Acts. With the dissolution of the Greater London Council the London Building Act was repealed and the Building Regulations now apply equally in London.

Apart from this, water authorities have regulations specifying, for instance, minimum strengths of pipes. To say 'that the pipes are to be of the strengths required by the Water Authority' is not good enough. Requirements as to strength often differ with the pressure at the particular point in the supply system: the architect should ascertain this information and not expect tenderers to make these investigations.

PROVISIONAL SUMS FOR STATUTORY CHARGES

In the same way as for specialists' work provisional sums have to be included in a specification for work which local authorities, nationalised and privatised boards etc. have to carry out under statutory or similar powers, for example the connections of drains to public sewers, the bringing in to the premises of water, gas or electric services. The extent of work done by such authorities varies. A gas board will probably bring their main up to a meter within the building, whereas a water authority may only bring their supply up to the boundary of the property. A local authority may allow a contractor to break up a road or they may insist on doing this themselves. The specification must be quite clear as to how much work is covered by the provisional sum and how much the contractor has to do. Enquiry should be made of the authority concerned and an estimate obtained, any doubt as to the extent of the work being clarified with them. Payment for charges of these authorities is due to the contractor and these provisional sums do not include for cash discount, as do those for nominated sub-contractors. If however the same authorities undertake further work such as the electric wiring of the building they will do so as nominated sub-contractors on the same basis as private firms.

USE OF STANDARD SPECIFICATIONS AND CODES OF PRACTICE

The extent to which standard documents should be referred to in a specification is sometimes difficult to decide. The British Standards Institution (BSI) publishes standards for a very wide range of materials. Many of them are prepared with a view to use for local authority housing contracts, the standard for which may not be high enough for some other types of building. Codes of practice relating to workmanship in various trades are also published by the BSI. A thorough knowledge of these standards and codes would need almost a life's work and is not a practical possibility for the ordinary professional: moreover, few contractors will have a complete range of them in their office. On the other hand, merchants supplying, say, clayware drainage goods or specialists such as asphalt laying firms would be familiar with the standards and codes applicable to their particular line.

Generally speaking it is advisable to refer to British Standards for the more usual materials and Codes of Practice can be incorporated where considered useful, but it is most important in either case to be conversant with the terms of the document quoted. It is not sufficient to say 'All materials are to be in accordance with the relevant British Standard' nor 'that all work is to comply with the Codes of Practice issued by the BSI', without stating the reference number of the particular standard or code. Even where reference numbers are quoted a BS may have subdivisions for different designs or qualities of the same article. Provided that in such cases reference is made to the appropriate subdivision, a BS number may help considerably in shortening descriptions. Similarly an examination of one of the Codes of Practice may reveal alternatives or some requirement with which there may be reason to disagree. Many of the Codes of Practice have now been converted to standards and it is the policy of the BSI to continue this process. The use of references to these documents must not be allowed to encourage laziness.

BUILDING OWNER'S OWN MATERIALS

It sometimes happens that the contract is to include for the fixing of materials supplied by the building owner. This may be on a large scale when public authorities take advantage of the cheaper rates obtainable by bulk buying, or just because the building owner has picked up some old panelling or a fireplace which he wants incorporated in the building. In specifying such work it must be made clear that this material will be supplied to the contractor and delivered to the job without expense.

PRELIMINARY ITEMS

The preliminary items will cover all matters of a general nature or applying to the sections generally. They include any definitions and refer to such matters as conditions of contract, working rules, general organisation of the job, provision of plant, sheds, water, temporary lighting, temporary roads, insurances etc. The detailed Common Arrangement list will be found in Chapter 5. The conditions of contract being incorporated in the specification by reference, need not be repeated in full, but it may be necessary to amplify them by specific provisions, such as accommodation for the clerk of works, administrative directions as to preparation of daywork sheets or price adjustment statements etc.

MATERIALS AND WORKMANSHIP CLAUSES

These will come at the head of each section and deal with all general matters relating to the section: being thereafter implied, they will shorten the descriptions following. They will comprise descriptions of the materials to be used, defining the quality where there might be more than one, and of the workmanship, i.e. the technical procedure in carrying out each process in the erection of the building. For example in 'Masonry' the materials – bricks, cement, sand etc. – will be described. Such descriptions will be followed by such instructions as may be thought advisable to give for the carrying out of the work, e.g. the stacking of bricks, composition and mixing of mortar, protection from frost etc. A clause in the preliminary items that the materials shall be the best of their respective kinds or that workmanship is to be of the best quality can only be indicative; the kinds of material must be identified and opinions may differ as to standards of workmanship.

KEEPING UP TO DATE

It is most important for specification writers to keep up to date with developments. Not only must changes in such matters as the various forms of contract and the working rules of the industry be watched, but it is necessary to keep abreast with the use of new materials and study their methods of fixing, and not forgetting to take note when materials become obsolete. The development of modern materials has done away with many of the old traditional materials such as hand made stopping and white lead.

Whether architects or surveyors, the specification writers will probably have noticed in their professional journal new publications of interest and changes in practice either obligatory or recommended. Subscription services are available which provide helpful information. It may mean some duplication but this does no harm: in fact repetition helps to impress.

It is physically impossible to master the flood of literature on technical matters which comes from the presses, but to watch the titles with only a vague idea of the subject-matter may at some later date prove valuable. It is much better to spend time noting what information is available than trying to master the detail of something which may never be wanted. Persons who know where to put their finger on information when it is wanted are of much more value (except unfortunately in examinations) than the ones who try to cram facts into their heads.

3 Form of the specification

AUTHORSHIP

The specification, if not written by the architect, should be written by an assistant who has worked on the drawings and knows the conditions and requirements of the job. Ideally, a specification should be prepared in parallel with the preparation of the drawings. Some offices operate a system whereby a 'specification writer' is employed solely to write specifications. Such a person, to do the job properly, has to spend much time in studying the drawings to get a grasp of each scheme and probably take up the time of the architect or assistants in asking questions which have already been settled, but answers to which may not be obvious from the drawings. There is therefore much to be said for the person who has seen the thrashing out of the scheme through all its stages in preparation of the drawings writing the specification.

COMPOSITION AND STYLE

A specification is not literature to be written in the flowing periods of Macaulay: it is essentially a document for quick reference and quick understanding. This being so, it should be divided into short paragraphs with clear headings and arranged in a systematic order. Nevertheless, the writing must be grammatical and properly punctuated (for good punctuation does much to help clear understanding). If it is written in a hurry without time spared for reading through, it is bound to show defects and uncertainties which the writer could easily have avoided. It may be contended that bad style or spelling are not important if the words are understood, but they do show slovenliness which reflects on the writer.

The writer should use language that is plain and straightforward, bearing in mind that he is giving orders which must not be misunderstood. Every word should mean something, and the phrasing must be definite, except only where alternatives are intentionally left to the contractor.

The best rule for punctuation is to imagine reading the document aloud and putting the stops where natural pauses occur either for breath or for a change of tone (e.g. to indicate parenthesis or a conditional clause). Sentences should be short: separate sentences with a colon or semicolon between are clearer than a long sentence divided by commas into clauses the construction of which becomes involved.

Shortcomings in drafting, together with a lack of compatibility with other project information, has been highlighted (Project Information Group report – 1979) as one of the prime causes of disruption of building operations on site leading as they do so often to delays and subsequent loss and expense claims. In order to improve the situation the Co-ordinating Committee for Project Information (CCPI) was set up and after consultation with all interested parties produced a Common Arrangement of Work Sections for Building Works (CAWS) and a Code for Project Specification writing. They also produced a similar code for Production Drawings and worked in conjunction with the producers of SMM 7. The last two of these documents are mainly outside the scope of this book: the first two are of prime relevance.

COMMON ARRANGEMENT OF WORK SECTIONS

The purpose of CAWS, as set out in its introduction, is to define an efficient and generally acceptable identical arrangement for specifications and bills of quantities.

The main advantages are:

- Easier distribution of information, particularly in the dissemination of information to sub-contractors: one of the prime objectives in structuring the sections was that the requirements of sub-contractors should not only be recognised but kept together in relatively small tight packages
- More effective reading together of documents. Use of CAWS coding allows the specification to be directly linked to the bill of quantities descriptions: cutting down the descriptions in the latter whilst still giving all the information contained within the former
- Greater consistency achieved by implementation of the above advantages. The site agent and clerk of works can be confident that when they compare the drawings with the specification and both with the bill of quantities no longer will they have to ask the question 'Which is right?'

CAWS is a system based on the concept of work sections. To avoid boundary problems between similar or related work sections, CAWS gives, for each section, a list of what is included and what is excluded, stating the appropriate sections where the excluded items can be found.

CAWS has a hierarchical arrangement in three levels for instance:

- Level 1 R Disposal systems
- Level 2 R1 Drainage
- Level 3 R10 Rainwater pipework/gutters

There are 24 Level 1 group headings, 150 work sections for building fabric and 120 work sections for services. Although very much depending on size and complexity, no single project will need more than a fraction of this number – perhaps as a very general average 25–30%. Only level 1 and level 3 are normally used in specifications and bills of quantities. Level 2 indicates the structure, and helps with the management of the notation. New work sections can be inserted quite simply without the need for extensive renumbering.

PROJECT SPECIFICATION CODE

The purpose of this code is to help the practitioner review and upgrade their specification practice. The code is divided into three parts:

- Part A General principles of specification writing
- Part B Guidance on coverage
- Part C Libraries of specification clauses.

Each part sets out detailed guidance on how the main aims of the code can be achieved in practice. As with Chapter 5 of this book CAWS sections are looked at and lists of items for consideration are given. Chapter 5 however goes one stage further by listing the relevant British Standards.

The remainder of this chapter covers matters such as terminology, consistency, procedure in drafting, underlining, headings, numbering, indexing and standardisation.

TERMINOLOGY

The principles of terminology are described in BS 3669 but of more practical importance is BS 6100: Glossary of Building and Civil Engineering Terms, a detailed specification in six parts covering all sections of building and civil engineering construction. Terminology required for retrieval purposes is best found in a thesaurus. This is a document which indicates terms authorised for use in indexing. Terms that may not be used are cross-referenced to those which may be, and relationships between indexing terms are shown. Most thesauri are designed for use within a particular subject field or discipline, or are associated with a particular data base. The Construction Industry Thesaurus published by the Department of Environment, has been produced for use in the UK construction industry and contains some 9000 entries. There are two listings – an alphabetical list of terms and their synonyms and a classified display showing the more important relationships between terms. Ten categories are used:

(1) physical forms of presentation (for indicating whether a document is a report, a book, a piece of trade literature)
(2) peripheral subjects (i.e. those not unique to the construction industry, e.g. training, environment)
(3) time
(4) place (i.e. physical location)
(5) properties and measures
(6) agents of construction (i.e. people, plant)
(7) operations and processes
(8) materials
(9) parts of construction works
(10) construction works

PROCEDURE IN DRAFTING

It may be that the drafting of 'Preliminary Items' will be left to the end: they are almost independent of work in the sections and can be deferred. Within each section the materials and workmanship clauses are also best left until the actual work has been specified. Until this is done it is difficult to see exactly what materials and workmanship require description.

In thinking of the building by sections the ground works will be the first items to be put down and all the varieties must be thought of. Similarly with concrete, there will be foundations, beds, lintels and other items all to be considered.

Working according to sections of the building, the first matter will be the clearing of the site: then will follow foundations i.e. everything up to damp proof course level or equivalent including the concrete and masonry as well as the groundworks. Plenty of space will be allowed for coming back to these sections when dealing with the superstructure.

In the final form of the specification the clauses should generally be in the CAWS order given in Chapter 5 and any necessary transfers can be made on the draft by the use of symbols and cross references to make quite clear what alterations in order are to be made.

UNDERLINING

Caution is needed in the use of underlining. Tenderers are apt to skip the Preliminary Items and descriptions of materials and workmanship to save time, assuming that requirements are normal, a dangerous practice putting themselves very much at risk, but being done just the same. To underline important clauses almost makes matters worse, as, though attracted to the underlined words, the estimator tends to pay less attention to the rest. Underlining adds nothing legally to words which are already binding.

CLAUSE HEADINGS AND NUMBERING

Quick reference is very important in specifications. Clauses should, therefore, have headings to indicate their contents. A clause may well have more than one paragraph and unless there is a close relation between paragraphs it is better that they are each made a separate clause with its own number.

CONSISTENCY

It is important that the specification should be consistent in all its parts, though this may sometimes be difficult when the writing is spread over any length of time. It is obvious that in interpretation of a document confusion is bound to arise if the same word or expression is used in different senses in different parts of the document, and conversely if different words or expressions are used with the same meaning.

Another example of inconsistency often met with is in the order in which dimensions are given. There is a generally accepted custom in the building trade that they shall be given in this order: length, width (or depth on bed), height. This is not universal, for it is pedantic or eccentric to talk of 3″ × 6″ steel joists when one has always heard them spoken of as 6″ × 3″, and, though one will hear of 50 × 100 mm wood rafter, they are more usually called 100 × 50 mm. Still there are occasions when a well understood order is important. A 450 × 150 mm lintel fair on both faces has 0.30 sq. m of fair face per lineal metre, whereas a 150 × 450 mm lintel (being 150 mm on bed and 450 mm high) would have 0.90 sq. m per lineal metre. The steel window manufacturers, however, always give the height followed by the width, and care is necessary when asking for quotations from such firms to state clearly which is width and which is height.

Lack of consistency also appears when timber is described sometimes as 'finished' and sometimes by basic sizes. If finished sizes are given a calculation will have to be made by the estimator each time to find the commercial sizes to which the buying is necessarily restricted. The most usual thicknesses for softwood are 25 mm, 30 mm, 40 mm, 50 mm and 75 mm. The recognised allowances from these nominal thicknesses for thickness as finished on the job are 3.20 mm for each sawn face. The quantity surveyor usually describes the items in the bill by basic sizes and confusion is likely to arise if the architect in specification and drawings given to the surveyor does not do the same.

The draft of a specification whether hand-written or produced from a word-processor will have to be prepared in a form suitable for final reproduction. To dictate it, except as a rough first draft to be put into shape later, is practically impossible. In preparing a draft plenty of space should be taken: it is easy enough to close up a draft, but to decipher an overcrowded one having alterations in wording and order can be extremely difficult. The specification has traditionally been arranged in trade order, now with the introduction of common arrangement, the order will be that adopted in CAWS.

Clauses should be numbered serially using the CAWS reference number with subservient sectional numbering. A liberal use should be made of cross referencing when it facilitates explanation. Given sectional numbering, e.g. A50.01, additional clauses can be inserted at the end of each section without disturbing the numbering throughout. The form in which it is suggested a specification should be written will be seen in the Appendix.

SCHEDULES

Schedules are an invaluable part of a specification for all but the smallest buildings, particularly in drawing distinctions between different treatments, e.g. to show which rooms have papered walls and which painted, where the different types of doors are and what ironmongery is to be fitted to each. They give, so to speak, a bird's-eye view of the requirements, fuller detail of which can be found by looking up the relevant paragraphs. They are especially useful as a supplement to bills of quantities when their items are referenced to the items in the bill. Instead of having to read long paragraphs of descriptions one can quickly trace the location of the items in the bill.

Schedules are sometimes prepared in the same way as drawings so that prints can be taken from them and included with the set of drawings. They are not then required in the specification but will be read with it, and, of course, the matter they contain need not be repeated.

INDEX

An index is a very useful adjunct to a specification. If nothing else is done a table of contents setting out the sections and sub-divisions thereof should precede page 1 so that quick reference can at any rate be made to the section of the specification required. A detailed index of all items means time and is hardly justified for a small specification; where each section has only two or three pages a quick glance down the headings will fairly certainly find the required item. In an extensive specification, however, a detailed index is undoubtedly useful and may be thought worthy of the trouble but in today's competitive climate it could be said to be an expensive luxury. A specimen table of contents is included in the example specification in the Appendix.

STANDARDISATION OF FORM

Standard specifications can take two forms. There are specifications like the National Building Specification (see Chapter 4) which are primarily a library of clauses which, as will be described, can be used to build up a working specification. Careful coding of the clauses permits the writer to refer simply to the relevant number or code and the typist or computer operator can do the rest. The second form of standard specification is one prepared with plenty of space for alterations and additions and a number of copies can be duplicated to serve as the draft for each project. This is done by government departments and local authorities where, for instance crown offices or schools were to be built in substantial numbers over several years, with similar methods of construction and general features. In using this form of standard specification great care must be taken. It is easy for something inapplicable to be left in, which, when duplicated into the individual specifications leaves no indication of being an oversight and looks silly. There may be alternatives in the standard draft one of which is to be deleted. The authors have before now come across such clauses as 'Facings to be in English/Flemish bond' or 'Plastering to be two/three coat work', showing lack of care and perhaps excessive haste in drafting. It is easy, too, when drafting from a standard specification like this to overlook items specially applicable to the particular project, which the standard specification never contemplated.

The disadvantage of a standard specification to be adapted for each project is that it tends to make the brain lazy. Instead of the writer's mind being alert on the job as it should be, a semi-mechanical process tends to be carried out based on other people's thinking.

LINK WITH BILLS OF QUANTITIES

As was mentioned in Chapter 1, the bill of quantities prepared by the quantity surveyor will have incorporated the specification matter, which, therefore, need not be repeated in a separate document. With the introduction of bills of quantities prepared in Common Arrangement order, cross-referencing to the specification is so straightforward that even more repetition is avoided. If the drawings, too, have CAWS references the ultimate aim of the information contained in the specification equalling that shown on the drawings equalling the bill descriptions will become much nearer to being achieved. The supplementary information as to location etc., required by the site agent and the clerk of works can be set out in a form of notes referenced to the items in the bill of quantities.

REPRODUCTION

Traditionally specifications were produced by type and carbon when only one or two contractors were tendering or by typing stencils and duplicating when larger numbers were involved. Today however the more usual form is for the information to be stored in a computerised retrieval system and called off for each project, drafted, adapted or amended as the case may be and then printed either 'one off', for subsequent photo-copying, or by a multiple printing process contained within the retrieval system.

Protective covers should be used as the specification will receive a good deal of handling, especially if there are no bills of quantities and it is the sole document to interpret the drawings. It is more convenient to handle if stapled or bound down the left-hand side so that it can be opened like a book. A pin through the top left-hand corner makes pages awkward to turn over and inclined to tear.

PROOFS

Reproduction by any means must be examined before making up or binding, so that corrections can be made; obvious mistakes, pages out of order or upside down etc., give a very poor impression. Reading a draft, particularly if it has been produced by hand, the non-technical typist or printer can easily make mistakes which will look foolish, e.g. 'core' for 'cove' or 'hoarding' for 'boarding'. Though readers may guess the error, there might be cases where they will be puzzled by the misprint.

4 The National Building Specification

The National Building Specification (NBS) is not a standard specification. It is a large library of specification clauses all of which are optional, many are direct alternatives, and which in many cases require the insertion of additional information. NBS thus facilitates the production of specification text specific to each project, including all relevant matters and excluding text which does not apply.

The existence of NBS underlines the ambiguity that the building industry gives to the word specification. To most architects it means the description of the quality of materials and workmanship required in the building. To most quantity surveyors it has come to mean the description of the work required (Schedule of Works) where quantities are not provided. NBS could be thought of by the quantity surveyor as National Building Preambles.

NBS is available only as a subscription service, and in this way it is kept up to date by issue of new material, several times a year, for insertion into the loose-leaf ring binders. Until 1988 NBS was available in either CI/Sfb or SMM6 order, but these have now been discontinued. NBS is now available only in the Common Arrangement of Work Sections, matching SMM7. It has also been revised to comply fully with the recommendations of the CPI Code of Procedure for Project Specification referred to in Chapter 3. There are two versions of the text, the 'Standard Version' and an abridged 'Intermediate Version'. A third 'Minor Works Version' is due to be published in 1990.

The range of specification clauses offered is wide and care must be taken to include only those which are required for the particular project. It is important to specify only those requirements which one is prepared to enforce, for as in traditional specification writing the inclusion of clauses which will not, or cannot, be enforced undermines the authority of the document as a whole.

NBS includes against most of its clauses guidance notes which draw attention to other documents, alternative and related clauses, and indicate the type of additional information which is required to complete a clause. Expendable copies of NBS work sections are available for 'mark up' and these do not have the guidance notes. Also, the NBS clauses are available on disc to suit all types of word processor or computer in common use.

The recommended procedure for using NBS is as described in the CPI Project Specification Code. A list of the specification sections should be prepared and an expendable copy of those sections obtained. The responsibility for writing particular sections should be decided, essentially according to who has responsibility for detailed design.

For each section of the specification the NBS clauses and accompanying guidance notes should be read, together with the appropriate parts of the relevant Code of Practice and other reference documents. This can be time consuming, but it is essential if the finished specification is to be technically sound.

The expendable copy of the NBS clauses can then be marked up, building up the project specification clause by clause. This is best done concurrently with preparation of the drawings and may be an intermittent activity spread over several days or even weeks. At the end of the process it is important to check through to ensure that the specification has been completed. Each section should be reviewed in general terms to ensure that it is appropriate to the needs of the job – it should be abbreviated or elaborated as thought fit.

The specification should be prepared before bills of quantities are prepared, and in parallel with the production drawings so that, for example, the drawings for the main structure and the specification sections for concrete and brickwork are given to the quantity surveyors at the same time. It is usually advantageous to supply the specification to the quantity surveyors in draft form, because as the first users of the document they will therefore be likely to discover any omissions or discrepancies. These can then be corrected before the specification is issued to tenderers as part of the bill of quantities.

A specimen example of NBS Intermediate Version text is given on pages 25–33 showing how it is marked up for a particular project. It will be clear from this that the system encourages standardisation, but at the same time provides a framework for describing the particular requirements of each project. The completed specification is set out in a way which facilitates cross-reference from the drawings and bill descriptions, e.g. 'Clay facing brickwork F10/110.'

NBS is a very useful tool for the preparation of a job specification. At best it is used by the design team to produce comprehensive, concise and constructive specifications to support the information given on drawings and in measured items. At worst it provides the quantity surveyor with a comprehensive query list for the architect whose only specification is 'You know the sort of thing I like.'

F10

BRICK/BLOCK WALLING

SCOPE

This section covers the laying of bricks and blocks of clay, concrete and calcium silicate in courses on a mortar bed to form walls, partitions, chimneys, plinths, boiler seatings, etc.
Other sections related to brick/block walling are:

- F20 Natural stone rubble walling
- F21 Natural stone ashlar walling/dressings
- F30 Accessories/Sundry items for brick/block/stone walling
- F31 Precast concrete sills/lintels/copings/features
- Z21 Mortars.

Simple reconstructed stonework, of the type intended for laying by general bricklayers, is included in this section (see clause 290).

GENERAL GUIDANCE ON SPECIFYING TYPES OF WALLING

USING THE TYPES OF WALLING CLAUSES

A range of clauses is included on pages 4–6 which can be used to specify the various types of brick/block walling, e.g.

110 FACING BRICKWORK *ABOVE DPC*:
- Bricks: Clay to BS 3921.
  Manufacturer and reference: *Birtley Brick Co: Blue Multis.*
  ~~Special shapes:~~ . . . .
- Mortar: As section Z21.
  Mix: *1:1:6 cement:lime:sand.*
  ~~Special colour:~~ . . . .
- Bond: *Half lap stretcher.*
- Joints: *Bucket handle.*
- Features: *Soldier course over openings.*

Complete a clause for each variation of brickwork and blockwork. Delete items which do not apply. Clauses can be repeated and renumbered as necessary. If a clause is repeated it is quite likely that this is because of a change in brick or mortar or bond or joints, but unlikely that all of these elements will change. In other words, repetition of the same information is likely. This can be overcome by referring to a previous clause, e.g.

111 FACING BRICKWORK *BELOW DPC*:
- Bricks: Clay to BS 3921.
  Manufacturer and reference: *As F10/110.*
  ~~Special shapes:~~ . . . .
- Mortar: As section Z21.
  Mix: *1:½:4 with sulphate resisting cement.*
  ~~Special colour:~~ . . . .
- Bond: *As F10/110.*
- Joints: *As F10/110.*
- Features: *As F10/110.*

The guidance notes below apply generally and are arranged in the same sequence as the clause subheadings.

Clause headings

These should describe the use and/or location of the brick/blockwork in a way which will be helpful to the Contractor, e.g.
CLAY FACING BRICKWORK *ABOVE DPC ON HOUSE TYPES 5 AND 7*:
CONCRETE FACING BLOCKWORK *IN CORRIDORS*:
ENGINEERING BRICKWORK *FOR MANHOLES*:
The Contractor will normally be looking first for information on the drawings and then to the specification. Therefore the clause heading need not be absolutely precise but more a confirmation that the clause is the correct one for the situation identified and/or cross-referenced on the drawings.

CLAY BRICKS

The last revision of BS 3921 introduced major changes in the classification of clay bricks, which are now designated according to frost resistance and soluble salt content:

Designation	Frost resistance	Soluble salt content
FL	Resistant	Low
FN	Resistant	Normal
ML	Moderately resistant	Low
MN	Moderately resistant	Normal
OL	Not resistant	Low
ON	Not resistant	Normal

BS 3921 specifies a maximum limit for soluble salt content in bricks designated 'Low' but not for 'Normal'.

Apart from engineering and damp proof course bricks, BS 3921 no longer classifies bricks by compressive strength.

CALCIUM SILICATE BRICKS

These should comply with BS 187. Apart from strength Class 2 (which, anyway, is not generally available) calcium silicate bricks have good frost resistance, even when saturated. Classes 3 to 7 can be used in any situation with the exceptions of:
- Cappings, copings and sills: Minimum class 4
- Earth retaining walls not waterproofed on the retaining side: Minimum class 4
- Manholes with foul effluent in continuous contact with the brickwork: Minimum class 7
- Damp proof courses: Not suitable.

See BS 5628:Part 3, table 13.

CONCRETE BRICKS

These should comply with BS 6073:Part 1. This standard does not give recommendations as to suitability for particular purposes.

Concrete bricks should not be used in contact with ground from which there is danger of sulphate attack unless the units are protected or have been specifically made for this purpose, i.e. with sulphate resisting cement and/or a certain minimum cement content – see BRE Digest 250.

Durability is related to the strength of the bricks and requirements vary greatly, depending on the situation. For detailed guidance see BS 5628:Part 3, table 13.

CONCRETE BLOCKS

BS 6073:Part 1 does not specify functional properties such as thermal conductivity, nor does it give recommendations as to suitability for particular purposes – for guidance on selection of blocks see:
- 'Products in practice supplement: Concrete blocks', The Architects' Journal 27.2.85.
- 'Concrete masonry for the designer', C&CA.

BS 6073:Part 1 specifies a minimum average crushing strength of 2.8 N/mm² for concrete blocks 75 mm thick or greater. Blocks less than 75 mm thick are classed as nonloadbearing. ➜

Concrete blocks are not suitable as damp proof courses or in foul drainage manholes. For guidance on use in other situations see BS 5628:Part 3, table 13. They should not be used in contact with ground from which there is danger of sulphate attack unless the units are protected or have been specifically selected for this purpose, i.e. made with sulphate resisting cement and/or with a certain minimum cement content – see BRE Digest 250.

SPECIAL SHAPES – BRICKS

BS 4729 specifies the dimensions of 'standard' special bricks. Many manufacturers will also make other shapes to order, but these may be limited by the manufacturing process and the nature of the material.

BS 4729 includes brick slips as special shapes. They should be manufactured with a keyed back to ensure a mechanical bond with the mortar.

For guidance on the use of special bricks see 'Products in Practice, Bricks', Architects' Journal 27.2.85.

Insert, e.g.

45° external and internal angles as BS 4729.
To BS 4729, type as shown on drawings.
As detailed on drawing number FC(2)1-7.

SPECIAL SHAPES – BLOCKS

There is no British Standard for special shapes of blocks and therefore availability must be checked with manufacturers. As with bricks, purpose designed blocks and matching components can be produced. The special blocks required should be listed, e.g.

Half sizes	Corner blocks
Reveal blocks	Horizontal duct blocks
Half reveal blocks	Pier blocks
Coursing bricks	Lintels
Return blocks	Lintel trough blocks.

MORTAR

The basic ingredients and making of mortar generally are specified in section Z21. Mortar mixes for brickwork and blockwork may be specified in the types of walling clauses at the beginning of the section, by stating the constituents and proportions, e.g.

1:1:6 cement:lime:sand
1:5-6 white cement:selected white sand.

For guidance on selection of mortars for use in various situations see BS 5628:Part 3, table 13 or BRE Digest 160. In general, stronger bricks and blocks and more severe exposure conditions require mortars of higher strength, i.e. with a greater cement content. The weaker the mortar, the more will it accommodate small movements in the brickwork or blockwork, this being more critical in blockwork because there are fewer joints. Richer mortars are desirable for use with exposed bricks that contain appreciable quantities of soluble sulphates (See BRE Digest 89).

SPECIAL COLOUR OF MORTAR

When coloured mortar is required, it is better to specify coloured ready mixed material. Site mixing using pigments is rarely satisfactory, adequate control of colour consistency being very difficult. Remember that the colour of mortar changes whilst drying and setting. Insert, e.g.

Mortaco Ltd. brown reference BR44.
Red, using ready-mixed lime:sand and subject to approval of sample panel.

If mortar is not required to be coloured, delete the item.

BOND

Use the following terminology:

Half lap stretcher	*Flemish*
Third lap stretcher	*Flemish garden wall*
English	*Heading*
English garden wall	*Honeycomb*

JOINTS

This item applies only to facework including fair face work. Wall faces which are to be plastered or which will not be visible in the finished work are covered by general clauses (see 610, 620 and 630). Delete the item accordingly.

Insert details of joint profile and method of execution, e.g. *Flush pointed as clause 860.* Profiles may be specified as:

Flush
Weathered
Bucket handle
Recessed square
Approved.

FEATURES

Insert brief details of any unusual requirements, cross referring where appropriate to sections F21, F30 and F31 and/or drawings, e.g.

Reinforcement over openings as section F30.
Brick corbelling to support joists.
Soldier course over openings.
Ashlar stone dressings as section F32.
Internal walls on suspended slabs reinforced in every bed joint as section F30.
Round windows formed with tapered blocks.

GUIDANCE NOTES

TYPES OF WALLING
See general guidance on pages 2 and 3.

Clauses are included for the following:
CLAY FACING BRICKWORK
CALCIUM SILICATE FACING BRICKWORK
CONCRETE FACING BRICKWORK
SECOND HAND FACING BRICKWORK
CONCRETE FACING BLOCKWORK
RECONSTRUCTED STONEWORK
CLAY COMMON BRICKWORK
CALCIUM SILICATE COMMON BRICKWORK
CONCRETE COMMON BRICKWORK
CONCRETE COMMON BLOCKWORK
ENGINEERING BRICKWORK
DAMP PROOF COURSE BRICKWORK

230
For advice on the use of second hand bricks see 'Salvage and reuse of bricks', Architects' Journal.

SPECIFICATION CLAUSES

F10 BRICK/BLOCK WALLING
To be read with Preliminaries/General conditions.

TYPE(S) OF WALLING

(110) CLAY FACING BRICKWORK EXTERNALLY:
- Bricks: To BS 3921.
  Manufacturer and reference: Smith's Bricks, Russet
  ~~Special shapes:~~ ....
- Mortar: As section Z21.
  Mix: 1:1:6 cement : lime : sand
  ~~Special colour:~~ ....
- Bond: Stretching
- Joints: Flush
- Features: Soldier course to parapet

~~170~~ CALCIUM SILICATE FACING BRICKWORK ....
- Bricks: To BS 187, Class ......
  Manufacturer and reference: ....
  Special shapes: ....
- Mortar: As section Z21.
  Mix: ....
  Special colour: ....
- Bond: ....
- Joints: ....
- Features: ....

~~210~~ CONCRETE FACING BRICKWORK ....
- Bricks: To BS 6073:Part 1.
  Manufacturer and reference: ....
  Minimum average compressive strength: ...... N/mm²
  Work size(s): ....
  Special shapes: ....
- Mortar: As section Z21.
  Mix: ....
  Special colour: ....
- Bond: ....
- Joints: ....
- Features: ....

~~230~~ SECOND HAND FACING BRICKWORK ....
- Bricks: Second hand bricks free from deleterious matter such as mortar, plaster, paint, bituminous materials and organic growths. Bricks to be sound, clean and reasonably free from cracks and chipped arrisses.
  Supplier/source: ....
- Mortar: As section Z21.
  Mix: ....
  Special colour: ....
- Bond: ....
- Joints: ....
- Features: ....

250
Blocks: Insert type of concrete, e.g.
Dense aggregate.
Aerated.
Work size(s): Specify face dimensions and the range of thicknesses required, e.g. *440 x 215 x 100, 140 and 190 mm.*
Finish/colour: Insert, e.g.
Shot blasted white.
Riven.
Exposed limestone aggregate.
Fluted charcoal.

290
This clause is for proprietary blocks which simulate natural stone rubble walling and normally form the outer leaf of a cavity wall. This type of reconstructed stonework is really only a variant of concrete blockwork and is quite distinct from simulated ashlar work.
Walling type: Insert, e.g.
Squared random rubble.
Rough hewn brought to courses.
Ashlar coursed and bonded as shown on drawings.
Finish/colour: Insert, e.g. *Weathered York.*
Special shapes: These include jumper blocks (to break up regular coursing), quoins, battered blocks, end blocks, capping blocks, sills, lintels, etc.
Mortar: See general guidance on page 3. Recommendations are normally the same as for concrete blockwork, but check with manufacturer.
Joints: Flush or concave (bucket handle) joints are usually the most suitable. Raked joints should not be used.
Features: Insert brief details of any unusual requirements, cross referring where appropriate to sections F30 and F31 and/or drawings, e.g.
Cast stone dressings as section F30.
Red brick quoins to corners and around openings, brickwork type F10/110.

310-340
Common brickwork is that which is not normally intended to be seen. If a brick marketed as a common is to be used as a facing or in fair faced work, specify in the appropriate facing brickwork clause.

250 CONCRETE FACING BLOCKWORK INTERNALLY:
- Blocks: Dense aggregate to BS 6073:Part 1.
  Manufacturer and reference: Fortihall Settle White
  ~~Minimum average compressive strength: . . . . . . N/mm²~~
  Work size(s): 440 x 225 x 100 and 140 mm
  Finish/~~colour~~: Smooth or shot blasted (drawing 101)
  Special shapes: Half blocks, lintel units
- Mortar: As section Z21.
  Mix: 1:1:6 cement : lime : sand
  ~~Special colour:~~ . . . .
- Bond: Stretching half lap
- Joints: Bucket handle
- Features: Reinforced block lintels

~~290~~ RECONSTRUCTED STONEWORK . . . .
- Walling type: . . . .
  Manufacturer and reference: . . . .
  Finish/colour: . . . .
  Special shapes: . . . .
- Mortar: As section Z21.
  Mix: . . . .
  Special colour: . . . .
- Bond: . . . .
- Joints: . . . .
- Features: . . . .

~~310~~ CLAY COMMON BRICKWORK . . . .
- Bricks: To BS 3921.
  Minimum average compressive strength: . . . . . . N/mm²
  Durability designation: . . . . . .
- Mortar: As section Z21.
  Mix: . . . .
- Bond: . . . .

~~330~~ CALCIUM SILICATE COMMON BRICKWORK . . . .
- Bricks: To BS 187, Class . . . . . .
- Mortar: As section Z21.
  Mix: . . . .
- Bond: . . . .

~~340~~ CONCRETE COMMON BRICKWORK . . . .
- Bricks: To BS 6073:Part 1.
  Minimum average compressive strength: . . . . . . N/mm²
  Work size(s): . . . .
- Mortar: As section Z21.
  Mix: . . . .
- Bond: . . . .

350
Common blockwork is that which is not normally intended to be seen. If a block marketed as a common is to be used as a facing or in fair faced work, specify in clause 250.
Blocks: Insert type of concrete, e.g. *Dense aggregate, Lightweight aggregate, Aerated.*
Thermal resistance: Thermal resistance need not be specified if blocks are not to be used as insulation, nor if a particular proprietary block is specified.
If the blocks are to be used, e.g. in the external leaf with a rendered coating, the thermal resistance should be specified at a moisture content of 5%. Amend the clause accordingly.
Work size(s): Specify face dimensions and the range of thickness required, e.g. 440 x 215 x 100, 140 and 190 mm.

380
If engineering bricks are to be used as facings they should be specified in clause 110. There are two classes of engineering brick specified in BS 3921:
A Compressive strength not less than 70 N/mm² and water absorption not more than 4.5% by mass.
B Compressive strength not less than 50 N/mm² and water absorption not more than 7% by mass.

390
Two classes are specified in BS 3921:
1 Water absorption not more than 4.5% by mass.
2 Water absorption not more than 7% by mass.

410
Delete either section if not included in the project specification. Insert any of the following if included in the project specification:
F20 Natural stone rubble walling
F21 Natural stone ashlar walling/dressings.

420
Concrete bricks and blocks are subject to drying shrinkage and should be protected from rain to ensure that they are dry when laid. Wetting of clay and concrete products will increase the possibility of efflorescence and/or staining.
Accuracy of construction is specified generally in Preliminaries section A33. Where higher levels of accuracy are required they can be specified using clause A33/360 or 361.

520
The limitation on rate of laying, i.e. 1.5 m height per day, is specified in BS 5628:Parts 1 and 2.

(350) CONCRETE COMMON BLOCKWORK BELOW DPC:
- Blocks: Dense aggregate to BS 6073:Part 1.
  Manufacturer ~~and reference~~: Fortihall
  Minimum average compressive strength: 20 N/mm²
  ~~Thermal resistance: Not less than .... m² °C/W at 3% moisture content.~~
  Work size(s): 440 x 215 x 100 and 140 mm
  ~~Special shapes:~~ ....
- Mortar: As section Z21.
  Mix: 1:¼:3 sulphate resisting cement: lime: sand
- Bond: Stretching

(380) ENGINEERING BRICKWORK FOR MANHOLES:
- Bricks: to BS 3921, Engineering Class B . . .
- Mortar: As section Z21.
  Mix: 1:¼:3 sulphate resisting cement: lime: sand
- Bond: English
- Joints: Flush.

~~390~~ DAMP PROOF COURSE BRICKWORK:
- Bricks: to BS 3921, Damp proof course Class ......
  Manufacturer and reference: ....
- Mortar: As section Z21.
  Mix: ....
  Special colour: ....
- Bond: ....
- Joints: ....

WORKMANSHIP GENERALLY

(410) RELATED WORK is specified in the following sections:
F30 Accessories/Sundry items for brick/block/stone walling.
F31 Precast concrete sills/lintels/copings/features.

(420) WORKMANSHIP GENERALLY:
- Store bricks and blocks in stable stacks, clear of the ground and clearly identified by type, strength, grade, etc. Protect from inclement weather and keep dry.
- Construct walling with all materials fully bonded or tied together to ensure compliance with design requirements for stability, strength, fire resistance, thermal and sound insulation as relevant.
- Lay bricks/blocks on a full bed of mortar and fill all cross joints and collar joints. Do not use shell bedding.
- Keep courses level, true to line and evenly spaced using gauge rods. Accurately plumb all wall faces, angles and features. Set out carefully to ensure satisfactory junctions and joints with adjoining or built-in elements and components.

(520) HEIGHT OF LIFTS: Carry up work with no portion more than 1.2 m above another at any time, racking back between levels. Do not carry up work higher than 1.5 m in one day.

540, 541
Alternative clauses.

540 COURSING: Gauge brick courses four to 300 mm including joints.

~~541~~ COURSING: Arrange brick courses to line up with existing work.

550
It is important that bricks are laid 'frog up' in cavity walls to ensure adequate bonding for ties.

550 FROGGED BRICKS: Lay single frog bricks with frog uppermost, double frog bricks with deeper frog uppermost.

~~560~~ BLOCKWORK: Lay cellular blocks with cavity downwards. Do not fill hollows in hollow blocks. Use cut or special shape blocks to make up courses and piece in; do not use bricks or other materials.

580 LINTEL BEARINGS: Carefully predetermine setting out so that full length bricks/blocks occur immediately beneath ends of lintels.

~~590~~ SUPPORT OF EXISTING WORK: Where new lintels or walling are to support existing structure, completely fill top joint with semidry mortar, hard packed and well rammed to ensure full load transfer after removal of temporary supports.

600
Insert appropriate details, e.g.
where shown on drawings.
unless specified otherwise.

~~600~~ BLOCK BOND new walls to existing, ..., by cutting pockets into existing walls, not less than 100 mm deep, the full thickness of the new wall, and vertically as follows:
Brick to brick: 4 courses high at 8 course centres.
Brick to block, block to brick or block to block: Every alternate block course.
Bond new walling into pockets with all voids filled solid with mortar.

610 JOINTS which are not to be visible in the finished work to be struck off with the trowel as the work proceeds.

~~620~~ BRICKWORK FOR PLASTER: Rake out mortar joints to a depth of not less than 13 mm in brickwork to be plastered or rendered.

630 BLOCKWORK FOR PLASTER: Strike off and leave rough joints in blockwork to be plastered or rendered.

640, 641
See BRE Digest 214 and the Building Regulations. Sealing around pipes and ducts is covered by section P31.

~~640~~ FIRE STOPPING: Fill joints around joist ends built into cavity walls with mortar to seal cavities from interior of building.

641
Use for brick cladding to timber frame construction.

~~641~~ FIRE STOPPING: Ensure a tight fit between brickwork and cavity barriers to prevent fire and smoke penetration.

650
Mortar that has failed to set after seven days may be assumed to be damaged by frost. Make sure mortar has set and is not unset but frozen. If the damage is merely to the surface of the joint the work can be made good by raking out and repointing. It is important that all defective material is raked out back to a face of hardened mortar. The depth of material needing to be removed should be not more than 20-25 mm. In cases of more severe damage the work should be rebuilt.

650 INCLEMENT WEATHER:
- Do not use frozen materials and do not lay on frozen surfaces.
- Do not lay bricks/blocks when air temperature is at or below 3 °C unless mortar has a minimum temperature of 4 °C when laid and walling is protected.
- Maintain temperature of the work above freezing until mortar has fully hardened.
- Adequately protect newly erected walling against rain and snow by covering when precipitation occurs and at the completion of each days work.
- Rake out and replace mortar damaged by frost and where instructed, rebuild damaged work.

ADDITIONAL REQUIREMENTS FOR FACEWORK

740 REFERENCE PANEL(S): Prepare panel(s) as set out below and, after drying out, obtain approval of appearance before proceeding. Construct panels with randomly sampled bricks/blocks in an approved location.

- Walling type F10/ 110
  Size of panel: 2 m²
  Including example of soldier course
- Walling type F10/ 250
  Size of panel: 4 m²
  Including example of lintel and control joint

740
BS 3921, Appendix F gives guidance on the visual assessment of facing bricks using panels of brickwork. The panels should be located in good natural light and, if possible, so that they can be seen in conjunction with the finished work. A viewing distance of 3 m will normally be satisfactory.
Walling type: Insert clause number.
Size of panel: 1.5 x 1.5 m is normal.
Including example of ... Any features which are to be included in the sample should be listed, e.g.
soldier course.
brick on edge coping.
cut bricks at reveals.

750 COLOUR MIXING:

- Agree with manufacturer and CA methods for ensuring that the supply of facing bricks/blocks is of a consistent, even colour range, batch to batch and within batches.
- Check each delivery for consistency of appearance with previous deliveries and do not use if variation is excessive.
- Mix different packs and deliveries which vary in colour to avoid horizontal stripes and racking back marks.
- Distribute bricks/blocks of varying colour evenly throughout the work so that no patches appear.

750
Manufacturers can help to overcome the problem of uneven colour distribution in several ways:

- Mixing bricks from different parts of the kiln before packaging.
- Reserving the whole of a kiln batch for the project when bricks are to be supplied over a long period.
- Mixing different kiln batches for very large projects.

760 APPEARANCE:

- Select bricks/blocks with unchipped arrisses and flat surfaces.
- Set out and lay carefully to give a satisfactory, uniform appearance with joints consistent in colour, width and profile and perpends vertically aligned in alternate courses.
- Cut bricks/blocks only where necessary at jambs and junctions. Cut with a masonry saw where cut edges will be exposed to view.
- Protect facework against damage and disfigurement during the course of the works, particularly arrisses of openings and corners.

760
Brickwork and blockwork should be dimensioned accordingly.

780 GROUND LEVEL: Facework to start not less than 150 mm below finished level of external paving or soil except where shown otherwise.

790 PUTLOG SCAFFOLDING to facework will not be permitted.

790
Putlog holes are made good as the scaffolding is taken down. This can be detrimental to the appearance of the wall if mortar colour or workmanship in pointing are markedly different. The alternative is to use independent scaffolding which will increase the cost of the job. Therefore use this clause only when very high quality facing work is required.

~~800~~ TOOTHED BOND: Except where a straight vertical joint is specified, new and existing facework in the same plane to be bonded together at every course to give a continuous appearance.

~~820~~ BRICK SILLS/CAPPINGS: Bed solidly in mortar with vertical joints completely filled. Press mortar firmly into exposed joints and finish neatly.

820
Brick sills and cappings are particularly vulnerable to frost action. A stronger mortar is required – see BS 5628:Part 3, table 13. It is preferable to joint the sills and cappings as the work proceeds rather than pointing separately. Sills and cappings should be specified as a separate type of brickwork.

830 CLEANLINESS: Keep facework clean during construction and until practical completion. Ensure that no mortar encroaches on face when laying. Turn back scaffold boards at night and during heavy rain. Rubbing to remove marks or stains will not be permitted.

860, 870, 880
See BRE Digest 200 and BS 6270:Part 1. Repairs should not be attempted unless movement has ceased. If cracked joints are not causing leakage, the decision to repoint will depend on visual considerations – cracks of 1.5 mm or less may be preferable to partial repointing. Rebedding should normally be done using a 1:1:6 mortar.

Mortar for repointing must not be very different in strength to the existing mortar. As a general guide:

Existing mortar	**Mortar for repointing**
Strong	1:1:6
Normal	1:2:9
Weak	1:3:12

State if coloured mortar is required. See BDA Practical Note 7 and BS 6270: Part 1, clause 13.

~~860~~ CRACKED BRICKS in existing facework to be cut out and replaced with matching bricks bedded in . . . . cement: lime:sand . . . . mortar, before repointing adjacent cracked joints as specified.

~~870~~ CRACKED JOINTS in existing facework which is not to be repointed: joints with cracks wider than . . . . . . mm to be cut out to form a square recess of 15–20 mm depth. Remove dust, lightly wet and neatly point in . . . . cement: lime:sand mortar to match existing work.

~~880~~ REPOINTING: Where specified carefully rake out existing joints by hand to form a square recess of 15–20 mm depth. Remove dust, lightly wet and neatly point in . . . . cement:lime: sand mortar to a . . . . profile in a continuous operation.

5 Specification work sections

INTRODUCTION

The previous chapters have covered the general principles of specification writing, this chapter deals with the actual preparation of a specification. Each Common Arrangement work section from A–Y is listed and each list (with the exception of A Preliminaries) opens with the full list of Common Arrangement headings. Under each section is set out a selection of elements likely to be encountered in domestic work or work of a simple nature, the selected elements each being marked with an asterisk and a selection of matters requiring specification is given. No such list can be comprehensive but the selections made are intended as a general guide. The CAWS headings not covered are, in the main, specialist work requiring a more detailed specification outside the scope of this book. It does not follow that each point mentioned should correspond with a clause in the specification: it may need several clauses or may be altogether unnecessary. It is obvious, for instance, that 'materials and workmanship' clauses of concrete will be much fuller for a reinforced concrete building than for a small house of traditional construction; or in a job where there are no suspended concrete floors the relative clauses will be omitted. The lists are lists of headings, and the writer should take each in turn asking 'Does this apply?' If it does, the next question is: 'What do I want?' When the section or part of the section has been completed the final question must not be forgotten: 'What else is wanted that is not on this list?'

Reference is given, where possible, to relative British Standards and Codes of Practice published by the British Standards Institution. The year reference is not quoted here, as these documents are revised from time to time: then the year reference is altered but the classification number normally remains the same. Up-to-date information as to currency of these standards and codes of practice will be found in the latest list (BSI Standards Catalogue – BSI) or can be ascertained by enquiry from the BSI. In addition to the sectional list there are available in four loose-leaf volumes summaries of British Standards for Building (BS Handbook 3 – BSI) to which addendum packets are issued from time to time on a subscription basis.

A PRELIMINARIES/GENERAL CONDITIONS

Note: The JCT Standard Form of Contract is assumed and its Conditions are not repeated here. If any other form is used, it should be checked with the Standard Conditions and any clauses found missing should be dealt with in the specification added to the conditions of contract.

A10	Project particulars	Position of the site Access for inspection, keys etc. Names and addresses: employer consultants
A11	Drawings	Schedule of drawings
A12	The site/Existing buildings	Site boundaries Existing buildings on site Existing buildings adjacent to site Existing services under and over the ground Trial hole information
A13	Description of the work	Brief description Details of work by others
A20	The Contract/Sub-contract	See note above Define the form Completion of blanks Amplifications or amendments Employer's insurance responsibilities Performance bonds Under hand or under seal

A30–A37 of the Common Arrangement cover the employer's requirements, that is to say matters at the discretion of the employer which must be made known to the contractor. It is therefore important that the employer's wishes are known and if necessary advice will need to be given. Each project must be treated on its merits but the following list, extracted in the main from SMM7 gives a guide. To this list would need to be added other matters special to the project, hence 'others – details stated'.

A30	Tendering/Sub-letting/Supply	Tender requirements Restrictions on sub-letting Purchase of materials
A31	Provision, content and use of document	Extra drawings Operating manuals

A32	Management of the Works	Site agent Site meetings Instructions Programmes, records etc.
A33	Quality standards/control	Samples Testing Certificates British Standards/Codes of Practice Clerk of works
A34	Security/Safety/Protection	Noise and pollution control Maintain adjoining buildings Maintain public and private roads Maintain live services Security Protection of work in all sections Others – details stated
A35	Specific limitations on method/sequence/timing	Design constraints Limitations, restrictions etc. Others – details stated
A36	Employer's requirements for facilities/temporary works/services	Accommodation temporary fences, hoardings, screens and roofs Name boards Technical and surveying equipment Temperature and humidity Telephone/Facsimile installation and rental/maintenance Others – details stated
A37	Operation/Maintenance of the finished building	

A40–A44 of the Common Arrangement cover the contractor's general cost items. Here the items are just listed for the convenience of the contractor in pricing. No details can be given as these are matters for the contractor to decide upon, hence 'others' is solely a reminder for the contractor to think of anything else he needs to price. The only exception to this is where the employer intends to make facilities available, such as space for messrooms, water and power for the works etc. In these cases details do need to be given.

A40	Management and staff	Contractor to decide
A41	Site accommodation	This is for the contractor to decide. If space etc is being made available details must be stated setting out any restrictions on use, leaving clean etc.

A42 Services and facilities	Power Lighting Fuels Water Telephones and administration Safety, health, and welfare Storage of materials Rubbish disposal Cleaning Drying out Protection of work in all sections Security Maintain private and public roads Small plant and tools General attendance on nominated sub-contractors Others
A43 Mechanical plant	Cranes Hoists Personnel transport Earthmoving plant Concrete plant Piling plant Paving and surfacing plant Others
A44 Temporary works	Temporary roads Temporary walkways Access scaffolding Support scaffolding and propping Hoardings, fans, fencing etc. Hardstanding Traffic regulations Others

A50–A55 of the Common Arrangement cover the remaining general items under this section.

A50 Work/Materials by the Employer	Work by others directly employed by the employer – details stated Attendance on others directly employed by the employer – details stated Materials provided by the employer – details stated
A51 Nominated sub-contractors	Add to PC sums for: profit unloading and storing materials

A51 Nominated sub-contractors continued	distributing materials in the building returning empties supplying water allowing use of plant or fixed scaffolding any special scaffolding general attendance cleaning up examine specialists' estimates for any special conditions
A52 Nominated suppliers	Add to PC sums for: profit carriage to site unloading and storing fixing as described returning empties carriage paid
A53 Work by statutory authorities	Provisional sums for: Crossings to public footways Sewer connections Water mains Gas and/or electric mains Building control fees etc. Hoardings Tests of materials
A54 Provisional work	Provisional sums how adjusted Contingency provision
A55 Dayworks	Any restrictions on daywork Information required Provision for daywork

B COMPLETE BUILDINGS

B1 Proprietary buildings B10 Proprietary buildings*

Specification of proprietary buildings is a matter for detailed reference to the chosen supplier's technical information giving as much or as little detail as is required to adequately price, purchase and subsequently erect the building or buildings. The infra-structure required to accommodate such buildings would be specified either as a section on its own or as part of a main specification.

C DEMOLITION/ALTERATION/RENOVATION

C1 Demolition	C10 Demolishing structures*
C2 Alteration – composite items	C20 Alterations – spot items*
C3 Alteration – support	C30 Shoring
C4 Repairing/Renovating concrete/masonry	C40 Repairing/Renovating concrete/brick/block/stone C41 Chemical dpcs to existing walls
C5 Repairing/Renovating metal/timber	C50 Repairing/Renovating metal C51 Repairing/Renovating timber C52 Fungus/Beetle eradication

Note: This section of the specification covers those works for which the estimator will require to see the site. In the case of a new building and a clear site there may be nothing at all, or perhaps only some openings in fence or hedge.

C1 Demolition

C10 Demolishing structures

Demolition	Describe extent with reference to drawings Grubbing up or sealing off drains and services Felling of trees (identified on drawing) and general clearance of site
Clearing site	Clearing site of rubbish, debris, overgrowth, etc.
Boundaries of site	Alterations (e.g. openings through or moving of fences, etc.): each boundary in turn
Temporary cross-overs	Temporary cross-overs to public footpath for access to the works
Obstructions	Provisional sums for moving lamp posts, telephone poles, etc.
Disposal of old materials	State if reserved for employer, otherwise clearing away will imply that they become the contractors' property and they will allow credit accordingly

Removal of toxic materials	State if the presence of materials such as asbestos lagging is known or suspected to exist, including indicating any special measures and/or restrictions as to their removal

C2 Alteration – composite items

C20 Alterations – spot items

Alterations	Alterations in buildings, taken, so far as practicable, room by room. Systematic and consistent order in each room, e.g.: Ceiling Walls Floors Windows Doors Fittings Define in a preamble the full meaning of such short descriptions as may be constantly repeated, e.g. 'overhaul window', 'overhaul lock', 'make good plaster', etc.

D GROUNDWORK

D1	Ground investigation/ stabilisation/dewatering	D10	Ground investigation
		D11	Soil stabilisation
		D12	Site dewatering
D2	Excavation/Filling	D20	Excavating and filling*
D3	Piling	D30	Cast in place concrete piling
		D31	Preformed concrete piling
		D32	Steel piling
D4	Diaphragm walling	D40	Diaphragm walling
D5	Underpinning	D50	Underpinning*

D2 Excavation/filling

D20 Excavating and filling

BS 8004 : Code of practice for foundations

Surface soil — Stripping any turf, rolling and stacking
Stripping surface soil and vegetable matter

Reducing levels — Any further surface excavation necessary

Basement — Excavation for basement

Trenches — Excavation for wall foundations, stanchion bases, etc. Define the depths to be estimated for if not shown on drawings. 'Depth required' is not good enough for pricing

Consolidation — Trimming and consolidating ground under concrete or hardcore

Excess excavation — Excess excavation (e.g., dug too deep) to be filled with concrete

Disposal — Disposal of excavated material –
(a) vegetable soil;
(b) general material;
including filling and compacting around foundations and under floors; where disposed of on site, position of dumps to be defined as accurately as possible. (Where material is removed from the site the tip is a matter for the contractor)

Temporary dumps for vegetable soil reserved for flower beds, etc. (See Q30 and 31)

	Material to be reserved for making up levels around building on completion and any forming of banks, etc. (See Q30 and 31)
Earth support	Provide all necessary earth support
Pumping etc.	Keep excavations, basements, ducts, etc., free from water
Filling	Material to be used Maximum thickness of layers and method of consolidation Extent and thicknesses (after consolidation) of beds Blinding top surface to receive concrete Any finishing to falls

D5 Underpinning

D50 Underpinning

Underpinning	Define nature and extent of underpinning and limit of length to be carried out in one operation Contractor to allow for necessary shoring Cutting away of old concrete or brick footings and excavation

E IN SITU CONCRETE/LARGE PRECAST CONCRETE

E1 In situ concrete	E10 In situ concrete* E11 Gun applied concrete
E2 Formwork	E20 Formwork for in situ concrete*
E3 Reinforcement	E30 Reinforcement for in situ concrete* E31 Post tensioned reinforcement for in situ concrete
E4 In situ concrete sundries	E40 Designed joints in in situ concrete* E41 Worked finishes/Cutting to in situ concrete* E42 Accessories cast into in situ concrete*
E5 Precast concrete large units	E50 Precast concrete large units*
E6 Composite construction	E60 Precast/Composite concrete decking

E1 In situ concrete

E10 In situ concrete

Materials and Workmanship

BS 12 : Portland cement
877 : Foamed slag
882 : Aggregates
915 : High alumina cement
1047 : Air-cooled blast furnace aggregate
1165 : Clinker aggregate
1199–1200 : Building sands
1521 : Building paper
1881 : Testing concrete

Cement
Fine aggregate (sand)
Coarse aggregate:
  general concrete
  fine concrete
  any special
Waterproofing compound
Water
Composition of mixes with note of abbreviated references (e.g. 1:2:4)
Method of mixing (hand and machine)
Frost and temperature control
Depositing, including minimum cover to reinforcement

3148 : Tests for water	Compacting, vibrating etc.
3681 : Sampling and testing of lightweight aggregates	Protection and curing
	Tests of individual materials
	Tests of concrete cubes
	Loading tests on finished work
3797 : Lightweight aggregates	Building paper or polythene sheeting under concrete
4016 : Building paper (breather type)	
4027 : Sulphate resisting Portland cement	
5075 : Concrete admixtures	
5328 : Methods of specifying concrete including ready mixed concrete	
5835 : Testing aggregates	
8110 : Structural use of concrete	
CP 102 : Protection from ground water	
117 : Composite construction in structural steel and concrete	

Mass concrete foundations	Mix
BS 8004 : Code of practice for foundations	Widths and thicknesses (or reference to drawings if there shown)
	Laps at steps
	Any sloping tops, etc.

Mass concrete beds	Mix
Reinforced foundations and beds	Thickness
Suspended slabs	Construction joints
Walls	Any preparatory wrapping of steel joists
Column and stanchion casings	Compacting or vibrating
Beam and beam casings	
Staircases	

E2 Formwork

E20 Formwork for in situ concrete

Materials and workmanship		Formwork:
BS 3809	: Wood wool permanent formwork and infill units	Strength Hardboard lining, etc. Retarding liquid Time for removal Finish from wrot formwork Splays, chamfers, etc. Mortices, boxings, etc.

E3 Reinforcement

E30 Reinforcement for in situ concrete

Materials and workmanship		Reinforcement:
BS 4449 4461	: Steel bars	Steel bars Steel wire, fabric, giving laps, etc.
4466	: Bending dimensions of bars	Hooked ends to bars, cranking, placing, tying wire, etc.
4482–3	: Steel fabric	
4486	: H.T. bars for pre-stressed concrete	
5896	: Steel wire for pre-stressed concrete	
6722	: Recommendations for dimensions	

Reinforced foundations and beds Suspended slabs Walls Columns and stanchion casings Beam and beam casings Staircases	Reinforcement (e.g. reference number of wire fabric or reference to drawings or schedule of bars)

E4 In situ concrete sundries

E40 Designed joints in in situ concrete

Expansion and contraction joints		Type of joint
BS 6093	: Code of practice	

Design of joints and jointing in building construction

E41 Worked finishes/Cutting to in situ concrete

Preparation of concrete for key for plaster etc.
Description and extent of any worked finish to concrete surface
Floor channels etc.
Grouting to holes, bolts, stanchion bases, etc.
Surface hardeners

E42 Accessories cast into in situ concrete

Provision for holding down bolts
Fixing strips and anchors
Insert plugs
Ties

E5 Precast concrete large units

E50 Precast concrete large units

Hollow tile or pre-cast slab floors and roofs (for estimating by specialists)	Superimposed load per square metre which is to be assumed (a) for floors (any floors taking special loads, e.g. safes, tanks, etc., mentioned with the loads to be reckoned) (b) for roofs Drawings supplied showing extent of the work Any supporting beams, beam casings, etc., required to be included shown marked on drawings Minimum bearings (end and side) Specialist to supply specification of his system Conditions of sub-contract including cash discount to be included

Note: If the floors or roofs have been designed in detail, they can be specified as described above for suspended slabs or pre-cast work, with the addition of the description of hollow tiles or other special units

F MASONRY

F1	Brick/Block walling	F10	Brick/Block walling*
		F11	Glass block walling*
F2	Stone walling	F20	Natural stone rubble walling*
		F21	Natural stone/ashlar walling/ dressings*
		F22	Cast stone walling/dressings*
F3	Masonry accessories	F30	Accessories/sundry items for brick/block/stone walling*
		F31	Precast concrete sills/lintels/ copings/features*

F1 Brick/Block walling

F10 Brick/Block walling

Materials and workmanship

- BS 12 : Portland cement
- 187 : Calcium silicate bricks
- 890 : Limes
- 1199–1200 : Sand
- 3921 : Clay bricks
- 4551 : Testing mortars
- 4721 : Ready-mixed lime sand for mortar
- 4729 : Special bricks
- 4887 : Mortar admixtures
- 5628 : Structural recommendations
- 6073 : Precast masonry units
- 8004, 8103 : Code of practice for foundations
- CP 102 : Protection from ground water

General bricks –
- (a) below damp-proof course
- (b) above damp-proof course

Facing bricks
Cement
Lime
Sand
Admixtures

Mortar composition and mixing –
- (a) cement mortar
- (b) lime or cement – lime mortar
- (c) any special pointing mortar

Bond of general brickwork
General workmanship in bricklaying (filling joints solid, keeping perpends, carrying up walls evenly, wetting bricks, protection against frost, etc.)
Pointing of fair face to common brickwork
Pointing of facing bricks
Gauge for height of courses (e.g. 13 courses to 1.00 m, assuming a metric brick 65 mm thick and a 10 mm joint)

Walls	Building of walls Construction of hollow walls, giving width of cavity and spacing of ties If generally lime or cement-lime mortar, state extent of work to be in cement mortar Preparation of existing walls for raising Any thickening of existing walls with extent of bonding to old
Wall finish (internal)	Extent of fair face. Refer to Schedule if any Describe pointing and state if to be carried out as the work proceeds Grooved bricks on surfaces to be plastered (if applicable)
Wall finish (external)	Extent of facing bricks String courses, bands, quoins and other dressings
Block partitions and walling BS 3921 : Clay blocks 6073 : Precast concrete masonry units	Make, thickness and surface finish Mortar, erection and bonding Bonding to brickwork Pinning up top edges
Sundries and general labours	Rough and fair cutting Raking out joints of brickwork (if necessary) as key for plastering, etc. Raking out joints for asphalt damp-proof courses, skirtings, etc., and pointing top edge Raking out joint for turn in of lead or copper flashings and pointing Eaves filling Fixing bricks Cutting rebates, splays, chases, holes, etc. Cutting and pinning ends of timbers, lintels, etc. Cutting, toothing and bonding new walls to old Pinning up top of walls to underside of steel joists (not bearing on them) or to existing soffites Centring

F11 Glass block walling

Type and size of glass blocks
Mortar
Temporary works including formwork if required

F2 Stone walling

F20 Natural stone/rubble

Walling		Stone
BS 5390	: Stone masonry	Mortar
		Whether random rubble, coursed, etc.
		Finish to exposed face (if in the same stone)
		Facing (if of a different stone) with finish to face and spacing of bonders
		Type of quoins, lintels, etc.
		Coping
		Centring

F21 Natural stone/ashlar walling/dressings

Dressings		To be set on natural bed
BS 3826, 6477	: Water repellents	Finish to surface generally
5642	: Sills and copings	Mortar for setting and pointing
		Bonding to backing
		Treatment to back, e.g. coating of lime
CP 202	: Tile and slab flooring	Coating face with slurry and cleaning down
		Protection
		Extent of general stone facing and any special descriptions or instructions
		Where stonework is in dressings only, give positions and sizes:
		Plinth
		String courses
		Window and door surrounds
		Cornices
		Copings
		Pavings, steps, etc.
		Padstones
		Casing up and protection
		Centring

F22 Cast stone walling/dressings

BS 340 : Kerbs
368 : Paving slabs
1217 : General
3826, 6477 : Water repellents
5642 : Sills and copings
6457 : Reconstructed stone masonry units
CP 202 : Tile and slab flooring

Paving: thickness, size of slabs and surface finish
Corbel courses: width and thickness
Cover stones on girders: ditto
Padstones under ends of beams: sizes and positions with surface finish to exposed faces
Steps: sizes and positions
Landings: ditto
Hearths: ditto
Maker's name if a specialist's material is required
Colour and texture
Otherwise specify in the same way as a variety of natural stone

F3 Masonry Accessories

F30 Accessories/Sundry items for brick/stone walling

Proprietary sills, lintels, copings, windows boards, etc.

Type and size and proprietary reference
Hoisting
Bedding and pointing
Temporary supports

Reinforcement

Expanded metal or proprietary brand
Spacing of reinforcement in thickness and height of wall

Forming cavities
BS 6676 : Thermal insulation of cavity walls

Width of cavity
Rigid sheet insulation

Wall ties, anchors, cramps, dowels, etc.
BS 1243 : Wall ties

Type size and spacing

Damp proof courses
BS 743 : Generally
6398 : Bitumen
6515 : Polythene

General damp-proof course
Ditto to parapets
Ditto to chimney stacks
Ditto over openings in hollow walls or where hollow walls are built solid
Ditto to jambs of hollow walls

Flue linings, terminals
BS 41 : Cast iron flue pipes
567 : Asbestos cement flue pipes (light)

Fireplaces and their fixing
Firebrick linings
Chimney pots
Brick hearths

835 : Asbestos cement flue pipes (heavy)	
1181 : Clay flue linings and flue terminals	
1289 : Precast concrete flue blocks and terminals	
5440 : Flues	
5854 : Code of Practice for flues and flue structures in buildings	
6461 (Part 1) : Masonry chimneys and flue pipes	
(Part 2) : Factory made insulated chimneys for internal	
Air bricks and gratings BS 493 : Air bricks and gratings	Type, size and location (check these are shown on drawings or specify the number required) Flues through walls Inside finish (cross-reference if in another section)
Fixing items	Cutting and pinning ends of timbers, lintels, etc. Bedding plates, door frames, etc. Building in metal windows, doors, etc.

F31 Pre-cast concrete sills/lintels/ copings/features (Non proprietary)

BS 340 : Kerbs 368 : Paving slabs 5642 : Sills and copings 5977 : Lintels 6073 : Masonry units	Mix Sizes of members Reinforcement if any Surface finish Dowels Bedding and pointing Forming holes Hoisting Fixing slips, cramps, etc. Fixings Temporary supports

G STRUCTURAL/CARCASSING METAL/TIMBER

G1 Structural/Carcassing metal	G10 Structural steel framing* G11 Structural aluminium framing G12 Isolated structural metal members*
G2 Structural/Carcassing timber	G20 Carpentry/Timber framing/First fixing*
G3 Metal/Timber decking	G30 Metal profiled sheet decking G31 Prefabricated timber unit decking G32 Edge supported/Reinforced woodwool slab decking*

G1 Structural/Carcassing metal

G10 Structural steel framing

Structural Steel BS 4 : Hot-rolled sections 449, 5950 : Use of structural steel in building 2994 : Cold rolled steel sections 3139, 3294 : Bolts 4174 : Self-tapping screws 4604 : High strength friction grip bolts 4848 : Hot-rolled structural steel sections 6323 : Steel tubes	For simple construction fixed by the builder the sizes of members must be given according to the British Standard sections. It is important where the same size member is made with two different weights to specify the weight required, e.g. 203 × 133 mm × 25 kg per m or 203 × 133 m × 30 kg per m beams Cleated connections and their rivets and bolts Holding down bolts Reference to any padstones, etc., specified elsewhere In more elaborate construction the work will usually be entrusted to a specialist and a specification is a matter for a structural engineer

G12 Isolated structural metal members

Isolated members	Wall plates, bearing bars etc. Isolated structural beams, columns, etc.

G2 Structural/Carcassing timber

G20 Carpentry/Timber framing/First fixing

Materials and workmanship BS 144 : Creosote 589, 881 : Nomenclature of timber 913 : Pressure creosoting of timber 1202 : Nails 1494 : Fixing accessories 1579 : Connectors for timber 3051 : Coal tar creosotes for wood preservation 4072 : Wood preservation 4169 : Laminated structural members 4174 : Drive screws 4471 : Sizes of sawn and processed softwood 4978 : Timber grades for structural use 5268 : Code of practice for structural use of timber 6100 : Glossary of terms 6178 : Joist hangers CP 112 : Structural use of timber	Timber generally Softwood Hardwoods Basic sizes given: allowances or finished sizes (sawn and wrot) Spacing of structural timbers in timber floor, roof or partition construction Trimming of ditto; extra thickness of trimming joists and trimmers Metal connectors, straps, hangers, bolts, ties, etc. off site preservatives
Floors	Plates Joists Trimming Strutting Bridging pieces
Timber stud partitions	Heads Sills Studs Braces Noggings Trimming

Pitched roofs	Plates Rafters and collars Ridge Hips and dragon ties Valleys Purlins and their struts Trusses Sprockets and tilting fillets Ceiling joists Binders and hangers to ditto Dormers Trimming Battens Roof boarding Walking boards in roof space Trap door to ditto Any special features, e.g. turrets
Eaves and verge finishes	Wrot ends to rafters Fascia Soffit and bearers (size and spacing) Bedmould Barge boards Parapet and other boxed gutters Snow boards
Grounds, bearers, firrings, etc	Wall battens: size, spacing and fixing Noggings for joints of plaster board or similar coverings: ditto. (Where not an integral part of linings or partitions system) Grounds for fibrous plaster work: ditto Firrings
Fixings	Bearers (e.g. for cisterns) Fixing slips in joints Plugging

G3 Metal/Timber decking

G32 Edge supported/Reinforced woodwool slab decking

Flat roofs BS 1105 : Woodwool cement slabs	Decking Supports and fixings Angle fillets for turn-up of felt roofing etc.

H CLADDING/COVERING

H1	Glazed cladding/covering	H10	Patent glazing*
		H11	Curtain walling
		H12	Plastics glazed vaulting/walling
		H13	Structural glass assemblies
		H14	Concrete rooflights/pavement lights*
H2	Sheet/Board cladding	H20	Rigid sheet cladding
		H21	Timber weatherboarding
H3	Profiled sheet cladding/covering/siding	H30	Fibre cement profiled sheet cladding/covering/siding*
		H31	Metal profiled/flat sheet cladding/covering/siding*
		H32	Plastics profiled sheet cladding/covering/siding*
		H33	Bitumen and fibre profiled sheet cladding/covering*
H4	Profiled panel cladding	H40	Glass reinforced cement cladding/features
		H41	Glass reinforced plastics cladding/features
H5	Slab cladding	H50	Precast concrete slab cladding/features
		H51	Natural stone slab cladding/features
		H52	Cast stone slab cladding/features
H6	Slate/Tile cladding/covering	H60	Clay/Concrete roof tiling*
		H61	Fibre cement slating*
		H62	Natural slating*
		H63	Reconstructed stone slating/tiling*
		H64	Timber shingling
H7	Malleable sheet coverings/cladding	H70	Malleable metal sheet prebonded coverings/cladding
		H71	Lead sheet coverings/flashings*
		H72	Aluminium sheet coverings/flashings*
		H73	Copper sheet coverings/flashings*
		H74	Zinc sheet coverings/flashings*
		H75	Stainless steel coverings/flashings*

H76 Fibre bitumen thermoplastic sheet coverings/flashings

H1 Glazed cladding/covering

H10 Patent glazing

Patent roof glazing BS 5516 : Code of practice for patent glazing	Type of glass Type of bar, spacing and method of fixing Clips to foot of glazing Finish to ridges, hips and valleys

H14 Concrete rooflights/pavement lights

Concrete rooflights and pavement lights	Maker and type (or PC sum) Over-all sizes Size of lenses Method of fixing (with reference to kerb in Masonry) Any intermediate supports

H3 Profiled sheet cladding/covering/siding

H30 Fibre cement profiled sheet cladding/covering/siding
H31 Metal profiled/flat sheet cladding/covering/siding
H32 Plastics profiled sheet cladding/covering/siding
H33 Bitumen and fibre profiled sheet cladding/covering

Fibre cement sheet coverings (for slating see under H61) BS 690 : Asbestos-cement sheets 1494 : Fixing accessories 5247 : Sheet roof and wall coverings	Material Type of sheet, e.g. standard corrugated (small section), standard corrugated (large section), reinforced corrugated Side lap and end lap Method of fixing (drive screws, clip bolts and their washers) Ridge covering, Hip ditto, Valley ditto, Vertical angle covering } If in lead, aluminium, copper or zinc specify under H71–H74

Sheet metal coverings (other than flat lead, zinc, aluminium or copper)

BS 1494	: Fixing accessories
3083	: Hot-dip zinc coated corrugated steel sheets
4154	: Corrugated plastic translucent sheets
4842	: Liquid coatings
4868	: Profiled aluminium sheets
6496, 6497	: Powder coatings
CP 143 (Part 1)	: Aluminium corrugated and troughed sheets
(Part 10)	: Galvanised corrugated steel

Bitumen and fibre sheet

CP 143 (Part 16)	: Semi-rigid asbestos bitumen sheet

H6 Slate/Tile cladding/covering

H60 Clay/Concrete roof tiling
H61 Fibre cement slating
H62 Natural slating
H63 Reconstructed stone slating/tiling

Tiling

BS 402	: Clay plain tiles
473, 550	: Concrete ditto
5534	: Code of Practice

Type and size of tiles/slates
Lap
Nailing (any differences for steep slopes or vertical faces)
Double course to eaves
Verge (tile/slate-and-a-half width, undercloak and bedding)
Ridge covering, Hip ditto, Valley ditto, Vertical angle covering — If in lead, aluminium, copper or zinc specify under H71–H74
Hip irons
Fixing lead soakers
Glass tiles or similar items

Slating		
BS 680	: Roofing slates	
690	: Asbestos cement slates	
5534	: Code of Practice	
Roofing felt		
BS 747	: Roofing felts	Underfelt to flat roofs Ditto to sloping roofs

H7 Malleable sheet coverings/cladding

H71 Lead sheet coverings/flashings

Sheet lead work		Flats Gutters and cesspools Flashings and aprons Ridge, hip and valley coverings Dormer tops and cheeks Saddles, slates, etc. Soakers Shoots into rainwater heads Damp-proof courses, aprons through cavity walls, to edges of asphalt, etc. Wedging flashings	Thickness of lead and girths where not defined on drawings. Size and spacing of tacks.
BS 1178	: Milled lead sheet for building purposes		
6915	: Lead roof and wall coverings		
		Bedding edges or turning into grooves, welting, etc. Soldered angles and seams Bossing to rolls, mouldings, etc. Copper nailing Timber rolls	

H72 Aluminium sheet coverings/flashings
H73 Copper sheet coverings/flashings
H74 Zinc sheet coverings/flashings
H75 Stainless steel sheet coverings/flashings

Aluminium work (flat sheets: for corrugated sheeting see H31) CP 143 (Part 15) : Aluminium roof and wall coverings	Sub-divided as for lead sheet coverings (H71) Gauge Method of jointing
Copper work BS 2870 : Sheet copper CP 143 (Part 12) : Copper sheet roof and wall coverings	
Zinc work BS 849 : Zinc roofing CP 143 (Part 5) : Zinc sheet roof and wall coverings	

J WATERPROOFING

J1 Cementitious coatings	J10 Specialist waterproof rendering
J2 Asphalt coatings	J20 Mastic asphalt tanking/damp proof membranes* J21 Mastic asphalt roofing/insulation/ finishes* J22 Proprietary roof decking with asphalt finish
J3 Liquid applied coatings	J30 Liquid applied tanking/damp proof membranes J31 Liquid applied waterproof roof coatings J32 Sprayed vapour barriers J33 In situ glass reinforced plastics
J4 Felt/flexible sheets	J40 Flexible sheet tanking/damp proof membranes J41 Built up felt roof coverings* J42 Single layer plastics roof coverings J43 Proprietary roof decking with felt finish

J2 Asphalt coatings

J20 Mastic asphalt tanking/damp proof membranes

Tanking/Damp-proof courses
BS 6577 : (natural rock)
6925 : (limestone aggregate)

General damp-proof course (usually 2 coat)
Tanking (usually 3 coat)
Key of vertical into brick joints
Angle fillets

J21 Mastic asphalt roofing/insulation/finishes

Roofing
BS 747 : Felt
6577 : (natural rock)
6925 : (limestone aggregate)
CP 144 (Part 4) : Mastic asphalt

Felt underlay
Flat roof (usually 2 coat)
Vertical or sloping faces
Angle fillets
Arrises and nosings
Skirtings: thickness, height and finish to top edge

Tile paving	Size and thickness of tiles (colour if coloured) Bedding material Joints: thickness and whether broken Skirtings: section

J4 Felt/flexible sheets

J41 Built up felt roof coverings

Felt roofing BS 747 : Felt roofing CP 144 (Part 3) : Roof coverings	Name of material, if a particular brand If single layer, type and weight per roll If built-up roofing, number of layers and type and weight per roll of each layer Bedding composition Surface finish, e.g. gritting, insulating tiles, macadam, etc. Angle fillet Skirtings and turn-ups over angle fillet

K LININGS/SHEATHING/DRY PARTITIONING

K1 Rigid sheet sheathing/linings	K10 Plasterboard dry lining* K11 Rigid sheet flooring/sheathing/ linings/casings* K12 Under purlin/Inside rail panel linings K13 Rigid sheet fine linings/ panelling*
K2 Board/Strip sheathing/linings	K20 Timber board flooring/ sheathing/linings/casings* K21 Timber narrow strip flooring/ linings
K3 Dry partitions/linings	K30 Demountable partitions K31 Plasterboard fixed partitions/ inner walls/linings* K32 Framed panel cubicle partitions K33 Concrete/Terrazzo partitions
K4 False ceilings/floors	K40 Suspended ceilings* K41 Raised access floors

K1 Rigid sheet sheathing/linings

K10 Plasterboard dry lining

BS 1230	: Gypsum board	Type of board, thickness and method of fixing
4022	: Gypsum wall board panels	Scrimming of joints
6214	: Jointing materials	
6452	: Beads	

K11 Rigid sheet flooring/sheathing/linings/casings

BS 1105	: Woodwool cement boards	Boarding: Make and thickness
3444	: Blockboard and laminboard	Method of fixing, nails, etc. Bearers
5669	: Chipboard	Cover fillets
6566	: Plywood	Pipe casings

K13 Rigid sheet fine linings/panelling

Standard	Title	Items
BS 3757	: Rigid PVC sheets	Type and thickness of material
3794	: Decorative high pressure laminates based on thermosetting resins	Bearers Treatment of joints Applied mouldings Mouldings formed on the solid Carving Fixings
4965	: Decorative panels	

K2 Board/strip sheathing/linings

K20 Timber board flooring/sheathing/linings/casings

Standard	Title	Items
BS 8201	: Code of practice for timber flooring	Type and thickness of material Method of fixing Jointing Surface finish

K3 Dry partitioning/linings

K31 Plasterboard fixed partitions/inner walls/linings

Standard	Title	Items
BS 1230	: Gypsum board	Linings
4022	: Gypsum wall board panels	Studs, head and sole plates, etc. Insulation
5234	: Code of practice partitioning	Joint supports and treatment Intersection angles, abutments, etc.
6214	: Jointing materials	Cutting or fitting for doors and the like
6452	: Beads	Surface treatment

K4 False ceilings/floors

K40 Suspended ceilings

Standard	Title	Items
CP 290	: Suspended ceilings	Type and size of tile, boards, panels or strips Method of fixing Spacing of supports Perimeter detail Access panels

L WINDOWS/DOORS/STAIRS

L1 Windows/Rooflights/Screens/Louvres	L10 Timber windows/rooflights/screens/louvres* L11 Metal windows/rooflights/screens/louvres* L12 Plastics windows/rooflights/screens/louvres
L2 Doors/Shutters/Hatches	L20 Timber doors/shutters/hatches* L21 Metal doors/shutters/hatches* L22 Plastics/Rubber doors/shutters/hatches
L3 Stairs/Galleries/Balustrades	L30 Timber stairs/walkways/balustrades* L31 Metal stairs/walkways/balustrades*
L4 Glazing	L40 General glazing* L41 Lead light glazing L42 Infill panels/sheets

L1 Windows/Rooflights/Screens/Louvres

L10 Timber windows/rooflights/screens/louvres

BS 1186 : Workmanship in joinery

Casement windows/borrowed lights

BS 584 : Wood trim
1227 : Hinges

CP 153
(Part 1) : Cleaning and safety
(Part 2) : Durability and maintenance

Generally – material

Casements:
- thickness
- division into panes and size of bars
- glazing beads
- treatment of bottom rails and meeting stiles if not detailed
- hinges etc. (where supplied with the component)

Frames:
- section: jambs and head
- ditto: sill (state if different material)
- ditto: transome (ditto)
- beads to swing casements

Jamb linings

Architraves and cover fillets

Double-hung sash windows BS 584 : Wood trim 644 : Windows (Part 2) (Part 3 for Scotland) CP 153 (Part 1) : Cleaning and safety (Part 2) : Durability and maintenance	Generally – material Sashes: thickness division into panes and size of bars horns glazing beads treatment of bottom and meeting rails if not detailed Boxed frames inside and outside linings pulley stiles and pocket pieces back linings parting slip parting bead inside bead sill Boxed mullions (as for frames) Jamb linings Architraves and cover fillets
Frames for metal windows BS 1285	All as frames to casement windows

L11 Metal windows/rooflights/screens/louvres

Metal windows BS 5286 : Aluminium framed sliding glass doors 6510 : Steel windows, sills, window boards and doors CP 153 (Part 2)	As for timber windows (L10) except BS windows which can be specified by their reference number Make clear whether fixing is by contractor or specialist If by specialist, contractor probably must distribute to positions, cut and pin lugs and bed frames If in wood frames reference to them

L2 Doors/Shutters/Hatches

L20 Timber doors/shutters/hatches

BS 1186 : Workmanship in joinery
CP 151 (Part 1) : Timber doors

Ledged (and braced) doors BS 459 (Part 4) : Matchboarded doors 1227 : Hinges	Generally – material Each type of door: thickness of boarding width of boards and type of joint thickness of ledges fixings and fastenings (where supplied with the component)
Framed, ledged and braced doors	As for ledged and braced, but giving thickness of stiles and rails instead of ledges and stating whether boarding is filled in or sheathed (over middle and bottom rails)
Panelled doors, etc. BS 1227 : Hinges	Generally – material Each type of door thickness of stiles and rails number of panels thickness of panels design of panel on each face, e.g. square framed, moulded, etc. if part glazed, number of panes and size of bars glazing (where supplied with the component) fixings and fastenings (where supplied with the components)
Flush doors BS 1227 : Hinges 4787 : Door sets, door leaves and frames	Whether for internal or external use Thickness Ply facing – if for staining to be stated Core Glue Fixings and fastenings (where supplied with the component)
Door linings or frames and architraves BS 584 : Wood trim 1567 : Wood door frames and linings	Material Jamb linings: thickness Wood frames: section: jambs and head ditto: transome ditto: threshold (state if different material) Architraves and cover fillets

L21 Metal doors/shutters/hatches

Metal doors BS 5286 : Aluminium sliding doors	Generally – material Size and thickness Frame Fittings

Metal door frames BS 1245		Profile of frame Size of opening Method of fixing

L3 Stairs/Galleries/Balustrades

L30 Timber stairs/walkways/balustrades

Stairs		
BS 585	: Wood stairs	Treads, landings and risers: thickness of each
1186	: Workmanship in joinery	Construction Wall strings: thickness and finish to top edge Outer strings: ditto, also whether cut or close, finish to bottom edge and framing to newels Applied mouldings to strings Skirtings to landings Newels: size, design and finish to top Balusters: size, design, spacing and framing of ends Solid balustrades: as for panelled framings Hand-rail: material, section and framing Spandril framing: as for cupboard fronts Soffite of stairs, if matchboarded, ply covered, etc.

L31 Metal stairs/walkways/balustrades

Staircases		Metal staircases, e.g. external escape stairs or steps and landings in machine rooms, etc., will usually be erected by specialists and be the subject of a PC sum
Tubular rails, balustrades, etc.		Wrought iron or steel
BS 1387	: Steel tubes and tubulars	Size of members Fittings – e.g. bends, tees, crosses, etc.
1740	: Wrought steel pipe fittings	Jointing – e.g. screwed or welded Finish – e.g. black or galvanised
4127	: Stainless steel tubes	
4360	Weldable structural steels	

4604	Use of high strength friction grip bolts

L4 Glazing

L40 General glazing

Materials and workmanship BS 544 : Linseed oil putty 952 : Glass for glazing 6262 : Code of practice for glazing	Sheet glass General obscure glass Any other kinds of glass to be used Putty for glazing to wood Ditto for glazing to metal Sprigged and front and back puttied Where to be glazed with beads (reference to beads in L10 timber windows or as supplied with metal windows L11) Extent of bedding in washleather or velvet
General glazing	Each kind of glass in turn with extent of its use Polished edges, etching, etc. Bending

M SURFACE FINISHES

M1 Screeds/trowelled flooring	M10 Sand cement/Concrete/Granolithic screeds/flooring* M11 Mastic asphalt flooring* M12 Trowelled bitumen/resin/rubber-latex flooring*
M2 Plastered coatings	M20 Plastered/Rendered/Roughcast coatings* M21 Insulation with rendered finish M22 Sprayed mineral fibre coatings M23 Resin bound mineral coatings
M3 Work related to plastered coatings	M30 Metal mesh lathing/Anchored reinforcement for plastered coatings* M31 Fibrous plaster*
M4 Rigid tiles	M40 Stone/Concrete/Quarry/Ceramic tiling/Mosaic* M41 Terrazzo tiling/In situ terrazzo* M42 Wood block/Composition block/Parquet flooring*
M5 Flexible sheet/tile coverings	M50 Rubber/Plastics/Cork/Lino/Carpet tiling/sheeting* M51 Edge fixed carpeting* M52 Decorative papers/fabrics*
M6 Painting	M60 Painting/Clear finishing*

M1 Screeds/trowelled flooring

M10 Sand cement/Concrete/Granolithic screeds/flooring

Screeds BS 4551	: Testing screeds	Composition Floor screeds with thicknesses Ditto to steps and landings, etc. Wall screeds with ditto Mat spaces Angle fillets for turn-up of felt roofing Filling to access covers
Granolithic BS 882	: Aggregates for granolithic	Mix and colouring (if any) Pavings with thicknesses Ditto to steps and landings, etc.

concrete floors

Carborundum finish
Non-slip nosings
Finish to skirtings, strings, kerbs, etc.
Floor strips

M11 Mastic asphalt flooring

Paving
BS 6577 : (natural rock)
6925 : (limestone aggregate) (coloured)
CP 204 : In situ floor finishes

Floors: thickness and colour (if coloured)
Skirtings: thickness, height and finish to top edge

M12 Trowelled bitumen/resin/rubber-latex flooring

Jointless floors
CP 204

Type, thickness and colour
Skirtings and covers
Floor strips

M2 Plastered coatings

M20 Plastered/Rendered/Roughcast coatings

Materials and workmanship
BS 12 : Cement
890 : Limes
1191 : Gypsum building plasters
: Premixed lightweight plasters
1199–1200 : Sands for plastering and external rendering
1230 : Gypsum plaster board
4049 : Glossary of terms
4551 : Testing plaster/render
5262 : Code of Practice for external rendered finishes
5270 : Bonding agents

Portland cement
Coloured cements
Sand
Waterproofing compound
Lime
Plaster board
Hard plaster
Any special plaster (e.g. barium or vermiculite)
Fibre reinforcement, hair, etc.
Mix of each coat of plaster
Interval between coats, scoring undercoats, finish to face, etc.
Cement and sand mixes
Protection

5492 : Code of practice for internal plastering	
Generally	Temporary rules for plastering and for specialists work
Plastering to ceilings and beams	Type of plaster Number of coats Salient angles Describe any system for suspension, or provisional sum if by specialist
Coves and cornices	Cove or cornice between ceiling and wall finish Ditto between ceiling and beam finish Bracketing
Wall plastering	Type of plaster Number of coats Dado plastering (if different) Coves Salient angles Dado mouldings, etc.
Skirtings	Material Number of coats Joints with wall and floor finishes
Cement work	Plain face, etc., to walls } As for wall Roughcast, stucco, etc. } plastering Cornices, pilasters, quoins and other features

M3 Work related to plastered coatings

M30 Metal mesh lathing/Anchored reinforcement for plastered coatings

Metal lathing BS 1369	Gauge and mesh Method of fixing Ceilings and beams Pipe casings or other surfaces

M31 Fibrous plaster

Fibrous plaster	Method of fixing Grounds (references to G20) Girths of cornices, etc. Where work is more elaborate, a general description with reference to detail drawings

M4 Rigid tiles

M40 Stone/Concrete/Quarry/Ceramic tiling/Mosaic

Quarry tiles, etc.		Type and quality of tiles
BS 1197	: Concrete flooring tiles	Thickness, colour and size of tiles Bedding and jointing material
6431	: Ceramic ditto	Floors
CP 202	: Floor tiling	Steps, etc. Skirtings Window sills, cappings, offsets, etc. Coves and rounded edges
Glazed tiles		All as for quarry tiles (ordinary glazed surface will be assumed unless 'eggshell glaze' is specified)
BS 5385	: Code of Practice for wall tiling	
5980	: Adhesives	
6431	: Ceramic tiling	

M41 Terrazzo tiling/In situ terrazzo

Terrazzo		PC sum for specialist's work, including dividing strips
BS 4131	: Terrazzo tiles	
4357	: Terrazzo units	State if screed by specialist
CP 204	: In situ floor finishes	Protection

M42 Wood block/Composition block/Parquet flooring

Block floors		Materials	or PC sum
BS 1187	: Wood blocks for floor	Thickness Pattern and border	
CP 209	: Maintenance	Laying and cleaning off Polishing Protection	
Parquet floors		As for block floors with reference to the item specifying the sub-floor	
BS 4050	: Mosaic parquet panels		

M5 Flexible sheet/tile coverings

M50 Rubber/Plastics/Cork/Lino/Carpet tiling/sheeting

Tile floors		Thickness and size of tiles
BS 1711	: Rubber	Make and colour
2592	: Thermoplastic flooring tiles	Laying Protection
3260	: Semi-flexible PVC floor tiles	

5325	: Installation	
5442	: Adhesives	
6826	: Linoleum and cork carpet tiles	
8203	: Code of practice for installation of tile flooring	
Sheet floors		Thickness
BS 1711	: Rubber	Make and colour
3261	: Flexible PVC flooring	Laying Protection
5442	: Adhesives	
6826	: Linoleum and cork carpet sheet	
8203	: Code of practice for installation of sheet flooring	

M51 Edge fixed carpeting

BS 5325	: Installation	Type of manufacturer's reference or PC sum Underlay Method of fixing Jointing/cover strips Stair nosings

M52 Decorative papers/fabrics

Paperhanging		Lining paper
BS 3046	: Adhesives for hanging flexible wall coverings	Decorative paper/fabric or PC price per piece Borders, etc.
3357	: Glue, size	

M6 Painting

M60 Painting/Clear finishing

Materials and workmanship		Ceiling emulsion
BS 381C	: Colours for identification coding and special purposes	Wall emulsion Cement paint Flat wall paint (or other material for treatment of ceilings and walls)

BS 1070 : Black paint (tar base)
1282 : Guide to the choice, use and application of wood preservatives
1336 : Knotting
1710 : Colour identification of pipelines and services
2015 : Glossary of paint terms
2523 : Lead based priming paints
3698 : Calcium plumbate priming paints
3981 : Iron oxide pigments
4764 : Powder cement paints
4800 : Paint colours for building purposes
5082 : Water-borne priming paints for woodwork
5358 : Solvent-borne priming paints for woodwork
5707 : Solutions of wood preservatives in organic solvents
6044 : Pavement marking paints
6150 : Code of Practice Painting of buildings
6952 : Exterior wood coating

Bituminous paint
Priming, undercoats and finishing paint for wood and metal work
Wood stain
Polish
Delivery and dilution of paint
Tints and selection
Surfaces dry – rubbing down coats – interval between coats – stippling, etc.
Any prohibition of spraying

Finishes

If finishes are more or less uniform throughout they can be specified in turn under the following heads:
Ceilings
Walls
Dadoes
External cement work

Internal metalwork
External metalwork
Internal woodwork
External woodwork

If the specification becomes at all complicated it is more satisfactory to set it out in the form of a schedule with explanatory notes to simplify the references, e.g.:

ROOM	CEILING	WALLS	DADO	METALWORK	WOODWORK	REMARKS
1	Emul.	Emul.	–	Gloss	Gloss	
2	Emul.	Flat	Gloss	Gloss	Gloss	Dado 1.00 m high
etc., etc.,						

Notes: Emul. = twice emulsion
Flat = prime and paint two coats flat wall paint
Gloss = prime and paint two undercoats and one coat hard gloss finish
All dadoes to be finished with 10 mm painted line cut in on both edges
Heights are from top of skirting
Etc., etc.

Where walls of a room are to be in different colours, this should be made clear, just as painting of woodwork should be stated to be 'in multi-colours' when panels, etc., are to be picked out

N FURNITURE/EQUIPMENT

N1 General purpose fixtures/furnishings/equipment	N10 General fixtures/furnishings/equipment* N11 Domestic kitchen fittings* N12 Catering equipment N13 Sanitary appliances/fittings* N14 Interior landscape N15 Signs/Notices*
N2 Special purpose fixtures/furnishings/equipment	N20, N21, N22, N23 — Appropriate section titles for each project

N1 General purpose fixtures/furnishings/equipment

N10 General fixtures/furnishings/equipment

N11 Domestic kitchen fittings

Shelving		Thickness, widths and number of tiers in each position Open slatted shelving: size and spacing of battens and ledges Wall bearers Posts, legs, rails, etc. Brackets
Cupboards		Cupboard fronts or doors and frames
BS 1195	: Kitchen fitments	Tops, bottoms, ends and divisions
3444	: Blockboard and laminboard	Locks or other fastenings Shelving or other fitting up
5669	: Chipboard	Bearers or brackets in the case of wall cupboards
6222	: Domestic kitchen equipment	
Sundry joinery		The sizes of the various members and their method of assembly Pelmets Cloak rails Backboards Seats Dressers Counters Racks or special fitments

Sundry glass items	Thickness: size: finish Mirrors Shelves Window sills, etc. Splashbacks
Sundry items BS 792 : Dustbins	Sundry items e.g. Curtain tracks, door mats, lockers, dustbins, etc., will usually be specified by maker's or merchant's catalogue number or be the subject of PC prices

N13 Sanitary appliances/fittings

Sanitary fittings		
BS 1125	: Flushing cisterns for w.c's	Manufacturer's reference and fixing or PC sum with clear and sufficient description of each type of fitting to enable fixing to be priced (e.g. whether brackets are screwed or cantilever type, whether the sum includes flush pipes or not) Towel rails Toilet roll holders
1188	: Ceramic washbasins	
1189	: Cast iron baths	
1206	: Fireclay sinks	
1212	: Float operated valves	
1244	: Metal sinks	
1254	: W.c. seats (plastic)	
1329	: Metal lavatory basins	
1390	: Sheet steel baths	
1415	: Mixing valves	
1876	: Urinal cisterns	
1968	: Floats for ball valves (copper)	
2081	: Chemical closets	
2456	: Floats for ball valves (plastic)	
4305	: Cast acrylic sheet baths	
5388	: Pillar taps	
5503–6	: W.c. pans, bidets, wash basins	
5779	: Spray mixing taps	
6340	: Shower trays	
6465	: Code of practice	
6731	: Handrinse basins	

N15 Signs/Notices

Signwriting BS 5499	: Fire safety signs, etc.	Description of work, style and size of letters, etc., so far as not shown on drawings Notices: size, material, lettering, fixing

P BUILDING FABRIC SUNDRIES

P1 Sundry proofing/insulation	P10 Sundry insulation/proofing work/fire stops*
	P11 Foamed/Fibre/Bead cavity wall insulation
P2 Sundry finishes/fittings	P20 Unframed isolated trims/ skirtings/sundry items*
	P21 Ironmongery*
	P22 Sealant joints
P3 Sundry work in connection with engineering services	P30 Trenches/Pipeways/Pits for buried engineering services*
	P31 Holes/Chases/Covers/Supports for services*

P1 Sundry proofing/insulation

P10 Sundry insulation/proofing work/fire stops

Insulating quilts, boards, etc.	Material thickness and method of fixing
BS 3837 : Polystyrene board	Over ceiling joists in roof space
4841 : Rigid foam board	In wooden upper floors
5803 : Roof insulation	In partitions
	To tanks, cisterns, etc.

P2 Sundry finishes/fittings

P20 Unframed isolated trims/skirtings/sundry items

Skirtings	Material
BS 584 : Wood trim	Thickness and height
	Description of section, e.g. moulded, etc.
	Backings
	Floor fillets
Cornices, frieze rails and dado rails	Material
BS 584 : Wood trim	Dimensions
	Description of section
	Grounds
	Angle brackets for cornices
	Bracketing for plaster cornices, pilasters, etc.

P21 Ironmongery

BS 3827 : Glossary of terms for builders' hardware	
Door and window ironmongery	Hinges, latch or lock and other furniture to doors, e.g. bolts, push plates, kicking plates, cabin hooks, etc.
BS 1227 : Hinges	Hinges or pivots, fasteners and stays to casement windows
2911 : Letter plates	Sash lines or chains, pulleys, weights, fasteners, lifts and pulls to sash windows
4951, 5872 : Locks, latches	
5725 : Emergency exit devices	
6125 : Sash lines	
6459 : Door closers	
Sundry ironmongery, etc.	Water bars: material and section
BS 1161 : Aluminium alloy sections	Floor strips: ditto
	Dowels: ditto
1494 : Fixing accessories	Hand-rail brackets
	Shelf brackets

P3 Sundry work in connection with engineering services

P30 Trenches/Pipeways/Pits for buried engineering services

BS 5834 : Valve boxes	Trenches for water, electric or gas mains Excavation Beds Haunching Surrounds
	Meter pits, stop valve pits, etc.
	Stop valve boxes, access/inspection chambers including covers

P31 Holes/Chases/Covers/Supports for services

Forming/Cutting holes, mortices, chases, etc.	Size, material, making good
Trench covers	Material and design
	Material and sections of kerbs or bearers

Q PAVING/PLANTING/FENCING/SITE FURNITURE

Much of the specification for work in this section for groundworks to roads, pavings, etc. will be covered in other sections, leaving such matters as asphalt road covering, planting, turfing, seeding, fencing, site furniture, etc. to be covered in this section.

Q1 Edgings for pavings	Q10 Stone/Concrete/Brick kerbs/edgings/channels*
Q2 Pavings	Q20 Hardcore/Granular/Cement bound bases/sub-bases to roads/pavings*
	Q21 In situ concrete roads/pavings/bases
	Q22 Coated macadam/Asphalt roads/pavings*
	Q23 Gravel/Hoggin roads/pavings
	Q24 Interlocking brick/block roads/pavings
	Q25 Slab/Brick/Sett/Cobble pavings*
	Q26 Special surfacings/pavings for sport
Q3 Site planting	Q30 Seeding/Turfing*
	Q31 Planting*
Q4 Fencing	Q40 Fencing*
Q5 Site furniture	Q50 Site/Street furniture/equipment*

Q1 Edgings for pavings

Q10 Stone/Concrete/Brick kerbs/edgings/channels

BS 340	: Pre-cast concrete Kerbs, channels, etc.	Excavation or making up levels Foundation Kerbs/Edgings/Channels
435	: Dressed natural stone Kerbs, channels, etc.	

Q2 Pavings

Q20 Hardcore/Granular/Cement bound bases/sub-bases to roads/pavings

Sub-base	Levelling and compacting sub-grade Sub-base material to be used Maximum thickness of layers and method of consolidation Extent of thickness (after consolidation) Blinding top surface Any finishing to falls

Q22 Coated macadam/Asphalt roads/pavings

BS 594	: Rolled asphalt	Number of layers and finished thickness of each
1446 } 1447 }	: Mastic asphalt for roads and footways	Type of aggregate and grading of each layer
3690	: Bitumens	Falls, cambers and rolling
4987	: Coated macadam for roads and pavings	Gritting or other surface finish
5273	: Dense tar road surfacing	

Q25 Slab/Brick/Sett/Cobble pavings

BS 368	: Precast concrete flags	Type size and thickness of material Bedding
6677	: Clay and calcium-silicate paviers	Laying to falls Pattern Steps

Q3 Site planting

BS 4428	: Recommendations for general landscaping operations

Q30 Seeding/Turfing

Lawns and banks	Excavation or making up levels Trimming banks Soiling (giving thickness) Grass seed (quality and weight per metre) Turf

Q31 Planting

Flower beds	Moving of reserved vegetable soil, depositing, levelling, breaking up and raking over
Trees, shrubs and other planting	Usually a provisional sum for work by specialists As this work is normally carried out after the main building work is finished, it is often excluded from the building contract and dealt with direct by the Employer

Q4 Fencing

Q40 Fencing

Fencing BS 1485 : Galvanised wire netting 1722 : Fencing 4102 : Steel wire 4092 (Part 1) : Metal gates (Part 2) : Wooden gates	Type of fencing and spacing of posts with their stays Foundation to posts and stays In-filling between posts with method of fixing Gates, posts, hinges and furniture (where part of fencing)

Q5 Site furniture

Q50 Site/Street furniture/equipment

BS 4092 (Part 1) : Metal gates (Part 2) : Wooden gates 5649 : Lighting columns 5696 : Play equipment	Gate posts, hinges and furniture (where not part of fencing)

R DISPOSAL SYSTEMS

R1 Drainage	R10 Rainwater pipework/gutters* R11 Foul drainage above ground* R12 Drainage below ground* R13 Land drainage R14 Laboratory/Industrial waste drainage
R2 Sewerage	R20 Sewage pumping* R21 Sewage treatment/sterilisation
R3 Refuse disposal	R30 Centralised vacuum cleaning R31 Refuse chutes* R32 Compactors/Macerators R33 Incineration plant

R1 Drainage

R10 Rainwater pipework/gutters

Cast iron rainwater pipes BS 460 : Cast iron rainwater goods	Quality, section (round or rectangular) and size of pipe Any surface finish, e.g. galvanised Method of jointing and fixing (e.g. ears or holderbats) Bends, branches, etc. Shoes or connection to drain Outlet pieces from flat roofs Connection pieces through roofs Hopper heads and gratings
Cast iron eaves gutters BS 460 : Cast iron rainwater goods	Quality, section and size Any surface finish, e.g. galvanised Method of jointing and fixing (e.g. brackets) Angles, stop ends, etc. Outlets and gratings
Pressed steel, asbestos-cement, aluminium, zinc or plastic rainwater goods BS 569 : Asbestos-cement rainwater goods 1091 : Pressed steel rainwater goods	As for cast iron

1431 : Wrought copper and wrought zinc rainwater goods 2997 : Aluminium rainwater goods 4576 : UPVC rainwater goods	
Sheet steel gutters	Thickness of metal and method of fixing Any surface finish, e.g. galvanised Angles, stop ends, outlets, etc.
Lead outlet or rainwater pipes	Outlet pipes from flats or gutters: diameter, weight per metre, connection to flat and connection (if any) to other end Weep pipes, etc. Gratings
Lead rainwater heads	PC price and describe method of fixing Gratings

R11 Foul drainage above ground

Wastes to basins, baths, sinks, etc. BS 1184 : Copper traps 3380 : Wastes and overflows 5255 : Plastic waste pipes and fittings	Traps Material and size of branch and main waste pipes in each case Joint to fitting; Joint of branch to main pipe — With any brass ferrules, etc. Joint of main pipe to gulley or drain Grating to top of main pipe Hopper head, where open wastes, and grating
Ventilating pipes to wastes	Material and size Point of and method of connection to branch wastes and to main stack (or if separate anti-syphon pipes)
Soil and ventilating pipes BS 416 : Cast iron pipes 3868 : Prefabricated galvanised steel stack units 5572 : Sanitary pipework	Material and size of branch and main soil pipes Joint to fitting; Joint of branch to main pipe — With any brass ferrules, thimbles, etc. Joint of main pipe to drain Grating to top of main pipe Vent pipes to top of drains

Overflow pipes BS 3380 : Wastes and overflows	Size and material Method of jointing Hinged flap or splay cut ends, etc.
Anti-siphon pipes	Material and size Connection to soil pipe or arm of pan with ferrule Point and method of connection to main stack
Testing	Requirements for testing

R12 Drainage below ground

Materials and workmanship BS 65 : Clay drain and sewer pipes 437 : Cast iron socketed drain pipes 534 : Steel drain pipes 1194 : Concrete porous pipes 1196 : Clayware field drain pipes 1211 : Spun iron pipes 3656 : Asbestos-cement pipes 4660 : Unplasticised PVC 4772 : Ductile iron pipes 5911 : Precast concrete pipes and fittings for drainage and sewerage 5995 : Code of practice PVC 6087 : Flexible joints for cast iron 8301 : Code of practice for building drainage	Cement Sand Mortar Clayware or concrete field pipes Salt-glazed ware pipes Concrete pipes Cast iron pipes Pitch-fibre pipes Drain excavation and backfilling Concrete beds: width thickness haunching or surrounding Laying and jointing of pipes
Runs of drains	Excavating and backfilling Any distinction in pipes to be used, e.g. soil, surface water, inside building, etc. Extent of concrete beds (if not to all) Minimum cover and extent of entirely surrounding with concrete Any backfilling required in concrete

		Suspended drains (hangers, etc.) Bends, junctions, etc. Any special items, such as cleaning eyes Subsoil drainage
Gullies, etc. BS 65 437	 : Clay ware : Cast iron	Types of rainwater shoes and gullies and their respective positions and uses Gully kerbs
Manholes BS 497 1247	 : Cast iron and cast steel covers and frames : Step irons	Sizes (if not indicated on drawings or in a schedule) Excavation Concrete bottoms: mix thickness spread beyond walls Walls: bricks mortar thickness internal finish Covers: concrete slabs iron covers, identified as to type and weight margins to covers Pipes: main channels branch channels iron bolted inspection pipes with branches concrete benching and finish intercepting trap Step irons: minimum depth of manhole spacing
Fresh air inlet		F.A.I. fitting Any length of vertical pipe and its fixing Pipe connection from manhole
Petrol interceptors		As for manholes including ventilation
Soakaways		Size and depth Bottom and walls (if any) Nature and extent of filling
Testing		Requirements for testing, e.g. water, smoke, etc. Make good defects

Note: In works of alteration the contractor cannot be expected to estimate the cost of making good defects to existing drains. Insert a provisional sum for this

R2 Sewerage

R20 Sewage pumping

Sewage disposal plant BS 6297 : Code of practice for design and installation of small sewage treatment works and cesspools	Construction of tanks, filter beds, etc., on the same lines as manholes Filtering media Special ironwork: PC sum or makers' catalogue references Pump (if any) and piping Disposal of effluent

R3 Refuse disposal

R31 Refuse chutes

BS 1703	Type and size of pipe Method of fixing Intake access Outlet access

S PIPED SUPPLY SYSTEMS

S1	Water supply	S10	Cold water
		S11	Hot water
		S12	Hot and cold water (small scale)*
		S13	Pressurised water
		S14	Irrigation
		S15	Fountains/Water features
S2	Treated water supply	S20	Treated/Deionised/Distilled water
		S21	Swimming pool water treatment
S3	Gas supply	S30	Compressed air
		S31	Instrument air
		S32	Natural gas*
		S33	Liquid petroleum gas
		S34	Medical/Laboratory gas
S4	Petrol/Oil storage	S40	Petrol/Oil – lubrication
		S41	Fuel oil storage/distribution
S5	Other supply systems	S50	Vacuum
		S51	Steam
S6	Fire fighting – water	S60	Fire hose reels
		S61	Dry risers
		S62	Wet risers
		S63	Sprinklers
		S64	Deluge
		S65	Fire hydrants
S7	Fire fighting – gas/foam	S70	Gas fire fighting
		S71	Foam fire fighting

As an example of specification information required, Sections S12 and S32, representing as they do supplies for a domestic building, are chosen. The remaining sections are of a special nature and would in all probability be matters dealt with by a specialist engineer.

S1 Water supply

S12 Hot and cold water (small scale)

Materials and workmanship		Steel tubing:
BS 486	: Asbestos-cement pressure pipes	quality and if galvanised jointing and fixing

BS 534 : Steel spigot and socket pipes
1010 : Draw-off valves and stop valves
1211 : Spun iron pipes
1387 : Steel tubes
1740 : Wrought steel pipe fittings
3284 : Polythene pipe for cold water services
2871 : Copper and copper alloy tubes
2879 : Draining taps
3380 : Wastes
3457 : Materials for tap washers
3505 : UPVC pipes for cold potable water supply
3974 (Part 1) : Rod type hangers
4118 : Glossary of terms
4346 : Joints and fittings for use with unplasticised PVC pipes
5292 : Jointing materials and compounds
5556 : Thermoplastic pipes
5572 : Sanitary pipe work
6700 : Water supply for domestic use
6730 : Polythene pipes

bends, tees, etc.
Copper tubing:
quality and gauges
jointing and fixing
type of couplings, etc.
Polythene tubing:
quality and gauges
butt welding and bending
compression fittings or other method of jointing
Joints of different materials

Connection to water main	Refer to A53 with a definition of extent of Authority's work
Service up to building	Material and size Stopcock Wrapping pipe Stopcock inside building
Rising main and branches	Material and size Connection to cistern and ball valve Branches to drinking water or other points requiring main service Branch to heating engineers' feed cistern Stopcocks controlling branches

Storage cistern BS 417 : Galvanised steel cisterns 1563 : Cast iron sectional tanks 1564 : Pressed steel ditto 4213 : Cold water storage cisterns	Material and size of cistern Ditto of overflow pipe Coupling up (if more than one)
Down services	Material and size at outlet of cistern Stopcock near cistern Sizes and general arrangement of distribution pipes or reference to lay-out shown on drawings Stopcocks controlling branches Provision for draining any dips below draw-off level Service to hot water system: material and size and controlling stopcock
Hot water service BS 699 : Copper cylinders 1565 : Galvanised steel indirect cylinders 1566 : Copper ditto 3198 : Copper storage units 5422 : Specification for insulation materials	Sizes and general arrangement of distribution pipes etc., as for down services Cylinder Controls Heated towel rails
Chlorinating	Chlorinating water pipes
Testing	Requirements for testing plumbing work

S3 Gas supply

S32 Natural gas

Service up to meter CP 331 : Service pipes	Refer to A 53 with a definition of extent of Authority's work
Distribution pipes CP 331 : Installation pipes	Material and size of pipes, general arrangement or reference to lay-out on drawings Gas cocks

Fittings		Any special connections to be supplied, e.g. brass tube, flexible connections, etc.
BS 669	: Flexible metallic tubing	
746	: Gas meter unions	
1552	: Control plug cocks	
CP 331	: Installation of pipes and meters	
Testing		Requirements for testing pipework

T MECHANICAL HEATING/COOLING/ REFRIGERATION SYSTEMS

T1	Heat source	T10	Gas/Oil fired boilers
		T11	Coal fired boilers
		T12	Electrode/Direct electric boilers
		T13	Packaged steam generators
		T14	Heat pumps
		T15	Solar collectors
		T16	Alternative fuel boilers
T2	Primary heat distribution	T20	Primary heat distribution
T3	Heat distribution/utilisation – water	T30	Medium temperature hot water heating
		T31	Low temperature hot water heating
		T32	Low temperature hot water heating (small scale)*
		T33	Steam heating
T4	Heat distribution/utilisation – air	T40	Warm air heating
		T41	Warm air heating (small scale)
		T42	Local heating units
T5	Heat recovery	T50	Heat recovery
T6	Central refrigeration/ Distribution	T60	Central refrigeration plant
		T61	Primary/Secondary cooling distribution
T7	Local cooling/Refrigeration	T70	Local cooling units
		T71	Cold rooms
		T72	Ice pads

As an example of specification information required section T32 represents a domestic heating system. The remaining sections are of a special nature and would in all probability be matters dealt with by a specialist engineer.

T3 Heat distribution/utilisation – water

T32 Low temperature hot water heating (small scale)

BS 779	: Boilers for heating and hot water	Heat source Pipework and fixing

3377	:	Back boilers (domestic)	Radiators, heaters, etc.
			Pumps, fittings, etc.
4433	:	Solid smokeless fuel boilers	Insulation
			Testing
5449	:	Code of Practice for central heating	
CP 341.300–7	:	Central heating by low pressure hot water	

Sections U–X are for specialist work generally outside the scope of this book and the entries that follow under these headings are restricted to the CAWS first and third levels to indicate the grouping of the detailed specification for these items which will be required.

U VENTILATION/AIR CONDITIONING SYSTEMS

First level	Third level
U1 Ventilation/Fume extract	U10 General supply/extract U11 Toilet extract U12 Kitchen extract U13 Car parking extract U14 Smoke extract/Smoke control U15 Safety cabinet/Fume cupboard extract U16 Fume extract U17 Anaesthetic gas extract
U2 Industrial extract	U20 Dust collection
U3 Air conditioning – all air	U30 Low velocity air conditioning U31 VAV air conditioning U32 Dual-duct air conditioning U33 Multi-zone air conditioning
U4 Air conditioning – air/water	U40 Induction air conditioning U41 Fan-coil air conditioning U42 Terminal re-heat air conditioning U43 Terminal heat pump air conditioning
U5 Air conditioning – hybrid	U50 Hybrid system air conditioning
U6 Air conditioning – local	U60 Free standing air conditioning units U61 Window/Wall air conditioning units
U7 Other air systems	U70 Air curtains

V ELECTRICAL SUPPLY/POWER/ LIGHTING SYSTEMS

V1 Generation/Supply/HV distribution	V10 Electricity generation plant V11 HV supply/distribution/public utility supply V12 LV supply/public utility supply
V2 General LV distribution/ lighting/power	V20 LV distribution V21 General lighting V22 General LV power
V3 Special types of supply/ distribution	V30 Extra low voltage supply V31 DC supply V32 Uninterrupted power supply
V4 Special lighting	V40 Emergency lighting V41 Street/Area/Floor lighting V42 Studio/Auditorium/Arena lighting
V5 Electric heating	V50 Electric underfloor heating V51 Local electric heating units
V9 General/Other electrical work	V90 General lighting and power (small scale)

W COMMUNICATIONS/SECURITY/ CONTROL SYSTEMS

W1	Communications – speech/audio	W10	Telecommunications
		W11	Staff paging/location
		W12	Public address/Sound amplification
		W13	Centralised dictation
W2	Communications – audio-visual	W20	Radio/TV/CCTV
		W21	Projection
		W22	Advertising display
		W23	Clocks
W3	Communications – data	W30	Data transmission
W4	Security	W40	Access control
		W41	Security detection and alarm
W5	Protection	W50	Fire detection and alarm
		W51	Earthing and bonding
		W52	Lightning protection
		W53	Electromagnetic screening
W6	Control	W60	Monitoring
		W61	Central control
		W62	Building automation

X TRANSPORT SYSTEMS

X1 People/Goods

- X10 Lifts
- X11 Escalators
- X12 Moving pavements

X2 Goods/Maintenance

- X20 Hoists
- X21 Cranes
- X22 Travelling cradles
- X23 Goods distribution/Mechanised warehousing

X3 Documents

- X30 Mechanical document conveying
- X31 Pneumatic document conveying
- X32 Automatic document filing and retrieval

Sections Y and Z are for general reference specification for services and building items. Cross reference in other sections to the general reference specification will avoid inconsistencies and unnecessary repetition of like items. The CAWS levels are given to indicate the nature of such general reference items.

Y SERVICES REFERENCE SPECIFICATION

Y1	Pipelines and ancillaries	Y10	Pipelines
		Y11	Pipeline ancillaries
Y2	General pipeline equipment	Y20	Pumps
		Y21	Water tanks/cisterns
		Y22	Heat exchangers
		Y23	Storage cylinders/calorifiers
		Y24	Trace heating
		Y25	Cleaning and chemical treatment
Y3	Air ductlines and ancillaries	Y30	Air ductlines
		Y31	Air ductline ancillaries
Y4	General air ductline equipment	Y40	Air handling units
		Y41	Fans
		Y42	Air filtration
		Y43	Heating/Cooling coils
		Y44	Humidifiers
		Y45	Silencers/Acoustic treatment
		Y46	Grilles/Diffusers/Louvres
Y5	Other common mechanical items	Y50	Thermal insulation
		Y51	Testing and commissioning of mechanical services
		Y52	Vibration isolation mountings
		Y53	Control components – mechanical
		Y54	Identification – mechanical
		Y59	Sundry common mechanical items
Y6	Cables and wiring	Y60	Conduit and cable trunking
		Y61	HV/LV cables and wiring
		Y62	Busbar trunking
		Y63	Support components – cables

Y7	General electrical equipment	Y70	HV switchgear
		Y71	LV switchgear and distribution boards
		Y72	Contactors and starters
		Y73	Luminaires and lamps
		Y74	Accessories for electrical services
Y8	Other common electrical items	Y80	Earthing and bonding components
		Y81	Testing and commissioning of electrical services
		Y82	Identification – electrical
		Y89	Sundry common electrical items
Y9	Other common mechanical and/or electrical items	Y90	Fixing to building fabric
		Y91	Off-site painting/Anti-corrosion treatments
		Y92	Motor drives – electric

Z BUILDING FABRIC REFERENCE SPECIFICATION

Z1 Fabricating
- Z10 Purpose made joinery
- Z11 Purpose made metalwork

Z2 Fixing/Jointing
- Z20 Fixings/Adhesives
- Z21 Mortars
- Z22 Sealants

Z3 Finishing
- Z30 Off-site painting

Appendix Example specification

The example specification chosen is that for a small extension to a typical suburban semi-detached house, an extension rendered necessary by a growing family who preferred to extend the house they liked rather than move to a larger house possibly in a strange district. This is a set of circumstances that often faces young architects and surveyors and the specification that follows is the kind of thing that might well be written.

This example specification has been adapted to Common Arrangement but the drawing has not been annotated as suggested in the Code for Production Drawings as it is not envisaged building work of such a simple nature warrants full cross referencing. The drawing and the subsequent text have not been amended to comply with current building regulations. The specification is therefore an example of the relationship between the drawing and a resulting specification; it is not an example of current building construction practice.

The example specification and drawing, except for the conversion to CAWS and to metric, minor alterations to delete trade names and the introduction of the wash basin to give some plumbing, are the specification and drawing from which the extension was built. Although the architect did not go on an extended holiday as suggested in Chapter 1, the extension was built with the minimum of variations and to the complete satisfaction of the client and, it is hoped, the architect and the builder!

SPECIFICATION OF WORKS

required to be done

in

CONSTRUCTION OF ADDITIONAL ACCOMMODATION

at

110 WHYTELEAFE HILL, WHYTELEAFE, SURREY

for

CHRISTOPHER J. WILLIS ESQ

David J. Juniper DipArch, RIBA
Chartered Architect
High Wold Gate
Woldingham, Surrey

Date:

TABLE OF CONTENTS

PRELIMINARIES/GENERAL CONDITIONS

A10 PROJECT PARTICULARS

01 The site of the works is 110 Whyteleafe Hill, Whyteleafe, Surrey.

02 The employer is Christopher J. Willis Esq., the owner/occupier of the premises. The architect is David J. Juniper DipArch, RIBA of High Wold Gate, Woldingham, Surrey.

A11 DRAWINGS

01 The scope of the work is shown on drawing CW 1.

A12 THE SITE/EXISTING BUILDINGS

01 The site of the works is the present car port area between No's 110 and 112 Whyteleafe Hill.

A13 DESCRIPTION OF THE WORKS

01 The work comprises taking down and clearing away the existing lean-to car port and constructing a new car port with a bedroom over being connected to the existing building at half landing level on the internal staircase.

A20 THE CONTRACT

01 The articles of agreement and conditions of contract will be those contained in the Agreement for Minor Building Works (Revised March 1988) issued by the Joint Contracts Tribunal. The conditions thereof, the headings of which are set out below, are to be read as incorporated herein.

Recital

1 The work – Delete 'contract administrator'
2 The specification – Delete 'or the schedules'
3 The contract documents
4 The quantity surveyor – Delete
5 The guarantee/warranty scheme – Delete

Article

1 Contractor's obligations
2 Contract sum
3 Architect/Contract administrator
4 Arbitration

01 Condition

1 Intentions of the parties
1.1 Contractor's obligations
1.2 Architect's/Contract Administrator's duties
2 Commencement and completion
2.1 Commencement and completion
Commencement date x 19 y
Completion date z 19 y
2.2 Extension of contract period
2.3 Damages for non-completion
Damages £100.00 per week
2.4 Completion date
2.5 Defects liability
3 Control of the works
3.1 Assignment
3.2 Sub-contracting
3.3 Contractor's representative
3.4 Exclusion from the works
3.5 Architect's/Contract Administrator's instructions
3.6 Variations
Delete 'schedules'
3.7 Provisional sums
4 Payment
4.1 Correction of inconsistencies
Delete 'schedules'
4.2 Progress payments and retention
4.3 Penultimate certificate
4.4 Final certificate
4.5 Contribution, levy and taxes
Nil %
4.6 Fixed price
5 Statutory obligations
5.1 Statutory obligations, notices, fees and charges
5.2 Value added tax
5.3 Statutory tax deduction scheme
5.4 Fair wages – delete
5.5 Prevention of corruption – delete
6 Injury, damage and insurance
6.1 Injury to or death of person
6.2 Injury or damage to property
6.3A Insurance of the works – Fire etc. – New works – delete
6.3B Insurance of the works – Fire etc. – Existing structures
6.4 Evidence of insurance
7 Determination
7.1 Determination by Employer
7.2 Determination by Contractor
8 Supplementary Memorandum
Part A Contribution, levy and tax changes

Part B Value added tax
Part C Statutory tax deduction scheme
Part D Fair wages clause – delete
Part E Guarantee Warranty scheme – delete

A33 EMPLOYER'S REQUIREMENTS: QUALITY STANDARDS CONTROL

01 Materials and workmanship to be best of their respective kinds to the satisfaction of the architect. Materials shall apply where applicable with the current British Standards.

A35 EMPLOYER'S REQUIREMENTS: SPECIFIC LIMITATIONS ON METHOD/SEQUENCE/TIMING

01 The contractor shall have access to the side of the house as necessary and to that part of the garden required for the extension. Care shall be taken to cause as little damage to the rest of the garden and as little inconvenience to the occupants as is possible.

02 The connection to the old building shall be left as late as possible and shall be made at a time mutually convenient to the contractor and the occupants.

03 Use of the garage space will be required by the occupants as much as is compatible with the building operations.

A42 CONTRACTOR'S GENERAL COST ITEMS: SERVICES AND FACILITIES

01 Water for the works and electricity supply for temporary lighting and small tools will be available from the house supply and providing the contractor is economical in the use thereof, no charge will be made.

02 Remove all debris as it accumulates, clean all glass outside and in and generally leave all clean and tidy at the completion of the works.

A44 CONTRACTOR'S GENERAL COST ITEMS: TEMPORARY WORKS

01 Supply all tools, brushes, scaffolding, hoisting tackle, ladders and other plant for the proper carrying out of the works.

A54 PROVISIONAL WORK

01 Provide the sum of £x for contingencies to be used as directed and deducted in whole or in part if not required.

DEMOLITIONS AND ALTERATIONS

C10 DEMOLISHING STRUCTURES

01 The contractor shall include both labour and materials (unless otherwise particularly stated) for any shoring, strutting, scaffolding or temporary works in connection therewith, for making good all work disturbed in all trades and for removing all rubbish from the site.

02 Pulling down, taking out and cutting away shall be carefully performed and every precaution shall be taken to ensure the safety of the works.

03 Supply, erect and maintain during the cutting of openings, etc., all necessary protection to the existing premises against damage by weather or other causes.

04 Allow for laying the dust as far as possible during the demolition by watering with a hose or other means.

05 Materials arising from the pulling down, except where described as 'set aside', shall become the property of the contractor who shall remove them from the site with reasonable speed.

06 Take down and clear away the glass lean-to roof over the existing car port together with timber plates, purlins, rafters, glazing bars, eaves gutter, downpipe and flashing.

07 Take off and set aside the existing garage doors, take down and clear away the frame and fascia and framing over.

08 Pull down and clear away the half brick boundary wall between No's 110 and 112 together with piers therein, the rendered one brick wall and foundations under and prepare for erection of new party wall. Allow for all necessary protection of adjoining property as above mentioned including adequate support of retained earth. After erection of new wall make good existing garage floor and adjoining path. This work shall be carried out with the minimum of inconvenience to the occupant of the adjoining property.

C20 ALTERATIONS – SPOT ITEMS

01 All making good of brickwork, building up of openings, etc., shall be in solid brickwork, unless otherwise described, in cement lime mortar (1:1:6) properly cut, toothed and bonded and pinned up to existing work and pointed where necessary.

02 The prices for cutting openings shall include for all necessary temporary strutting, easing and striking.

03 Move the metal storage bins clear of the boundary wall prior to demolition and adapt the timber top to the main coal bunkers including providing temporary back during rebuilding. Access to and full use of these bunkers will be required throughout the winter months.

GROUNDWORK

D20 EXCAVATING

01 The term excavate shall include for getting out. The price shall include for keeping the excavations free from water by pumping, baling or otherwise and grubbing up any roots or normal obstructions met with.

02 The contractor will be held responsible for any slips or falls of earth, etc., and must clear away any falls and make good all damage to the works arising therefrom without charge.

03 Excavate trenches 600 mm deep for foundations to 250 mm hollow walls and pier holes for brick piers as shown on the drawing, support the sides as required, level and consolidate the bottoms and on completion of foundation concrete and brickwork return and fill excavated material and remove surplus from site.

IN SITU CONCRETE

E10 IN SITU CONCRETE

01 The cement shall be British Portland cement to comply with BS 12 and shall be stored in a dry place and not in contact with the ground.

02 The sand for concrete shall be clean, sharp washed sand, a mixture of fine and coarse grains from 9.52 mm gauge downwards, free from loam and other impurities to comply with BS 882 Table 2.

03 The aggregate for general concrete shall be clean gravel from an approved source, free from loam, clay or organic matter and shall be graded for general concrete 38.10 mm down to 4.76 mm and for fine concrete 9.52 mm down to 4.76 mm as set out in Table 1 of BS 882.

04 All cement shall be measured by weight and other materials in approved gauge boxes.

05 The general concrete (1:3:6) shall be composed of one part of cement, three parts of sand and six parts of 38.10 mm aggregate.

06 The fine concrete (1:2:4) shall be composed of one part of cement, two parts of sand and four parts of 9.52 mm aggregate.

07 If concrete is hand mixed, the materials for each batch shall be mixed on a clean boarded or other approved platform and turned over twice while dry and twice while water is added through a fine rose in sufficient quantity to form a stiff workable mixture, the whole mass again turned over twice until thoroughly mixed.

08 If concrete is mixed by machine, the mixers shall be of approved type and mixing shall continue until there is a thorough distribution of the materials and the mass is of uniform consistency.

09 The contractor may use ready-mixed concrete from a supplier approved by the architect. The same conditions shall apply as for site-mixed concrete. In all other respects the ready-mixed concrete shall comply with BS 5328.

10 The concrete shall be deposited as quickly as possible after mixing and shall proceed continuously, so that, as far as possible, complete sections of the work are done in one operation. All levelling pegs are to be withdrawn as the concrete is laid.

11 Immediately on every cessation of work the platforms, mixers and wheelbarrows used shall be emptied and cleaned.

12 Concrete work shall not be carried out when the temperature is less than 3° centigrade. In dry weather the sand and aggregate shall be kept in a damp condition by watering. All concrete shall be protected from frost, inclement weather, damage or too rapid drying with sacking or other suitable material.

13 Place foundations in concrete (1:3:6) to trenches and pier bases to the widths and depths shown on the drawing.

14 Construct 250 × 230 mm beam in concrete (1:2:4) reinforced with two 12 mm diameter bars, the concrete filled into fair formwork and well packed around reinforcement and interstices made good after striking of formwork.

15 For cavity fill see F10.17.

E20 FORMWORK

01 All formwork is to be substantial, sufficiently tight to prevent leakage and supported to maintain its position and shape without deflection. No formwork is to be struck before the concrete has reached a suitable strength.

02 The description of fair formwork shall include for rubbing down the surface of concrete smooth after removal.

E30 REINFORCEMENT

01 Bar reinforcement shall be of mild steel bars to comply with BS 4449, free from anything likely to interfere with the adhesion of steel and concrete.

MASONRY

F10 BRICK WALLING

01 All bricks shall be approved white sand lime bricks.

02 Bricks for use below ground shall have a compressive strength of not less than 20.5 N/mm^2 and those for use in both internal and external work above ground shall have a compressive strength of not less than 14N/mm^2. All bricks shall be the best of their respective kinds, hard, square, sound and free from other cracks and defects and even in size to the architect's approval.

03 The cement shall be British Portland cement as Item E10 01.

04 The sand shall be clean, sharp, washed sand to comply with BS 1200.

05 The lime shall be approved hydrated lime to comply with BS 890.

06 Materials for mortar shall be measured by volume in approved gauge boxes: platforms and mechanical mixers shall be of approved type.

07 The cement mortar shall be composed of one part of cement and three parts of sand mixed dry with water added afterwards.

08 The cement-lime mortar shall be composed of one part of cement, one part of lime and six parts of sand mixed dry with water added afterwards.

09 Brickwork of half brick walls shall be built in stretcher bond and that to piers in English bond. Brickwork shall be carried up in level courses, each course well flushed up with mortar and all interstices properly filled in. Thirteen courses shall rise 1.00 m and all perpends shall be truly kept. Brickwork shall be in whole bricks as far as possible, bats and closers only being used where required for bond.

10 All bricks shall be soaked with clean water before laying.

11 All bricklaying shall be suspended when the temperature is less than 3° Centigrade.

12 No portion of brickwork shall be carried up more than 1.00 m above adjoining work. Differences in levels shall be raked back not toothed.

13 Construct the 215 mm and 250 mm hollow walls and the brick piers, to the lengths and heights shown on the drawing, of special purpose sand-lime bricks in cement mortar below ground and sand lime bricks in cement lime mortar above ground.

14 Finish all brickwork externally and in the car port with a neat struck joint as the work proceeds. Plumb all angles. Rake out joints internally on new and old walls as key for plaster.

15 Construct brick-on-edge coping 215 mm wide as shown bedded and pointed in cement lime mortar including single course tile creasing.

16 Build ends of hollow walls solid with brickwork half brick thick, point ends of walls to match remainder and plumb angles of faced work.

17 Fill cavities of hollow walls below ground level with fine concrete (1:2:4), the top edge splayed off.

F30 SUNDRY ITEMS

01 The ties for hollow walls shall be 200 mm galvanised steel twisted pattern to comply with BS 1243 type 'a' built in at 1000 mm centres horizontally and 500 mm centres vertically and staggered. Extra ties shall be incorporated at corners and cavities shall be kept clear of mortar droppings at all times.

02 The damp-proof courses shall be lead lined felt to comply with BS 743 and 6398. Provide and fix lead lined felt damp-proof course across cavity of hollow walls and at ends of walls built solid as indicated on detail drawing.

03 Provide and fix lead lined felt damp proof course across cavities of hollow walls built solid as indicated on detailed drawing.

04 Cut pocket in existing wall for new reinforced concrete beam and for ends of timber bearers under front and back screens. Cut and pin joist hangers (K20 03) to existing and new walls.

05 Provide and build in 225 × 150 × 100 mm square hole pattern air bricks to match sand lime facings in the positions indicated.

STRUCTURAL CARCASSING TIMBER

G20 CARPENTRY/TIMBER FRAMING/FIRST FIXINGS
K20 TIMBER BOARD FLOORING/SHEATHING
M52 DECORATIVE FABRICS

01 All timber shall be the best of its kind, perfectly dry, thoroughly well seasoned, sawn die square, free from sap, shakes, cracks, waney edges, loose and dead knots over 32 mm diameter and any other defects which in the opinion of the architect render the timber unsuitable for its purpose.

02 The softwood for carcassing shall be approved quality imported Baltic Red Deal selected from 5th grade.

03 The joist hangers shall be 10 gauge black bitumen finish mild steel with ragged top flanges size 150 × 40 mm for floor joists and 100 × 40 mm for roof joists.

04 The word 'framed' as applied to woodwork shall be understood as including all the best known methods of jointing, by mortice, tenon, dovetail or otherwise. All joints in joinery shall be glued and cross-tongued with hardwood tongues. Arrises and mouldings to be run clean true and sharp.

05 Construct the floor of 150 × 40 mm sawn softwood joists at 300 mm centres trimmed as required to stairwell, cover the underside with 10 mm fire-resisting insulation material on 25 mm battens and finish with 22 mm tongued and grooved softwood boarding in about 90 mm (face) widths, each board well cramped up and nailed to each joist, the boarding cleaned off on completion.

06 Construct the roof of 100 × 40 mm sawn softwood joists at 300 mm centres with 25 mm glass fibre quilt strip insulation between the joists, covered with 25 mm boarding firred to falls and traversed for felt roofing with fascias over the window screen and triangular timber fillets to parapet walls. Line the underside of the joists with 3 mm hardboard and fix 300 × 300 mm white chamfered polystyrene ceiling tiles fixed with approved adhesive, carefully set out to pattern all in accordance with the manufacturer's instructions.

07 Fix 150 × 25 mm softwood fascias to front and back edges of roof and 185 × 19 mm external quality plywood fascia to back screen at first floor level.

CLADDING

H71 LEAD SHEET FLASHINGS

01 The sheet lead shall be the best British milled lead of the full weights per square metre and sizes specified, neatly laid and dressed down in the best workmanlike manner with proper provision for expansion and contraction.

02 Supply and fix Code 4 lead flashings as follows:
at joint of new roof and existing walls, the top edge turned into groove in wall and the bottom edge dressed over turn up of felt roofing
over head of screen windows back and front, open copper nailed to joist and dressed over frame
under sills of front and back screen windows, open copper nailed to bearer, that to front screen dressed across glass and glazing bars.

H74 ZINC FLASHINGS
01 Fit 14 gauge zinc drips to transomes of front and rear screens.

WATERPROOFING

J41 BUILT UP FELT ROOF COVERINGS
01 The felt roofing shall comply with BS 747 Class I and shall consist of three layers of 20.4 kg/m^2 (Type 1b) The bottom layer shall be nailed to roof boarding at 50 mm centres along laps 20 mm from the exposed edge. The layers shall be laid to break joint and bonded with hot bitumen compound. The top surface shall be dressed with similar bitumen compound and finished with white spar chippings closely packed (not less than 50 kg/3 m^2).
02 Cover the high level roof with three layer felt roofing as described. Dress over fascia with welted edges at screens, over angle fillet and under coping to 250 mm walls and over angle fillet and under flashing to existing walls.

WINDOWS/DOORS/STAIRS

GENERALLY
01 The softwood for joinery shall be approved good class unsorted Baltic Redwood. Timber for frames and framing shall be selected for straightness of grain.
02 The hardwood shall be prime selected oak, free from all defects, selected for straightness of grain and approved by the architect.
03 All hardwood shall be kept clean and finished for polishing and properly protected.
04 Plywood shall conform to BS 6566, the bonding MR for external quality and INT for internal quality. Plywood shall be faced both sides with birch or Gaboon mahogany veneer and shall be grade 2 quality.
05 Softwood and hardwood for joinery shall comply with the requirements of BS 1186.

06 All of the dimensions and thicknesses of timber are basic unless otherwise stated. Allow for all necessary planing margins in accordance with BS 4471 having due regard to the end use of the timber.

07 Allow for protecting by casing all hardwood and softwood joinery fitted as first fixing.

08 All surfaces of joinery to be painted shall be primed as specified (M60).

09 The description for fixing softwood joinery, unless described as screwed or bolted shall include for punching nails well in and neatly stopping.

L10 TIMBER WINDOW/SCREENS

01 Provide and fix the two window screens to the detail drawing, form lower panel of 50 × 50 mm softwood framing faced one side with 19 mm external quality plywood and the other side with 12 mm approved insulation board. Form and hang side-opening casement windows each fitted with aluminium cockspur stay and fastener, with side lights for direct glazing and cover fillets, packings, timber bearers and double joist beams under all to detail.

L20 TIMBER DOORS

01 Fix 50 × 32 mm twice rebated softwood glazing bars framed in and prepared for glazing. Fix 100 × 50 mm rebated softwood door frame plugged to new and old walls, adapt as necessary old garage doors, including supplying and fixing two pairs of new Collinges hinges to match existing.

L22 PLASTIC DOORS

01 To the opening formed put 40 mm crosstongued softwood lining tongued at angles with transome for fanlight. Fit 25 × 16 mm softwood glazing beads to fanlight both sides and finish with 75 × 25 mm splayed softwood architrave both sides.

02 Provide and fix approved plastic covered folding door with aluminium frame and fittings PC £x the set to suit opening 690 × 2000 mm all in accordance with the manufacturer's instructions.

L30 TIMBER STAIRS

01 Construct the steps in softwood with 32 mm treads and risers glued, blocked and bracketed and framed to 40 mm strings, the sides and soffites covered with 10 mm insulation on bearers.

L40 GENERAL GLAZING

01 All glass shall be SQ quality, the best of its kind free from bubbles, specks and other defects, of approved manufacture to comply with BS 952.

02 Glazing to wood shall be bedded in linseed oil putty to comply with BS 544, back puttied, well sprigged and front puttied, with all superfluous putty neatly trimmed off.

03 Glaze the sashes and fixed lights of the front and back screens and the roof over the front of the car port with 4 mm clear sheet glass.
04 Glaze the door fanlight with 5 mm group 2 obscure glass.

SURFACE FINISHES

M20 PLASTERED COATINGS
01 The plaster on walls shall be lightweight aggregate gypsum plaster of approved manufacture and shall be applied in accordance with the manufacturer's instructions.
02 The rendering coat on brick walls shall be perlited browning plaster ruled to an even surface and lightly scratched to form a key. The finishing coat shall be neat plaster. The plaster shall finish a total of 12 mm thick.
03 The finishing coat shall be trowelled off to a fine smooth finish and shall not be applied until the rendering coat is adequately dried out.
04 Supply, fix and clear away all temporary rules or battens required. Dub out existing walls as necessary to give a true and even surface.
05 Render and set walls of first floor room internally as described.
06 Form V joint in plaster finish at junction of new and old walls as shown.

M60 PAINTING/CLEAR FINISHING
01 The whole of the paints shall, unless otherwise described, be obtained from ...
02 The paints shall be delivered in the maker's sealed containers. No dilution of painting materials will be allowed except strictly as detailed by the manufacturer's directions either on their containers or in their literature, or by permission of their representative.
03 The patent knotting shall be of the best quality, consisting only of shellac dissolved in methylated spirit, and free from resin or naphtha.
04 The wax polish shall be bleached beeswax and genuine American turpentine applied in two coats and polished to an approved finish. Surfaces to be wax polished shall be thoroughly and evenly rubbed down with fine abrasive paper and stopped as necessary.
05 Painting and sealing internally shall not be carried out unless the room is free from dust.
06 An interval of at least twenty four hours shall be allowed between each coat. Each coat when dry shall be properly rubbed down with fine abrasive paper and stopped as necessary.
07 All cracks, crevices and nail holes shall be stopped with a prepared stopping of approved make.
08 All ironwork shall be free from rust and scale before painting.
09 No painting of exterior work shall be carried out in wet, frosty or foggy weather or upon surfaces that are not properly dry.

10 No decorations are to be proceeded with on plaster until it is perfectly dry.
11 Prepare as described, prime and paint two undercoats and one finishing coat on surfaces of eaves gutters and downpipes in black oil paint.
12 Wash down surfaces of existing doors, touch up and apply one coat of oil paint to match existing.
13 Knot, prime, stop and paint one undercoat and two finishing coats on all new woodwork internally and externally, including priming only all backs of frames, linings, etc.
14 Apply two coats of emulsion paint on all surfaces of new plaster internally.
15 Wax polish general surfaces of oak shelving (a provisional area of 6 m^2).

FURNITURE

N10 GENERAL FIXTURES
01 Allow the provisional sum of £x for bunk and storage fittings to be used as directed or deducted in whole or part if not required.

N13 SANITARY APPLIANCES
01 Supply and fix as shown 559 × 406 mm white glazed vitreous china washbasin with pair of 13 mm chromium plated raised nose pillar valves with star heads red and green indexed, 38 mm chromium plated waste fitting, plug and chain with concealed screw to wall brackets.

BUILDING FABRIC SUNDRIES

P20 UNFRAMED ISOLATED/SKIRTINGS/SUNDRY ITEMS
01 Provide and fix 125 × 25 mm softwood splayed skirting fixed to window screens, plugged to outside wall and plugged to old inside wall, mitred at corners and stopped at stairwell.
02 Allow a provisional quantity of 3 m^2 of 19 mm oak shelving to be supplied and fixed as directed on and including a provisional amount of 6 m of 50 × 50 mm oak framing.

PAVING

Q22 COATED MACADAM PAVINGS

01 Include the provisional sum of £x for surfacing of drive and provision of kerb edging.

DISPOSAL SYSTEMS

R10 RAINWATER PIPEWORK/GUTTERS

01 Cast iron gutters, pipes and fittings shall be lightweight to comply with BS 460.

02 Provide and fix 76 mm half round eaves gutter to front fascias at high and low levels fixed with approved brackets, with stop end one end and stop end with outlet for 51 mm and 76 mm rainwater pipes respectively the other end.

03 To last gutters provide and fix 51 mm and 76 mm down pipes jointed to existing pipe or gulley as shown.

R11 FOUL DRAINAGE ABOVE GROUND

01 Provide and fix to basin 38 mm chromium plated deep seal bottle trap with black vulcathene base, jointed to 38 mm waste pipe carried through existing wall and connected to adjoining bedroom basin waste including cutting in and inserting 38 mm tee fitting jointed to new and old pipes.

MECHANICAL AND ELECTRICAL SERVICES

S12 HOT AND COLD WATER

01 The copper tubing shall be light gauge to comply with BS 2871.

02 No bend or curve in any pipe shall be made so as to diminish the waterway or alter the internal diameter of the pipe in any part.

03 Brass fittings shall be approved capillary fittings suitable for service pipes.

04 The copper tubing shall be chromium plated where exposed and fixed with chromium plated brass strip pipe brackets screwed.

05 Cut into existing 13 mm copper services under floor of adjoining bedroom, insert 13 mm tees in hot and cold water service pipes and run 13 mm copper services through existing wall, jointed to basin valves with 13 mm tap connectors and make good all work disturbed.

V9 GENERAL LIGHTING AND POWER

01 Provide the provisional sum of £x for electrical installation.

02 Cut away for and make good after electricians in all trades to the following with their switchgear, local switches, fittings and lamps. The conduit to lighting points will be embedded in the plaster, except in the car port where the conduit will be exposed.

Lighting points	Nr 3
Socket outlet points	Nr 2
Fitting outlet points	Nr 1

Index

Italics refer to Common Arrangement of Work Sections for building works – references restricted to first and second levels.